APPLICATIONS OF
AUTOMATA THEORY AND ALGEBRA
Via the Mathematical Theory of Complexity to Biology, Physics, Psychology, Philosophy, and Games

T0350196

APPLICATIONS OF
AUTOMATA THEORY AND ALGEBRA
Via the Mathematical Theory of Complexity to Biology, Physics, Psychology, Philosophy, and Games

John Rhodes
University of California at Berkeley, USA

Edited by
Chrystopher L. Nehaniv
University of Hertfordshire, UK

Foreword by
Morris W. Hirsch
University of California at Berkeley, USA

 **World Scientific**

NEW JERSEY · LONDON · SINGAPORE · BEIJING · SHANGHAI · HONG KONG · TAIPEI · CHENNAI

Published by

World Scientific Publishing Co. Pte. Ltd.
5 Toh Tuck Link, Singapore 596224
USA office: 27 Warren Street, Suite 401-402, Hackensack, NJ 07601
UK office: 57 Shelton Street, Covent Garden, London WC2H 9HE

British Library Cataloguing-in-Publication Data
A catalogue record for this book is available from the British Library.

Front cover image: "Illuminated Dragons" by Anita Chowdry, collection of Najma Kazi.
Back cover image and page (ii): "Nautilus" by Anita Chowdry, collection of Lionel de Rothschild.

APPLICATIONS OF AUTOMATA THEORY AND ALGEBRA
Via the Mathematical Theory of Complexity to Biology, Physics, Psychology,
Philosophy, and Games

ISBN-13 978-981-283-696-0
ISBN-10 981-283-696-9
ISBN-13 978-981-283-697-7 (pbk)
ISBN-10 981-283-697-7 (pbk)

Printed in Singapore.

Contents

v

Foreword to Rhodes' *Applications of Automata Theory and Algebra*

John Rhodes came to Berkeley in 1963 as assistant professor of mathematics and has been here ever since. In addition to full time teaching and research, he has been a real estate entrepreneur (I made a little money with him), and a therapist, and campus activist. In the sixties he and Krohn ran their own consulting firm, which supported their graduate students and ran conferences. Eventually it was sold at a profit.

I recall vividly my enchantment upon first seeing, in 1969, the mimeographed notes that are almost identical with this book. I knew nothing of the algebra and automata theory behind it, and what little I know I've learned from John. The book is still strikingly original, but in those days it was astonishing. Computers were practically unknown, not only to the public but to the vast majority of scientists, mathematicians and engineers. Many subjects intimately connected to the book were likewise either nonexistent or known only to very small groups: Complexity Theory, Neural Networks, Cellular Automata, Catastrophe Theory, Chaotic Dynamics, Genetic Algorithms .... Pure mathematics was just beginning to come out of a period of fruitful but esoteric rigorous axiomatic development, typified by Bourbaki. Applied mathematics, science and engineering had advanced sufficiently that it could begin to appreciate and use some of the more abstruse mathematical theories.

Rhodes' 1962 doctoral thesis at MIT was not only greatly original in its content, but also in the fact that it was a joint work with Kenneth Krohn. When the outside examiner at Rhodes' thesis defense complained that a Harvard student had written the same thesis, there was no little consternation. Rhodes and his coauthor Krohn had in fact informed their supervisors of their collaboration, saying "It is worth four theses but we only want two". When Krohn handed in his thesis he merely crossed out Rhodes' name and typed in his own.

In his thesis Rhodes (with Krohn) looked at semigroups as finite state machines. Mathematically there is a complete equivalence, but the perspective is different in the two approaches. To tinker with the multiplication table of a group or a semigroup is still a rather strange idea to an algebraist; but nothing could be more natural than to adjust the workings of a machine. Certain constructions are difficult to handle algebraically, but correspond to simple operations on machines. The wreath product of two semigroups, for example, translates into the cascade of two machines, where the output of the first is fed as input into the second.

On the other hand, the algebraic viewpoint unifies many subjects in which a machine approach seems at first sight to be inappropriate, such as the topics covered in this book. Anything with states, inputs and outputs can be usefully looked at in semigroup terms.

The Krohn-Rhodes theory represents any finite semigroup S as the homomorphic image of a subsemigroup of the wreath product of a finite sequence of semigroups acting on finite sets, in which permutation groups and combinatorial semigroups (of noninjective maps) alternate, beginning and ending with combinatorial semigroups. The minimal number of groups in such a representation is the *complexity* $\theta(S)$ of the semigroup S.

The complexity of a finite state machine is the minimal number of reversible computations which, in any cascade representation, must be performed successively, separated by a irreversible redirection of output. Here the reversible computations are done by groups, and the irreversible ones by combinatorial semigroups.

Rhodes outlines the rich and elegant axiomatic theory of complexity in the first chapters. What does it have to do with Biology, Physics, Psychology and Games?

The basic philosophical outlook in the rest of the book is this: Any system which can be investigated scientifically can be – or perhaps *must* be – usefully approximated as a finite state machine. To take only some easy cases: A game such as chess has only finitely many positions. The state of an organism is describable by finitely many chemicals, whose concentrations can be approximated by a finite set of vectors. These concentrations change in response to inputs from the environment, which can be similarly approximated.

Rhodes argues that many scientific and philosophical questions can be formulated in terms of the complexity of suitable semigroups, and that to do so leads to interesting theories and conjectures. Often these are based on computations of complexity whose outcomes are by no means obvious.

The simplest application is to the Theory of Games. The complexity of a game is defined as the minimum complexity of machines which can play the game perfectly (e.g., so as to achieve the von Neumann value of the game). Rhodes proposes that the rules of board games have evolved so as to maximize complexity. Rough estimates lead him to conjecture that this maximum is of the order of the typical number of moves in a game between expert players. For 7×7 Hex, he proves that the complexity is no greater than 24. For Go he proves it is less than 200. Rhodes conjectures that the complexity of Go and Hex, but *not* of 3-dimensional Tic-Tac-Toe, tends to infinity with the size of the board.

Rhodes looks closely at a semigroup S representing the classical Krebs cycle in Biochemistry. Here he argues that it is not just the complexity that is significant, but also the maximal prime factor groups of subgroups of S. In a *tour de force* he calculates that these are just $\mathbb{Z}_2$ and $\mathbb{Z}_3$, that the complexity is 2.

More complex chemical cycles lead to other prime groups – prime order cyclic groups and Simple Nonabelian Groups (SNAGs). Rhodes considers the oxidative pentose phosphate cycle, and suggests that highly transitive permutation groups, close to SNAGs, appear. He points out that the computation cannot be carried out owing to the failure of the Principle of Superposition: that all reactions that occur are derived from the basic ones by independent summation. He proposes that "the groups which actually do appear are large subgroups of the alternating or symmetric groups which theoretically appear under the superposition principle." The latter groups include the 4-transitive Mathieu group on 12 letters. These SNAGs "measure the intricacy of the interconnection of the reactions."

These computations are based on a labeled directed graph derived from the chemistry. Similar analyses can be done, Rhodes suggests, on flow charts of computer programs, or on graphs arising from the configurations in Conway's Game of Life. He asks the interesting question, "How does the complexity of the configurations change with time?"

In Evolution the application of semigroup theory is necessarily more speculative, but also more comprehensible. Here the objective is not precise computations of complexity or SNAGs, but rather general principles influencing Evolution.

Highly evolved organisms, he suggests, are in "perfect harmony" with their environments— otherwise they would either die out or evolve further:

> Either the organism uses effectively more of the possibilities, or
> the possible configurations diminish with lack of use.
>
> Thus if an organism has stopped evolving (e.g., the cell) then it
> must be stable under the forces of evolution and thus the com-
> plexity must be close to the possible number of configurations.

Rhodes then derives the following highly nontrivial and surprising conclusions from this principle:

- Enzymes catalyze (in almost all cases) specific reactions.
- A Mendelian genetic theory must hold for all evolved organisms.
- No simplification of the basic (but immense) data of an evolved
  organism is possible.
- Evolved organisms have internal relations of great depth.
- In the cell, the control of the genes over metabolism need not be
  absolutely complete but must be very good.

Rhodes discusses R. Thom's controversial models of morphogenesis.
These are at the opposite end of the mathematical spectrum, based as
they are on continuous state spaces and differential equations. Yet Thom's
approach is based on the *finite* number of elementary catastrophes. More-
over Rhodes, like Thom, assumes that biological development maximizes
some potential-like function on the state space.

Rhodes points out what many biologists consider a glaring weakness
in Thom's theory: it offers no way to identify physical variables with the
potential function. (Thom would no doubt say that that is a task for
biologists.) Rhodes claims his principle of evolution can be of service here,
as the possible elementary catastrophes will be intimately related to the
states of the machine representing the organism. Rhodes suggests that a
certain generalized Lagrangian function that he describes

> is the function that all living organisms are attempting to maxi-
> mize. But its definition requires modeling organisms with finite
> state machines plus a lot of algebra ... So we will need Thom's
> detailed results extended to arbitrary diffeomorphisms ... plus
> detailed modeling of organisms by finite state machines, plus
> algebraic theory of complexity ...
>
> One interesting thing appears on both the finite side and on the
> differentiable side, namely Lie groups ... most (all?) SNAGs
> come from Lie algebras (plus some 'twists') via the Chevalley
> method.

Recently the theory of '*punctuated equilibrium*' has received a good deal of attention, especially in the popular writings of Stephen Jay Gould. Rhodes does not refer to it by name (it may not have existed when this book was written), but he gives precise quantitative principle concerning "jumps in evolution":

> [I]f the actual path of evolution of the organism is refined into irreducible jumps, some jumps may double the complexity plus one, but never increase it more than this.

The final section, on Emotion, Neurosis and Schizophrenia, is long and complex. Rhodes introduces a "Lagrangian of individual emotional development", defined in terms of interacting semigroups representing relevant aspects of the individual and the environment, and discusses its relation to neurosis and schizophrenia. In his introduction he summarizes it as follows:

> The Lagrangian (true goal) of emotional life for the individual is to contact the environment maximally subject to reasonable understanding and ability to act.

The full audacity, originality, fecundity and rigor of Rhodes' ideas can be appreciated only by reading the book.

Morris W. Hirsch
Department of Mathematics
University of California at Berkeley

Editorial Preface

The book you are reading is commonly referred to as the *Wild Book*. The proof of the Krohn-Rhodes Prime Decomposition Theorem in the 1960s revealed a deep but completely unexpected connection between algebra (semigroups) and computation of all sorts (especially computation requiring only bounded memory). The book in draft manuscript form was first completed by John Rhodes around 1969 and quickly became an underground classic, potentially destined to become the source for fundamental advances in many branches of science. It contains radical ideas on the application of automata and semigroup theory via a rigorously developed theory of mathematical complexity to traditional and non-traditional areas for applications of mathematics. The book introduces a highly insightful approach to create a general, new applied mathematics of finite systems. It is "wild" in that it invades areas as diverse as philosophy (epistemology and the purpose of life), psychology (mathematical theory of psychoanalysis), physics (finite phase-space systems), biology (including metabolism, development and evolution), and games (complexity of games), with a completely original unifying viewpoint and rigorous mathematical methods.

The draft manuscript was used in courses on algebra and automata at the University of California at Berkeley, and was known to students of Rhodes and those lucky enough to find it in the Berkeley Mathematics Library in a simple spiral binding (dated 1971). Many physicists and mathematicians around the world, from Japan to Santa Fe, who became aware of it through academic rumor and word of mouth made photocopies for themselves and disseminated them to colleagues. By the 1990s, many scientists including Professor Morris Hirsch at Berkeley, who has kindly written the foreword, were encouraging Rhodes to publish the *Wild Book*. They realized that the approach of introducing sequential coordinates into the understanding of any phenomenon that can be described by a finite-state (or more general) automaton is a fundamental and general contribution to science whose time is coming, even if its full exploitation may require generations of scientists. As with many crucial ideas in mathematics and science that require time to mature, given the depth of these methods and the publication of this book only now even after 47 years since the Krohn-Rhodes Theorem was discovered, it is still only shortly after the first illuminating light-rays of dawn. Let there be light.

Much more has been done and published by John Rhodes and followers of his school, in particular in developing the abstract mathematical theory of finite semigroups and automata, since the first draft of this book was written. Yet due to the limited availability of the material to other scientists and interdisciplinary researchers, most of the suggested applications still need to be developed much further. Applications are also notoriously unfashionable (sometimes even considered scandalous) in the prevalent culture of many pure mathematicians, therefore many of those exposed to and inspired by the material were not being motivated to pursue these aspects.* If nothing else, even casual readers with a traditional view toward applications will likely find their prejudices about the confines of what comprises applied mathematics vigorously shaken by this book, and with just a little further effort may even awaken to substantially broadened horizons of what applied mathematics and applied algebra could look like in the future.

Happily, algorithmic and computational tools implementing Rhodes' ideas in order to automatically generate sequential coordinate systems are finally becoming available to mathematicians and interdisciplinary scientists thanks to the special abilities of a new generation versed not only in the mathematics and its potential application areas, but capable also of efficiently harnessing the computer algebra and high-performance parallel

---

*Indeed, a negative attitude toward applications is common in the orientation of many narrowly enculturated or specialized mathematicians (as the present editor knows only too well from his own intellectual development), but such a viewpoint is either an error of mere prejudice in the prevailing mathematical culture or is based on profound ignorance of the relationship of mathematics to its applications areas. Pure and applied mathematics can drive each other's creative development, as examples from the work of Archimedes, Isaac Newton, John von Neumann, or John Rhodes (e.g. the present volume) make plain.

computation necessary to achieve the full potential of these ideas.[†] These recent developments are also now approaching the point where some of the techniques described here can be fruitfully applied to natural systems in biology and to artificial intelligence to reveal insights into the hidden algebraic structures and symmetries of these systems in a manner that goes well beyond what is computable by an unaided human being.[‡]

For this first published edition, John Rhodes has written a new philosophical section included here as a prologue, outlining background of the overall viewpoint. Many of the references have been updated in this new edited version of the book while retaining its essential character, errors and typos have been corrected, and an index has been introduced. An almost completely successful effort has been made to retain the numbering of mathematical equations and assertions to agree with the original version. Editorial footnotes are indicated by non-numerical symbols, while the author's own footnotes are indicated with numerals.

Readers of the book, whether from mathematics or other fields, including physicists, biologists of various stripes, computer scientists, psychologists, philosophers, and game theorists, will find much original thought to stimulate the development of their own ideas here. The non-mathematical reader is advised to skip most of the sections developing the mathemati-

---

[†]The first realizations of software implementations of the Krohn-Rhodes Theorem were achieved only at the dawn of the 21st century with the remarkable PhD work of Attila Egri-Nagy at the University of Hertfordshire (see A. Egri-Nagy & C. L. Nehaniv, "Algebraic Hierarchical Decomposition of Finite State Automata: Comparison of Implementations for Krohn-Rhodes Theory", *Springer Lecture Notes in Computer Science* 3317:315-316, 2005). Ongoing work is associated with our newer software implementations for mathematical synthesis of hierarchical coordinate systems (i.e. wreath product decompositions usable for prediction, manipulation, understanding and automated solution of problems in finitary discrete dynamical systems). These include Krohn-Rhodes coordinate systems on transformation semigroups (via the holonomy method) and, as a special case, Frobenius-Lagrange coordinate systems on permutation groups. The open-source software (implemented as a package for the GAP computer algebra system) is freely available at http://sourceforge.net/projects/sgpdec/ with some at present minimal documentation.

[‡]For some current and recent work in this direction, the reader is referred to A. Egri-Nagy & C. L. Nehaniv, "Hierarchical Coordinate Systems for Understanding Complexity and Its Evolution, with Applications to Genetic Regulatory Networks", *Artificial Life* (Special Issue on Evolution of Complexity), 14(3):299-312, 2008; A. Egri-Nagy, C. L. Nehaniv, J. L. Rhodes, & M. J. Schilstra, "Automatic Analysis of Computation in BioChemical Reactions", *BioSystems*, 94(1-2):126-134, 2008; A. Egri-Nagy & C. L. Nehaniv, "Algebraic Properties of Automata Associated to Petri Nets and Applications to Computation in Biological Systems", *BioSystems*, 94(1-2):135-144, 2008; as well as to subsequent and forthcoming papers by these authors and collaborators.

cal theory (Chapters 2, 3 and 5) on first reading and concentrate on the
applications sections (the prologue introducing the philosophical view of
semigroups as algebraic models of time, the brief overview of the book
in Chapter 1, and the very important and deeply insightful Chapter 4
on coordinate systems for understanding phenomena in science, physics,
and biochemistry, and then the four substantial Parts of Chapter 6 on
metabolism, biology, psychology, and games, with their wide ranging vi-
sion and more detailed developments of the viewpoint for various fields).
Most of these sections can be understood prior to developing a detailed
understanding of the underlying mathematics. Given these applications as
motivations, these readers can later pursue the more mathematical chapters. Chapters 2 and 3 explain and justify the research program for finite
semigroups from the mathematical viewpoint of finite algebra (especially
group theory) and complexity. The motivating material in these chapters
introduces a guiding viewpoint on finite semigroup theory as a generalization of finite group theory, with the natural development of the complexity
theory framing the study of finite semigroups (and therefore, as the reader
shall see in later Chapters, of finite state automata and, therefore, myriad
other topics). These parts will be most accessible to those with some background in abstract algebra. Chapter 5 explains the mathematical core of
the Krohn-Rhodes prime decomposition theory for finite state machines and
then develops the associated complexity theory. The appendix to Chapter 5
describes the connection between the cascade product and to series-parallel
product of circuits and provides mathematical proofs of many of the assertions of Chapter 5; this is the most technical section of the book, providing
valuable insights for more mathematically sophisticated readers.

The ideas presented in Chapter 6 on applications to biology and psychology (together with their philosophical interpretations), to games, and
in Chapter 4 to physics are, for the most part, a still untapped intellectual
gold mine for applications of (group-)complexity.

The editor is grateful to the University of Hertfordshire for support and encouragement during the preparation of this volume (especially to Professors Jill Hewitt, Bruce Christianson, John Senior, Kerstin
Dautenhahn, and Martin Loomes). In typesetting, copyediting, correcting,
putting together this material, and/or indexing, the editor depended on
the various skills and invaluable work of Deborah Craig, Attila Egri-Nagy,
Laura Morland Rhodes, and the always helpful staff at World Scientific
Publishing Co., although they cannot be blamed for any remaining errors,
gaffes or inconsistency in style, nor the decision to retain the author's unique

citation style (against advice to the contrary). Artist Anita Chowdry provided wonderfully appropriate mathematically wild artwork for the front and back covers and page ii. Many thanks to her, and also to Lionel de Rothschild and Najma Kazi for permission to reproduce these exquisite creations! The editor also thanks Dr. Maria Schilstra and again especially Dr. Attila Egri-Nagy, both of the Royal Society / Wolfson Foundation Bio-Computation Research Laboratory at the University of Hertfordshire, for their advice leading to some technical corrections and improvements. As could have been expected, although it is still far from perfect, the work of editing has taken many years longer than planned. But it would be unfair to delay any further and not make this book available to those with the gifts to pursue its richness, rigor, depth, and to explore and develop the new vistas it opens for science and mathematics.

The material is intellectually demanding, but the rewards will more than repay the effort. Depending on the reader, various bits of the contents and style of this book may come as inspiration, revelation, or shock, as those who have studied with Rhodes will know from his inimitable lectures which hold a special intellectual excitement. In lectures and half-day long café meetings with John Rhodes, many of us as young mathematicians (including 26 PhD students he has seen through) learned to really think creatively for ourselves (wordlessly being expected to fill in gaps as if being instructed by a gentle mathematical Zen master), to drop naive and narrow minded notions in mathematics, to be optimistically curious and adventurous, to lash out with creativity and fearless metaphor in mathematical exploration, to create our own mathematical language and concepts as needed, to come back to discipline and reflective judgment backed up by rigorous proofs, and to begin to see the algebra in everything.[§] A similar spirit pervades this book. May it help the reader find illumination!

Chrystopher L. Nehaniv
Welwyn Garden City, U.K.
First Day of Spring, 20 March 2009

[§] Attending at least one lecture by John Rhodes has been suggested as an educational requirement for all Berkeley PhD students in mathematics. There are many colorful tales to tell about Professor Rhodes' lectures, such as the invention of new Greek letters as needed, fluid co-opting of pictures to denote algebraic operations, proofs presented (and understood by the students) without language, relentless perseverance through calculations and in pushing viewpoints, enlightening breakthrough insights, repeated re-engagement with things already proved yielding new questions and perspectives, unfinished sentences, blank slides, etc., but these stories must be told another time.

Prologue: Birth, Death, Time, Space, Existence, Understanding, Science, and Religion
[Outline]

Aristotle wrote a paper lost in the library at Alexandria giving some classical axioms about time. One important axiom was:

> Time satisfies (if a, then b, then c) is the same as (if b, then c, but prior to both a).

Algebraically this can be written as the *associative law*. Namely,

$$(a \cdot b) \cdot c = a \cdot (b \cdot c).$$

A *semigroup* is a set with an associative multiplication. Thus every semigroup is a model of time.

The standard model of time for classical physics is $(\mathbb{R}^1, +)$, the real numbers under addition. Surprisingly $(\mathbb{R}^1, +)$ is isomorphic to (the same as) $(\mathbb{R}^+, \cdot)$, the positive real numbers under multiplication, under $x \mapsto e^x$ and $\ln y \longleftrightarrow y$.

Another standard model is $(A^+, \cdot)$, where A is a finite (non-empty) alphabet and A^+ is all finite strings over A (e.g. if $A = \{a, b\}$, then $abba = ab^2a \in A^+$, $ab^{13}abbabb \in A^+$, etc.) with multiplication by concatenation (e.g. $abba \cdot abb = abbaabb$, $a \cdot b = ab$, $aa \cdot bb = aabb$, etc.).

If we think of the members of A as basic events, then $(a_1, \ldots, a_n) \in A^+$, where $a_i \in A$, represents event a_1 happening at time 1, $\ldots$, event a_n happening at time n (where we are using the standard model of time $\mathbb{N}^+ = (\{1, 2, 3, \ldots\}, +)$) and

1

$$a_1, a_2, \ldots, a_n$$

$$\downarrow$$

$$a_1, a_1 \cdot a_2, \ldots, a_1 \cdot \ldots \cdot a_n \in S^+$$

where $(S, \cdot)$ is a semigroup generated by A (denoted $(S, A, \cdot)$, so each member of S is a finite product of members of A) and the *result* of $(a_1, \ldots, a_n)$ is $(a_1, a_1 \cdot a_2, \ldots, a_1 \cdot \ldots \cdot a_n)$. This is how models of time (e.g. $(S, A, \cdot)$) code movement of events in time.

Also given say *abbabba* $\in A^+$ the fact that a can be discerned at places 1, 4, and 7 in the sequence is the concept of space ("at a different time but the same thing" is the concept of space). Thus time and space are intimately interrelated.

Space is standardly modeled as $\mathbb{R}^3$ or $\mathbb{R}^n$ since Descartes.

Given two semigroups $(S, \cdot)$ and $(T, \times)$, a map $f : S \to T$ is called a *morphism* iff* for all $s_1, s_2 \in S$, $f(s_1 \cdot s_2) = f(s_1) \times f(s_2)$, (e.g. $x \mapsto e^x$ is a morphism from $(\mathbb{R}^1, +)$ to $(\mathbb{R}^+, \cdot)$, actually an isomorphism, i.e. f^{-1} exists and is a morphism with $f(x) = e^x$ and $f^{-1}(x) = \ln x$). A *surmorphism* is by definition an onto morphism.[†]

Now every surmorphism of a model of time (or semigroup) is another model of time (or semigroup).[‡] For example every finite semigroup is a surmorphism of $(A^+, A, \cdot)$ for some finite A.

Now let us consider existence and understanding (epistemology). Existence implies feedback and is prior to understanding. That is, things exist, like cells, children, massive computer programs using inductive loops, ecological systems with complex feedback, etc., but we may not or do not understand them. Birth implies feedback.

Understanding comes later in the form of *introducing coordinates*, i.e., science, describing the system in question with time and space movements, in *sequential form*:

---

*As is usual in mathematics, throughout this book, "iff" is short-hand for "if and only if", i.e. logical equivalence. -CLN

[†]Again following standard mathematical terminology, a map $f : S \to T$ is said to be "onto" (or "surjective") if for every $t \in T$ there is at least one $s \in S$ with $f(s) = t$.

[‡]If $f : S \to T$ is a surmorphism, we sometimes loosely say that T is a surmorphism of S.

Thus, if the coordinates $(\ldots, x_n, \ldots, x_1)$ (finite or infinite) describe a system at time and space $c(t, p)$, then if the input π is a change in time or space, then in coordinates

$$(\ldots, x_n, \ldots, x_1)\pi = (\ldots, y_n, \ldots, y_1).$$

Understanding implies *sequential form* which means that y_n depends only on $\pi, x_1, \ldots, x_n$ and not on $x_{n+1}, x_{n+2}, \ldots$. Thus existence is prior, and to understand the existing is more difficult!

So the whole idea of science is to introduce coordinates* that move sequentially in time.

Now if the semigroup $(S, A, \cdot)$ is a group $(G, A, \cdot)$, then no information is lost via

$$(a_1, a_2, \ldots, a_n)$$

$$\downarrow$$

$$(a_1, a_1 \cdot a_2, \ldots, a_1 \cdot \ldots \cdot a_n) \in S^+$$

$$\|$$

$$(b_1, b_2, \ldots, b_n)$$

since $b_i b_{i-1}^{-1} = a_i$, so $\downarrow$ is one-to-one; or, in other words, from $(b_1, \ldots, b_n)$, the sequence of basic events $(a_1, \ldots, a_n)$ can be recovered.[†]

In this case time can be run backwards as in classical Newtonian mechanics (e.g. the planets can run backwards). In this situation there is not "death" — no loss of information.

The same is true of $(A^+, A, \cdot)$:

$$(a_1, a_2, \ldots, a_n)$$

$$\downarrow$$

$$(a_1, a_1 a_2, \ldots, a_1 \cdot \ldots \cdot a_n)$$

However consider the simple model of time SL, i.e. $(\{0, 1\}, \{0, 1\}, \cdot)$. So

$$1\ 1\ 1\ 1\ 0\ 1\ 1\ 1\ 0$$

$$\downarrow$$

$$1\ 1\ 1\ 1\ 0\ 0\ 0\ 0\ 0$$

---

*For much more on coordinate systems in science and mathematics, see Chapter 4.

[†]A *group* $(G, A, \cdot)$ is a semigroup with an *identity* element $1 \in G$ (i.e. for all $g \in G$, $1 \cdot g = g \cdot 1 = g$) and inverses for all elements, i.e. for each $g \in G$ there is an *inverse* g^{-1} with $g \cdot g^{-1} = g^{-1} \cdot g = 1$. It is not difficult to prove that if there is an identity element then it must be unique, and each element having an inverse with respect the identity element has a unique inverse. -CLN

and in fact

or

$$a_1 \ldots\ldots\ldots a_n$$

$$\downarrow$$

$$\ldots\ldots\ldots b_n$$

with $b_n = 1$ if *all* $a_1, \ldots, a_n = 1$ and $b_n = 0$ otherwise (i.e. there exists i so that $a_i = 0$).

So there is a loss of information. All living things eventually become a pile of dust and ashes. For the pile of dust and ashes, we cannot reconstruct the entire life, thoughts and deeds of the dead person!

Thus group models of time are useful models for non-living things (e.g. practical physics, and the groups are usually symmetry groups.) Semigroups are the models of time useful for living things (e.g. mathematical biology).

On the other hand in general each group is a "rare jewel", but there are many more semigroups (billions of semigroups with 8 elements*, but only 5 groups). For example if all semigroups were equally probable then most are weeds satisfying $a \cdot b \cdot c = 0$ (almost immediate death)[†], so we need an algebraic theory of semigroups to find interesting ones (e.g. smallest semigroups having complexity 1, 2, 3, ..., n, ..., etc.).

How is all this related to religion? The Egyptians believed the body would live eternally, hence food, water, chair, etc., in the tombs. But what does this mean? Well, it means that body moves around in the standard model of space $\mathbb{R}^3$ forever. But after the advent of Riemannian theories of space from 1880 onwards (like spiders must walk on the surface of a donut, so what is the shortest path to the fly stuck on the surface?), one must ask *which model* of space would the body move around in forever ($\mathbb{R}^3$, the sphere, the donut, etc.?), and the religion loses much of its power. Also the Egyptian religion would be hostile to the ideas of Riemann and other new concepts of space beyond $\mathbb{R}^3$.

---

*S. Satoh, K. Yama, and M. Tokizawa, "Semigroups of Order 8", *Semigroup Forum*, 49(1):7-29, 1994. -Ed.

[†]D. J. Kleitman, B. R. Rothschild, and J. H. Spencer, "The Number of Semigroups of Order n", *Proceedings of the American Mathematical Society*, 55(1):227-232, 1976.

Christianity asserts the soul will live in time forever. But again with only one model of time $(\mathbb{R}^1, +)$ this statement makes sense, but with the introduction of many models of time (e.g. semigroup theory), *which model* of time, $(\mathbb{R}^1, +)$, $(A^+, A, \cdot)$, $(\{0, 1\}, \cdot)$, etc.?[1] This seems hostile to Christianity and thus semigroup theory is not popular or fashionable! But again, as many models of time are shown to exist, the power of the living soul forever in time diminishes in persuasion.

John Rhodes

---

[1]For recent applications of these ideas on semigroups and models of time to computation, see J. Rhodes and P. V. Silva, "Turing machines and bimachines", *Theoretical Computer Science*, 400(1-3):182-224, 2008.

Chapter 1

Introduction

One of our goals in this book is to relate the following diverse topics:

- a generalization of finite group theory which includes finite semigroups and finite state automata;
- a reasonable, precise and useful definition of *complexity* of organisms and machines;
- a mathematical theory of physics and/or biology for the case in which models are applicable with nonlinear but finite phase spaces. Our theory will give rise to the following: relations between the Darwinian theory of evolution, the complexity of the evolved organisms and the Mendelian genetics;
- a definition of a global (generalized) Lagrangian from which the philosophy of Thom's program for studying morphogenesis (i.e., the origin, development, and evolution of biological structures) via the topological methods of the theory of structural stability (à la Smale, Thom, etc.) might begin;
- a Mendeleev type table for the classification of biological reactions (e.g., bacterial intermediary metabolism) isomorphic to the classification scheme (*à la* Dickson, Chevalley, Thompson, etc.) for simple non-abelian groups (SNAGs);
- a mathematical theory of personality including a definition of emotion which synthesizes the ideas of S. Freud and R. D. Laing;
- a theory of complexity for certain games (in the sense of von Neumann) which includes Chess, Checkers, Go, etc.

We will proceed from standard mathematics to non-standard mathematics and then to applications, generalities and possible applications.

Chapter 2

What is Finite Group Theory?

We first consider some of the theorems. We might start our list with the Jordan-Hölder theorem.[1] This proves that each finite group can be built up from a fixed set of simple groups (called the Jordan-Hölder *factors*) and the building set is unique (including repetitions).

This leads one naturally to investigate and classify the SNAGs (= Simple Non-Abelian Groups). Of course, the simple Abelian groups are just the integers modulo a prime p, $\mathbb{Z}_p$.

The most famous result concerning SNAGs is the Burnside Conjecture, proved by Feit and Thompson in 1963 which states that there are *no* SNAGs of odd order. See [Feit and Thompson]. Due to Galois theory and for other natural reasons a finite group G is said to be *solvable* if and only if its Jordan-Hölder factors are all simple Abelian groups, i.e., all of the form $\mathbb{Z}_p$ for various primes p. Then Feit and Thompson's result can be restated as *all groups of odd order are solvable*.

In fact, Thompson (see [Thompson I]) has classified in an explicit list the minimal SNAGs (i.e., those SNAGs all of whose proper subgroups are solvable) and found them to be generated by two elements. Thus a group is solvable if and only if each subgroup generated by two elements is solvable.

However the classification of all SNAGs is very difficult and has been taking very long complete. New SNAGs were being found yearly well into the 1970s, and with the work of producing a complete and correct proof of the classification still ongoing.[*]

---

[1]We assume the reader is familiar with elementary group theory. For example, see [Hall] or [Burnside] of the references at the end of Chapter 2.

[*]The classification of SNAGs is now widely believed by most finite group theorists to be complete (see M. Aschenbacher, The Status of the Classification of the Finite Simple Groups, *Notices of the American Mathematical Society*, 51(7):736–740, 2004), with the largest sporadic SNAG, the Fischer-Griess Monster group M, only first being constructed

Next we might mention P. Hall and G. Higman's results concerning the p-length of p-solvable groups, one of the most important results of the 1950s. See [Hall-Higman].

Let p denote a fixed prime integer. Then P is a *p-group* iff the order of P is p^n for some n. Q is a *p'-group* iff p does not divide the order of Q. By definition G is *p-solvable* iff each Jordan-Hölder factor of G is either a p- or p'-group. If G is p-solvable, then $\ell_p(G)$, the *p-length* of G, is the smallest non-negative integer k such that there exists a normal series
$$\{1\} = P_0 \trianglelefteq N_0 \triangleleft P_1 \triangleleft \cdots \triangleleft N_{k-1} \triangleleft P_k \trianglelefteq N_k = G$$
with each term normal in G and P_i/N_{i-1} is a p-group and N_i/P_i is a p'-group. Clearly such a series for G exists iff G is p-solvable.

Equivalently, $\ell_p(G)$ can be defined as the smallest number of surmorphisms k of the type
$$G \xrightarrow{p'} G_0 \xrightarrow{p} G_1 \xrightarrow{p'} G_{10} \xrightarrow{p} G_2 \xrightarrow{p'} G_{20} \xrightarrow{p} \ldots \twoheadrightarrow G_k \xrightarrow{p} G_{k0} \xrightarrow{p'} \{1\}$$
where $A \xrightarrow{p'} B$ denotes a surmorphism with *kernel* a p'-group and $A \xrightarrow{p} B$ denotes a surmorphism with *kernel* a p-group.

Clearly, G is solvable iff G is p-solvable for all primes p. Also by solvability of groups of odd order, G is 2-solvable iff G is solvable.

If G is a finite group, let G_p denote a Sylow p-subgroup of G. Then one form of Hall and Higman's theorem is:

Theorem 2.0. *If G is p-solvable and $x^{p^e} = 1$ for each $x \in G_p$, then* $\ell_p(G) \leq 3e$.

For a proof see [Hall-Higman] or [Higman]. The statement above is not the strongest form of the theorem. Hall-Higman type theorems were used by Thompson in his original proof of the Frobenius conjecture (see [Thompson II, Thompson III, Thompson IV]) and they also have important applications to the weak Burnside problem; see [Higman] and [Hall, chapter 18].

Note that if $\mathbb{N}$ denotes the non-negative integers, then $\ell_p : \{p\text{-solvable groups}\} \to \mathbb{N}$, (and in fact ℓ_p is onto*).

We next turn to another important concept of finite group theory. If G is a solvable group, the *Fitting height* or *Fitting length* of G is defined as follows: Let $F(G)$, the *Fitting subgroup* of G, be the unique maximal nilpotent[2] normal subgroup of G (it exists!). Let $F_0(G) = \{1\}$ and

---

in 1980 by R. L. Griess. It is conceivable however that any remaining gaps in the proof of the classification might still allow for the discovery of new SNAGs. -CLN

*onto = surjective. To assert that a mapping is onto, it will be denoted with $\twoheadrightarrow$. -CLN

[2]A finite group G is nilpotent iff G is the direct product of its Sylow subgroups.

let $F_{n+1}(G)$ be $F(G/F_n(G))$ considered as a normal subgroup of G (i.e., $F_{n+1}(G) = \eta^{-1}(F(G/F_n(G)))$ where $\eta : G \twoheadrightarrow G/F_n(G)$). Then, by definition, the Fitting length of G is the smallest integer n so that $F_n(G) = G$.

Equivalently the Fitting length of solvable G can be defined as the smallest number k for which there are surmorphisms of the type

$$G \xrightarrow{\eta} G_1 \xrightarrow{\eta} G_2 \xrightarrow{\eta} \ldots \xrightarrow{\eta} G_{n-1} \xrightarrow{\eta} G_n = \{1\}$$

where $A \xrightarrow{\eta} B$ denotes a surmorphism such that the *kernel* is a nilpotent group.

Special groups of low Fitting height are important in group theory and flipping through the pages of research publications shows this concept appearing frequently in important ways.

This will complete our very brief and scant outline of the results and concepts of finite group theory. We next turn to the techniques.

We will consider first the *monomial map* of Frobenius, used so successfully by Burnside and Frobenius around 1900, together with its derivative construction the *transfer* used by Grün, P. Hall and Wielandt in the 1930s.

Let G be a finite group. Let (G, G) denote the *right regular representation* of G so that G acts on the right of G by $g_1 \cdot g = g_1 g$. Let H be a subgroup of G. Choose a set of representatives of the cosets of H in G with $\bar{g}$ the representative of Hg. Then by Lagrange's Theorem each $g \in G$ has a unique expression as $g = h\bar{g}$ with $h \in H$. Denote $\{Hg : g \in G\}$ by G/H (even if H is not necessarily normal in G) and let $L(g) = (h, Hg) = (h, H\bar{g})$. Then $L : G \to H \times G/H$, called the *Lagrange coordinate map*, is one-to-one and onto, so L^{-1} exists.

Now if in (G, G) the letters* are transformed by the Lagrange coordinate map we find from $g_1 \cdot g = g_1 g$ that $L(g_1) \cdot g = L(g_1 g)$ so $(H \times G/H, G)$ is given by

$$(2.1) \quad (h_1, Hg_1) \cdot g = (h_1 \bar{g}_1 g(\overline{(g_1 g)}))^{-1}, Hg_1 g) = (h_1 f_g(Hg_1), Hg_1 g)$$

where $g_1 = h_1 \bar{g}_1$ and $h_1 \in H$ so $L(g_1) = (h_1, Hg_1)$ and $f_g : G/H \to H$ is given by $f_g(Hg_1) = \bar{g}_1 g(\overline{(g_1 g)})^{-1}$. Since the action (G, G) is faithful the action $(H \times G/H, G)$ is also faithful.[†]

---

*i.e., the elements of the set acted upon. -CLN

[†]An action (X, S) is said to be *faithful* if distinct elements act distinctly, i.e. for all $s \neq s' \in S$, there exists some $x \in X$ such that $x \cdot s \neq x \cdot s'$. -CLN

We note two important properties of the action (2.1). First it is in *triangular form*[‡] since if we consider Hg_1 as the first coordinate and h_1 as the second coordinate, then the new first coordinate (Hg_1g) only depends on the old first coordinate (Hg_1) and the action $(\cdot g)$ and not on the old second coordinate (h_1).

Second, we notice the action (2.1) restricted to the first coordinate is the natural action of G on G/H ($Hg_1 \cdot g = Hg_1g$) and when the action $(\cdot g)$ and the first coordinate (Hg_1) are held fixed, then the action on the second coordinate becomes $h_1 \mapsto h_1 \cdot f_g(Hg_1)$ which is in,the right regular representation of H (namely, multiplication by $f_g(Hg_1)$).

We formalize this as follows: Let (X_k, S_k) be given for $k = 1, \ldots, n$ where X_k is a finite non-empty set and S_k is a collection of mappings on X_k closed under composition so S_k acts on the right of X_k by $x_k \cdot s_k = (x_k)s_k$. (We do not require the mappings to be permutations so as to include some generalizations considered later.) (X_k, S_k) is termed a *(right) mapping semigroup.*

Definition 2.2. $(X_2, S_2) \wr (X_1, S_1)$, read the *wreath product* of (X_1, S_1) by (X_2, S_2), equals by definition $(X_2 \times X_1, (X_2, S_2)$ w $(X_1, S_1) \equiv S)$ where $S = \{\Pi : X_2 \times X_1 \to X_2 \times X_1 : \Pi$ *satisfies conditions* (2.2a) *and* (2.2b) *listed below* $\}$.

2.2a. (triangular form) There exists $\Pi_1 : X_1 \to X_1$ and $\Pi_2 : X_2 \times X_1 \to X_2$ so that

$$(x_2, x_1) \cdot \Pi = ((x_2, x_1)\Pi_2, (x_1)\Pi_1) \ .$$

In this case we write $\Pi = \mathrm{w}(\Pi_2, \Pi_1)$.

2.2b. (k^{th} component action lies in S_k) If $\Pi = \mathrm{w}(\Pi_2, \Pi_1)$, then we require $\Pi_1 \in S_1$, and for each $\overline{x}_1 \in X_1$ we also require $x_2 \mapsto (x_2, \overline{x}_1)\Pi_2$ lies in S_2.

It is easy to verify that S is closed under composition. Also conditions (2.2a) and (2.2b) can be equivalently replaced by the single condition

(2.3)   There exists $s_1 \in S_1$ and $\beta_2 : X_1 \to S_2$ so that
$$(x_2, x_1) \cdot \Pi = (x_2 \cdot \beta_2(x_1), x_1 \cdot s_1)$$

In this case we write $\Pi = \mathrm{w}(\beta_2, s_1)$.

[‡]This is also called *sequential form* and sometimes now also called *global hierarchical coordinate form.* -CLN

Also in a natural way $[(X_3, S_3) \wr (X_2, S_2)] \wr (X_1, S_1)$ is isomorphic as a mapping semigroup with $(X_3, S_3) \wr [(X_2, S_2) \wr (X_1, S_1)]$, i.e., wreath product is an associative operation. Thus

$$(X_n, S_n) \wr \cdots \wr (X_1, S_n) = (X_n \times \cdots \times X_1, (X_n, S_n) \text{ w } \cdots \text{ w } (X_1, S_1))$$

is well-defined.

Now from (2.1) and (2.2) we obtain

$$(2.4) \qquad (G, G) \le (H, H) \wr (G/H, G) .$$

Now taking a Jordan-Hölder series for G with simple factors $S_1, \ldots, S_n$ (in that order) and using (2.4) and induction we obtain

$$(2.5) \qquad (G, G) \le (S_n, S_n) \wr \cdots \wr (S_1, S_1)$$

since if $H \lhd G$, then $(G/H, G)$ made faithful is $(G/H, G/H)$.

Now members of $(H, H) \wr (G/H, G)$ can be viewed as $G/H \times G/H$ row-monomial matrices with coefficients in H.[3] Namely, by using (2.1) and (2.3),

$$g \mapsto \text{w}(f_g, \cdot g) \mapsto M_g$$

with $M_g(Hg_1, Hg_2) = 0$ if $Hg_1 g \ne Hg_2$ and $M_g(Hg_1, Hg_1 g) = f_g(Hg_1)$. Then $g \mapsto M_g$ is a one-to-one homomorphism of G into the $G/H \times G/H$ row-monomial matrices with coefficients in H, and this monomorphism is called the *monomial map*.

Let H' denote the derived group of H so H/H' is the maximal Abelian homomorphic image of H. Let $\det_{H/H'}(M_g)$ be the determinant of the matrix M_g after the entries have been mapped into H/H' (and 0 into 0). Then

$$(2.6) \qquad g \mapsto \det\ _{H/H'}(M_g)$$

is a homomorphism of G into H/H' called the *transfer homomorphism*. Writing (2.6) in terms of the wreath product and using (2.1) yields

$$(2.7) \qquad g \mapsto \sum \{H' f_g(x) : x \in G/H\} \in H/H'$$

which also gives the transfer homomorphism.

---

[3] That is, in each row exactly one entry lies in H and all other row entries are 0 with $0 \notin H$ and $0h = h0 = 0$ for all $h \in H$.

Now let ρ be a matrix representation of $H \leq G$ by $d \times d$ matrices with coefficients in the field F. Then

$$(2.8) \qquad\qquad g \mapsto M_g \mapsto \rho^{\#}(M_g),$$

where $\rho^{\#}(M_g)$ is the $d|G/H| \times d|G/H|$ matrix (given by replacing each entry $M_g(a,b)$ of M_g by the matrix $\rho(M_g(a,b))$), is the *induced representation* of G from ρ on H.

Of course this just scratches the surface of the techniques of finite group theory but for the sake of brevity we must stop here. For an exposition of relevant methods in group theory, see [Gorenstein]. See (5.94) and (5.95) for some non-standard comments on SNAGs.

Our exposition presented here is similar to the viewpoint in [Huppert and Wielandt].

Bibliography

[Burnside] W. Burnside. *Theory of Groups.* (Reprint of 2nd Edition, 1911) Dover, New York, 1955.

[Feit and Thompson] W. Feit and J. G. Thompson. Solvability of groups of odd order. *Pacific Journal of Math*, 13:775–1029, 1963.

[Gorenstein] D. Gorenstein. *Finite Groups.* Harper and Row, 1968.

[Hall] M. Hall. *Theory of Groups.* Macmillan, New York, 1959.

[Hall-Higman] P. Hall and G. Higman. The p-length of a p-solvable group and reduction theorems for Burnside's problem. *Proceedings of the London Mathematical Society*, 7:1–42, 1956.

[Higman] G. Higman. p-length theorems. In *1962 Proc. Symp. Pure Math.*, volume 6, pages 1–16. American Mathematical Society, 1963.

[Huppert and Wielandt] B. Huppert and H. Wielandt. Arithmetical and normal structure of finite groups. In *1962 Proc. Symp. Pure Math.*, volume 6, pages 17–38. American Mathematical Society, 1963.

[Thompson I] J. G. Thompson. Nonsolvable finite groups all of whose local subgroups are solvable. *Bulletin of the American Mathematical Society*, 74:383–437, 1968.

[Thompson II] J. G. Thompson. Finite groups with fixed-point-free automorphisms of prime order. *Proc. Nat. Acad. Sci.*, 45:578–581, 1959.

[Thompson III] J. G. Thompson. Normal p-complements for finite groups. *Math. Zeit.*, 72:332–354, 1960.

[Thompson IV] J. G. Thompson. Normal p-complements for finite groups. *Journal of Algebra*, 1:43–46, 1964.

Chapter 3

A Generalization of Finite Group Theory to Finite Semigroups

We begin with an elementary definition.

Definition 3.1a. Let S and T be semigroups. Then we write $S\,|\,T$, read "S *divides* T", iff S is a homomorphic image of a subsemigroup of T (or $S \twoheadleftarrow\leq T$). Clearly 'divides' is transitive.

b. Let (X, S) and (Y, T) be (right) mapping semigroups. Then we write $(X, S)\,|\,(Y, T)$, read (X, S) *divides* (Y, T), iff there exists $Y' \subseteq Y$, $T' \leq T$, $Y' \cdot T' \subseteq Y'$ and a surmorphism $\rho : T' \twoheadrightarrow S$ and an onto map $\theta : Y' \twoheadrightarrow X$ so that

$$\theta(y' \cdot t') = \theta(y') \cdot \rho(t')$$

for all $y' \in Y'$, $t' \in T'$. Clearly 'divides' is transitive. Notice $(X, S)\,|\,(Y, T)$ implies $S\,|\,T$.

c. We write $(\theta, \rho) : (Y, T) \twoheadrightarrow (X, S)$ iff $\rho : T \twoheadrightarrow S$ is a surmorphism and $\theta : Y \twoheadrightarrow X$ is an onto map and

$$\theta(y \cdot t) = \theta(y) \cdot \rho(t)$$

holds for all $y \in Y$, $t \in T$. Clearly "$\twoheadrightarrow$" is transitive.

Now with the aid of Definition (3.1) we can 'translate' finite group theory as follows (with the proofs indicated later):

3.2a. G is a simple group iff

$$G\,|\,(X_2, G_2)\ \text{w}\ (X_1, G_1)$$

implies $G\,|\,G_2$ or $G\,|\,G_1$.

3.2b. Let G be a solvable group. Then the Fitting height of G equals the smallest integer n such that

$$G\,|\,(X_n, N_n)\ \text{w}\ \cdots\ \text{w}\ (X_1, N_1)$$

with each N_j a nilpotent group.

3.2c. Let G be a p-solvable group. Then $\ell_p(G)$, the p-length of G, equals the smallest integer n such that

$$G \,|\, (Y_n, P_n') \text{ w } (X_n, P_n) \text{ w } \cdots \text{ w } (Y_1, P_1') \text{ w } (X_1, P_1) \text{ w } (Y_0, P_0')$$

with P_j a p-group and P_j' a p'-group.

3.2d. Let G be a solvable group. Consider

$$(*) \quad (G, G) \,|\, (\mathbb{Z}_{p_n}, \mathbb{Z}_{p_n}) \wr \cdots \wr (\mathbb{Z}_{p_1}, \mathbb{Z}_{p_1})$$

for primes $p_1, \ldots, p_n$. Notice (*) is taken in the sense of division of mapping semigroups. Then if the Jordan-Hölder factors of G are $\mathbb{Z}_{q_1}, \ldots, \mathbb{Z}_{q_m}$ (including repetitions) it can be proved that $m \leq n$ and $q_1, \ldots, q_m$ lie among $p_1, \ldots, p_n$ (including repetitions). Further, by (2.5) we can solve (*) with $m = n$ and $q_j = p_j$ for $j = 1, \ldots, n = m$. Thus we can translate the Jordan-Hölder Theorem for solvable groups as follows: Let G be a fixed solvable group and let $\mathcal{P}$ denote the set of prime integers and $\mathbb{N}$ the set of non-negative integers. Given (*), $JH_{(*)} : \mathcal{P} \to \mathbb{N}$, read "the *Jordan-Hölder function of* (*)", is defined by $JH_{(*)}(p)$ equals the number of times $\mathbb{Z}_p$ occurs among $\mathbb{Z}_{p_1}, \ldots, \mathbb{Z}_{p_n}$. If $f, g : \mathcal{P} \to \mathbb{N}$, then by definition $f \leq g$ iff $f(p) \leq g(p)$ for all $p \in \mathcal{P}$.

Jordan-Hölder Theorem for Solvable Groups. Let G be a solvable group. Then there exists a unique function $JH : \mathcal{P} \to \mathbb{N}$ satisfying the following:

(**) Given any (*) for G, we have $JH \leq JH_{(*)}$.

(***) There exists a (*) such that $JH = JH_{(*)}$.

(****) The function JH, uniquely determined by (**) and (***), is: $JH(p)$ equals the number of times $\mathbb{Z}_p$ occurs as a Jordan-Hölder factor of G.

A similar statement holds if G is not solvable but $\mathcal{P}$ is replaced by all simple groups and (*) is replaced by

$$(G, G) \,|\, (S_n, S_n) \wr \cdots \wr (S_1, S_1)$$

with each S_j a simple group. However, things are messier to state since one simple group can properly divide another (e.g., $A_5 \,|\, A_6$), but the idea is to minimize the S_j's — leading to the usual Jordan-Hölder factors. We repress the details here.

We briefly indicate the proofs of assertions (3.2a–d).[1] First, (2.5) shows the sufficiency of the condition of (3.2a). For the necessity see Chapter 5 of [M-L-S] (in particular, Definition 5.1.4 and Lemma 5.3.6). To show (3.2b) first notice that the Fitting height equals the smallest integer n such that $G = N_n \triangleright N_{n-1} \triangleright \cdots \triangleright N_1 \triangleright N_0 = \{1\}$ with N_i/N_{i-1} nilpotent. This is easy to see since by induction $F_0(G) \geq \{1\}$, $F_1(G) \geq N_1$, $F_2(G) \geq N_2$, etc., because if $F_i(G) \geq N_i$ then $N_{i+1}/N_i \twoheadrightarrow N_{i+1}/(F_i(G) \cap N_{i+1}) \cong N_{i+1}F_i(G)/F_i(G)$ so all factors are nilpotent and $F_{i+1}(G) \geq N_{i+1}F_i(G) \geq N_{i+1}$. (Here and later we will use that nilpotent groups are closed under division.) Next we claim if G_1, G_2 are solvable groups and $G_1 \mid G_2$, then Fitting height$(G_1) \leq$ Fitting height(G_2). This follows since if N_{i-1}, N_i and N are normal in G, $N_{i-1} \triangleleft N_i$ and $H \leq G$, then $N_i/N_{i+1} \twoheadrightarrow N_iN/N_{i-1}N$ and $(N_i \cap H)/(N_{i-1} \cap H) \leq N_i/N_{i-1}$ and thus we can use the previous characterization of Fitting height to obtain the desired inequality. But the Fitting height of $(X_n, N_n) \wr \cdots \wr (X_1, N_1)$ is $\leq n$ since the projection homomorphisms $(X_j, N_j)\wr(X_{j-1}, N_{j-1})\wr \cdots \wr(X_1, N_1) \twoheadrightarrow (X_{j-1}, N_{j-1})\wr \cdots \wr(X_1, N_1)$ have nilpotent kernels (namely, direct products of N_j with itself). Thus letting $f(G)$ be the integer defined by (3.2b) it follows from all the above that Fitting height$(G) \leq f(G)$. But (2.5) implies the reverse inequality. This establishes (3.2b). The proof of (3.2c) is similar. See [Survey] and [p-length]. (3.2d) can be shown easily using the classical Jordan-Hölder Theorem for finite groups and the results of [Glue].

Now that we have 'translated' the concepts of simple group, Fitting height, p-length, and the Jordan-Hölder Theorem, we introduce the theorem which allows us to generalize this 'translation' to finite semigroup theory.

Let $a \neq b$ and consider $(\{a, b\}, \{C_a, C_b, \text{Id}\})$ where $x \cdot C_a = a$, $x \cdot C_b = b$ and $x \cdot \text{Id} = x$ for all $x \in \{a, b\}$. We also write $(\{a, b\}, \{C_a, C_b, \text{Id}\})$ as $(\{a, b\}, U_3)$. This innocuous three-element semigroup plays an important role in what follows and its multiplication table is

U_3	Id	C_a	C_b
Id	Id	C_a	C_b
C_a	C_a	C_a	C_b
C_b	C_b	C_a	C_b

(An electrical engineer would call U_3 a 'flip-flop' with C_a being the instruction 'destroy contents and replace by a', C_b being the instruction 'destroy contents and replace by b', and Id being the instruction 'do nothing'.)

---

[1] The reader may omit this long paragraph with no loss of continuity.

Definition 3.3. **UNITS** $= \{U : U \,|\, U_3\}$. Up to isomorphism there are four units, namely $U_0 \cong \{C_a\} \cong \{C_b\} \cong \{\text{ Id }\}$, $U_1 \cong \{C_a, C_b\}$, $U_2 \cong \{C_a, \text{ Id }\} \cong \{C_b, \text{ Id }\}$ and U_3.

By definition, **PRIMES** $= \{G : G \text{ is a finite simple group and } G \neq \{1\}\}$. Clearly **PRIMES**=SNAGs $\cup \{\mathbb{Z}_p : p \geq 2 \text{ is a prime integer}\}$.

Let S be a finite semigroup. Then

$$\mathbf{PRIMES}(S) = \{P \in \mathbf{PRIMES} : P \,|\, S\}.$$

If $\mathcal{S}$ is a collection of finite semigroups, then

$$\mathbf{PRIMES}(\mathcal{S}) = \bigcup \{\mathbf{PRIMES}(S) : S \in \mathcal{S}\}.$$

By definition, $S \in \mathbf{IRR}$, read "S is *irreducible*", iff

$$S \,|\, (X_2, S_2) \text{ w } (X_1, S_1) \text{ implies } S \,|\, S_2 \text{ or } S \,|\, S_1.$$

The following theorem, first stated and proved by Krohn and Rhodes (see [Trans]) is the fulcrum on which all the following turns.

Theorem 3.4 (Prime Decomposition Theorem for Finite Semigroups). *Let (X, S) be given. Then there exists $(X_1, S_1), \ldots, (X_n, S_n)$ such that*

$$(3.5) \qquad\qquad (X, S) \,|\, (X_n, S_n) \wr \cdots \wr (X_1, S_1)$$

and for each j with $1 \leq j \leq n$ either

$$(3.5a) \qquad\qquad \begin{array}{c} S_j \in \mathbf{PRIMES}(S) \text{ and } (X_j, S_j) \text{ is a faithful} \\ \text{transitive permutation group,} \end{array}$$

or

$$(3.5b) \qquad\qquad (X_j, S_j) = (\{a, b\}, U_3) \ .$$

Further,

$$(3.6) \qquad\qquad \mathbf{IRR} = \mathbf{PRIMES} \cup \mathbf{UNITS}.$$

For other proofs see [Gins] or [M-L-S], Chapter 5.

Noteworthy special cases of Theorem (3.4) are the following. Say, by definition, that the finite semigroup C is *combinatorial* iff the maximal subgroups of C are singletons or equivalently **PRIMES**(C) is empty.

Corollary 3.7. *C is combinatorial iff*

$$(3.8) \qquad\qquad C \,|\, (\{a, b\}, U_3) \text{ w } \cdots \text{ w } (\{a, b\}, U_3)$$

for some finite number of terms on the right-hand side.

Let $W(\mathcal{S})$ denote the closure of $\mathcal{S}$ under division and wreath products (w), i.e. $W(\mathcal{S}) = \bigcap\{\mathcal{S}' : \mathcal{S} \subseteq \mathcal{S}',$ and $S_1, S_2 \in \mathcal{S}'$ implies (X_2, S_2) w $(X_1, S_1) \in \mathcal{S}',$ and $S \in \mathcal{S}'$ and $T \mid S$ implies $T \in \mathcal{S}'\}$. Equivalently (see [M-L-S] Chapter 5, section 2, and Chapter 6, Theorem 1.5),

$$W(\mathcal{S}) = \{S : S \mid (X_n, S_n) \text{ w } \cdots \text{ w } (X_1, S_1) \text{ with } S_j \in \mathcal{S} \text{ for } j = 1, \ldots, n\}.$$

Note if $\mathcal{G}$ is a collection of finite groups, then $W(\mathcal{G})$ is also a collection of finite groups. Also $W(W(\mathcal{S})) = W(\mathcal{S})$.

Corollary 3.9. *(a)* $W(\textbf{PRIMES}) = $ *all finite groups.*
(b) $W(U_3) = $ *all finite combinatorial semigroups.*
(c) $W(\textbf{PRIMES} \cup U_3) = $ *all finite semigroups.*
(d) Let $\mathcal{S}$ be a collection of finite semigroups. Then

$$S \in W(\mathcal{S} \cup U_3) \textit{ iff } \textbf{PRIMES}(S) \subseteq \textbf{PRIMES}(\mathcal{S}).$$

Thus finite semigroup theory is finite group theory plus the 'flip-flop' U_3. We now generalize (3.2) from groups to semigroups via Theorem (3.4).

3.10a. Let S be a semigroup. Then S is a simple group or a unit semigroup iff

$$S \mid (X_2, S_2) \text{ w } (X_1, S_1)$$

implies $S \mid S_2$ or $S \mid S_1$. Thus comparing with (3.2a) we find the analogue of simple groups for semigroups to be the simple groups and U_3 and its divisors (i.e. **IRR**=**PRIMES** $\cup$ **UNITS**). Thus U_3 and its divisors are the sole new additions.

In the following any group-theoretic terminology applied to semigroups means, by definition, that each subgroup of the semigroup has the property (e.g. solvable semigroup means each subgroup of the semigroup is a solvable group).

3.10b. Let S be a solvable semigroup. Then the Fitting height of S equals the smallest integer n such that

$$S \mid (X_n, S_n) \text{ w } \cdots \text{ w } (X_1, S_1)$$

with each S_j a nilpotent semigroup. This generalizes Fitting height from solvable groups to solvable semigroups. See (3.2b).

3.10c. Let S be a p-solvable semigroup. Then $\ell_p(S)$ (read "the p-length of S") equals the smallest integer n such that

$$S \mid (Y_n, P'_n) \text{ w } (X_n, P_n) \text{ w } \cdots \text{ w } (Y_1, P'_1) \text{ w } (X_1, P_1) \text{ w } (Y_0, P'_0)$$

with P_j a p-*group* and P'_j a p'-*semigroup*. This extends p-length from p-solvable group to p-solvable semigroups. See (3.2c), [Survey] Chapter 5, and [p-length].

3.10d. Let S be a solvable semigroup. Consider (similar to (3.2d))

$$(\dagger) \quad (S, S) \mid (C_n, C_n) \wr (\mathbb{Z}_{p_n}, \mathbb{Z}_{p_n}) \wr \cdots \wr (C_1, C_1) \wr (\mathbb{Z}_{p_1}, \mathbb{Z}_{p_1}) \wr (C_0, C_0)$$

for prime integers $p_1, \ldots, p_n$ and combinatorial semigroups $C_0, C_1, \ldots, C_n$.

We say the Jordan-Hölder Theorem is true for S and S has Jordan-Hölder factors JH, where $JH : \mathcal{P} \to \mathbb{N}$ iff JH satisfies (**) and (***) of (3.2d), replacing (*) by ($\dagger$).

This is a *statement* of the Jordan-Hölder Theorem for finite solvable semigroups. It does not hold (in the above form) for all solvable semigroups but for wide classes of solvable semigroups and in modified forms for even larger classes. However, this is still an active area of research, and final results have not yet been obtained. Similar remarks hold for the non-solvable case.

However the above captures the spirit of the correct generalization of the Jordan-Hölder Theorem to finite semigroups. See [Survey], section 7.

We note here that if in (3.2) all the division signs '$\mid$' are replaced by '$\leq$', then equivalent definitions are obtained (essentially because $\leq$ occurs in (2.4)). However this is *not* the case in (3.10). Thus the 'translation' from finite groups to finite semigroups *requires* '$\leq$' to be replaced by '$\mid$' and U_3 to be added to the **PRIMES**.

Recapitulating, the important concepts of finite group theory, namely simple group, Fitting height, p-length and the Jordan-Hölder Theorem, all involve in their definition by (3.2) equations like

(3.11a) $\qquad\qquad S \mid (X_n, S_n) \text{ w } \cdots \text{ w } (X_1, S_1)$

or

(3.11b) $\qquad\qquad (X, S) \mid (X_n, S_n) \wr \cdots \wr (X_1, S_1)$

with restrictions on the S_j's (e.g. S_j nilpotent groups for Fitting height, S_j alternately p' and p-groups for p-length, S_j simple for Jordan-Hölder).

An important number relating to the structure of the group is given by the 'length' of the shortest equation for S subject to the restrictions on the S_j's where 'length' of the righthand side is defined in a convenient and natural way. See (3.2). (Even 'simple' means that the shortest length of (3.11a) such that S does not divide S_j for any $j = 1, \ldots, n$, is zero.) Next, using the Prime Decomposition Theorem for Finite Semigroups (Theorem (3.4)) we have immediately generalized all these concepts from finite group theory to finite semigroup theory in (3.10a–d).

The definition of Fitting height for groups (from (3.2b)) is based on the fact that from finite nilpotent groups, division, and finite wreath products we obtain exactly the finite solvable groups, i.e. W(nilpotent groups) = solvable groups. The generalization of Fitting height to finite semigroups, (3.10b), is based on W(nilpotent semigroups) = solvable semigroups, which follows from (3.9d).

Similarly, the definition of p-length for groups (3.2c) is based on W(p-groups $\cup$ p'-groups) = p-solvable groups and the generalization of p-length to finite semigroups (3.10c) is based on W(p-groups $\cup$ p'-semigroups) = p-solvable semigroups, which follows from (3.9d).

Finally, the formulation of the Jordan-Hölder Theorem given in (3.2d) is based on W(**PRIMES**) = groups, (3.9a) and the generalization of the statement of the Jordan-Hölder Theorem to finite semigroups is based on W(**PRIMES** $\cup U_3$) = W(**PRIMES** $\cup$ combinatorial semigroups) = semigroups (3.9d).

Now from (3.9),

3.12.

W(**PRIMES**) = groups
W(groups) = groups
$W(U_3)$ = combinatorial semigroups
W(combinatorial semigroups) = combinatorial semigroups
W(groups $\cup$ combinatorial semigroups) = semigroups

Now by complete analogy with the above ((3.2b–d) and (3.10b–d)), using (3.12), we arrive at the following natural definition.

Definition 3.13. Let S be a finite semigroup. Then by definition $\#_G(S)$, read "the *(group) complexity** of S", is the smallest non-negative integer n such that

$$(3.14)\ S\,|\,(Y_n,C_n)\ \text{w}\ (X_n,G_n)\ \text{w}\ \cdots\ \text{w}\ (Y_1,C_1)\ \text{w}\ (X_1,G_1)\ \text{w}\ (Y_0,C_0)$$

holds with $G_1,\ldots,G_n$ finite groups and $C_0,\ldots,C_n$ finite combinatorial semigroups. (For alternate definitions see (3.15) below.)

Remarks 3.15. (1) Using (3.12), Definition (3.13) can be equivalently reformulated as: $\#_G(S)$ is the smallest non-negative integer n such that

$$
\begin{aligned}
(3.15a)\ S\ \ |\ &(U_3,U_3)\ \text{w}\ \cdots\ \text{w}\ (U_3,U_3)\ \text{w}\ G_{\pi(n),n}\ \text{w}\ \cdots\ \text{w}\ G_{1,n}\ \text{w}\\
&(U_3,U_3)\ \text{w}\ \cdots\ \text{w}\ (U_3,U_3)\ \text{w}\ \cdots\ \text{w}\ (U_3,U_3)\ \text{w}\ \cdots\ \text{w}\ (U_3,U_3)\ \text{w}\\
&G_{\pi(1),1}\ \text{w}\ \cdots\ \text{w}\ G_{1,1}\ \text{w}\ (U_3,U_3)\ \text{w}\ \cdots\ \text{w}\ (U_3,U_3)
\end{aligned}
$$

where $G_{i,j}$ are simple groups for $j=1,\ldots,n$, $1\le i\le\pi(j)$.

Thus the complexity $\#_G(S)$ of S is the minimal number of alternations of blocks of simple groups and blocks of U_3's necessary to obtain (3.15a) for S or, equivalently, the minimal number of alternations of groups and combinatorial semigroups to obtain (3.14) for S.

(2) Let (X,S) be a right mapping semigroup. Suppose Definition (3.13) is changed by replacing (3.14) with

$$(X,S)\,|\,(Y_n,C_n)\,\wr\,(X_n,G_n)\,\wr\,\cdots\,\wr\,(Y_1,C_1)\,\wr\,(X_1,G_1)\,\wr\,(Y_0,C_0)\ .$$

This then yields an equivalent definition, i.e. X plays no role. This follows since (S^1,S) divides $(X,S)\times\cdots\times(X,S)$ ($|X|$-times) and $\wr$ behaves well with respect to direct products, i.e., $((X_2,S_2)\wr(X_1,S_1))\times((Y_2,R_2)\wr(Y_1,R_1))$ divides $(X_2\times Y_2,S_2\times R_2)\wr(X_1\times Y_1,S_1\times R_1)$. For proofs see [M-L-S], Chapter 5, Fact 2.14, Chapter 6, Fact 2.2, and in general Chapter 6 of [M-L-S].

To show how closely the definition of complexity follows group theory we next show that the standard group-theoretic concept of π-length (defined below) includes complexity.

Let π denote a collection of prime integers and π' the complementary set of prime integers, e.g. if $\pi=\{2,3\}$, $\pi'=\{5,7,11,13,\ldots\}$. Then, as in standard group theory, G is a π-group iff the prime p divides the order of G implies $p\in\pi$. Now in all our previous treatment of p-length (for both groups and semigroups) we can replace p everywhere by π (and

---

*In this volume $\#_G(S)$ will be called the *complexity* of S or the *group-complexity* of S. Elsewhere it is sometimes called the *Krohn-Rhodes complexity* of S. -CLN

so p' everywhere by π') and obtain the concepts of π-solvable and define $\ell_\pi(S)$, the π-*length* of S. When S is a group G, $\ell_\pi(G)$ is the standard group-theoretic concept of π-length of G introduced by [Hall-Higman] and others. Now when $\pi = \{p\}$, $\ell_{\{p\}}$ (which we write as ℓ_p) is just the usual p-length. However, notice $\ell_\pi(S)$ is only defined for π-solvable semigroups, and π-solvable semigroups = all semigroups iff π = all prime integers. Thus $\ell_{\text{all prime integers}\}}$ is defined for all semigroups S (and the only such ℓ_π). But

$$\#_G = \ell_{\{\text{all prime integers}\}} \ .$$

Thus complexity arises naturally from group theory; first by close analogy with the concepts of simple, Fitting height, p-length, and the Jordan-Hölder Theorem as formulated in (3.2) and (3.10); and second, as ℓ_π with $\pi = \{\text{all prime integers}\}$.

Further, methods powerful enough to give a theory for ℓ_π for some specific π, or for Fitting height, etc., will undoubtedly yield a theory for the others. Further, $\#_G$ is defined for all finite semigroups while the others are not. Also by (3.9) and (3.12), $\#_G$ is the most natural.

See [Survey] for another exposition of material similar to the foregoing.

We devote the remainder of this section to the mathematical theory of complexity (which can be easily adapted to p-length) of finite semigroups.

Let $\mathcal{S}_{Fin}$ denote the collection of all finite semigroups and let $\mathbb{N} = \{0, 1, 2, \dots\}$ denote the non-negative integers. Then $\#_G : \mathcal{S}_{Fin} \to \mathbb{N}$. We would like to compute $\#_G(S)$ given S, and one modern approach is to give axioms characterizing the function $\#_G : \mathcal{S}_{Fin} \to \mathbb{N}$. However before turning to this we list some elementary properties of complexity and give some examples.

Remark 3.16. (a) $\#_G(S) = 0$ if S is a combinatorial semigroup.

(b) $\#_G(S) = 1$ if $S \neq \{1\}$ is a group.

(c) $S \mid T$ implies $\#_G(S) \leq \#_G(T)$.

(d) (Stiffler) If U_3 does *not* divide S, then $\#_G(S) \leq 1$. For a proof, see [Stiffler]. In particular, $\#_G(\text{Abelian semigroup}) \leq 1$.

(e) Let $G \neq 1$ be a group. Let $(2^G, \cdot)$ be the set of all subsets of G under the usual multiplication of subsets, $X \cdot Y = \{x \cdot y : x \in X, \ y \in Y\}$. Then,

$$\#_G(2^G) = 1 \ .$$

This follows from (d) since it can be verified that U_3 does not divide 2^G.

(f) Let $\mathcal{S}$ be a collection of finite semigroups and suppose $W(\mathcal{S}) = \mathcal{S}$. Let $\#_G(\mathcal{S}) = \max\{\#_G(S) : S \in \mathcal{S}\}$ where we assign $+\infty$ when $\{\#_G(S) : S \in \mathcal{S}\}$ is unbounded. Then *exactly one* of the following three cases occurs:

(1) $\#_G(\mathcal{S}) = 0$ or, equivalently, each member of $\mathcal{S}$ is combinatorial.
(2) $\#_G(\mathcal{S}) = 1$ or, equivalently, U_3 divides no member of $\mathcal{S}$ and some member is non-combinatorial.
(3) $\#_G(\mathcal{S}) = +\infty$ or, equivalently, U_3 divides some member of $\mathcal{S}$ and some member of $\mathcal{S}$ is non-combinatorial.

The proof follows from (3.16d) and (3.9d) and (i) below.

(g) Let $X_n = \{1, \dots, n\}$ and let $F_R(X_n)$ denote the semigroup of all functions of X_n into itself under the composition $(f \cdot g)(n) = g(f(n))$, i.e., the functions act on the right of X_n forming $(X_n, F_R(X_n))$. Then

$$\#_G(F_R(X_n)) = n - 1 .$$

For a proof, see [Results], section 5. In particular, semigroups of each complexity exist so $\#_G : \mathcal{S}_{Fin} \twoheadrightarrow \mathbb{N}$ is onto $\mathbb{N}$.

(h) Let F be a finite field. Let $M(n, F)$ denote the semigroup of all $n \times n$ matrices with coefficients in F under matrix multiplication. Then if $|F| > 2$,

$$\#_G(M(n, F)) = n .$$

For a proof, see [Survey], section 4, (4.3)e.

(i) Let $G_1, \dots, G_n$ be non-trivial groups. Then

$$\#_G((Y_{n+1}, U_3) \, \mathrm{w} \, (X_n, G_n) \, \mathrm{w} \, (Y_n, U_3) \, \mathrm{w} \, \cdots \, \mathrm{w} \, (Y_1, U_3) \, \mathrm{w} \, (X_1, G_1) \, \mathrm{w} \, (Y_0, U_3)) = n.$$

For a proof, see [M-L-S], Chapter 6, Theorem 2.10.

(j) $\#_G((X_2, S_2) \, \mathrm{w} \, (X_1, S_1)) \leq \#_G(S_1) + \#_G(S_2)$. For the proof use 3.15(2) and if $(X_2, S_2) \,|\, (Y_2, T_2)$ and $(X_1, S_1) \,|\, (Y_1, T_1)$ then $(X_2, S_2) \wr (X_1, S_1) \,|\, (Y_2, T_2) \wr (Y_1, T_1)$ as proved in [M-L-S], Chapter 5, Fact 2.14.

We now give the Axioms characterizing $\#_G : \mathcal{S}_{Fin} \to \mathbb{N}$. We write $S_1 \times \cdots \times S_n$ for the *direct product* of the semigroups $S_1, \dots, S_n$. We write $S \leq\leq S_1 \times \cdots \times S_n$, read "$S$ is a *subdirect product* of $S_1, \dots, S_n$" iff S is (isomorphic to) a subsemigroup of $S_1 \times \cdots \times S_n$ which projects *onto* each S_j for all $1 \leq j \leq n$.

In the following, θ (possibly with various subscripts) always denotes a function of $\mathcal{S}_{Fin}$ into $\mathbb{N}$.

3.17a. Axiom I. $S \mid T$ *implies* $\theta(S) \leq \theta(T)$. *Also* $\theta(S_1 \times S_2) =$ $\max\{\theta(S_1), \theta(S_2)\}$.

It is easy to verify that an equivalent formulation is

3.17b. Axiom I. $S \mid T$ *implies* $\theta(S) \leq \theta(T)$. *Also if* $S \leq\leq S_1 \times \cdots \times S_n$, *then* $\theta(S) = \max\{\theta(S_i) : i = 1, \ldots, n\}$.

It is not difficult to verify that $\#_G = \theta$ satisfies Axiom I (3.17). See [M-L-S], Chapter 6. Notice that an interpretation of the last line of (3.17b) is as follows: given homomorphisms $\rho_1, \ldots, \rho_n$ which separate the points of S (i.e., given $s_1, s_2 \in S$, $s_1 \neq s_2$, there exists a j such that $\rho_j(s_1) \neq \rho_j(s_2)$) with $\rho_j(S) = S_j$, then if θ satisfies Axiom I, $\theta(S) = \max\{\theta(S_j), j = 1, \ldots, n\}$. Also if θ satisfies Axiom I (e.g., $\#_G = \theta$), it suffices to compute $\theta(S)$ only for those S which are *subdirectly indecomposable* (i.e., $S \leq\leq S_1 \times \cdots \times S_n$ implies S is isomorphic to S_j for some j). The subdirectly indecomposable semigroups have some rather special and useful properties. See [M-L-S], Chapter 8, Lemma 2.19.

A large number of functions θ (other than just $\#_G$) satisfy Axiom I. In fact we have the following old theorem in universal algebra of G. Birkhoff.[2] See [Cohn].

Theorem 3.18. *The following statements are equivalent.*

(a) For every function $\theta : \mathcal{S}_{Fin} \to \mathbb{N}$ *satisfying Axiom I, we have*

$$\theta(S) = \theta(T).$$

(b) $S \mid \prod T$ *and* $T \mid \prod S$, *where* $\prod X$ *denotes some finite direct product of copies of* X.

We write $S \twoheadrightarrow T$ iff there exists a *surmorphism* (equals 'onto homomorphism') of S onto T. We write $S \underset{\gamma}{\twoheadrightarrow} T$ iff there exists a surmorphism $\theta : S \twoheadrightarrow T$ such that θ restricted to *each subgroup* of S is one-to-one. For example, let C be a combinatorial semigroup and suppose (Y, C) and (X, T) are given. Let $S = (Y, C)$ w (X, T) and let $(Y \times X, S) \twoheadrightarrow (X, T)$ be the natural surmorphism (given by restricting to the X-coordinate). Then it can be shown that this produces $S \underset{\gamma}{\twoheadrightarrow} T$. See [M-L-S], Chapter 8. Also note that the restriction of γ-surmorphisms to subsemigroups yield again γ-surmorphisms.

---

[2] This application was pointed out to me by J. Karnofsky.

Now we have the important Axiom II for complexity.

3.19. Axiom II. $S \underset{\gamma}{\twoheadrightarrow} T$ *implies* $\theta(S) = \theta(T)$. *Also* $\theta(\{0\}) = 0$.

Theorem 3.20 (Fundamental Lemma of Complexity). *Axiom II holds for* $\#_G = \theta$.

Rhodes first proved that Axiom II holds for $\#_G$. The existing proof is long and difficult. See [BuII], [AxII Weak] and [AxII Strong].*

We write $S \underset{\mathcal{L}'}{\twoheadrightarrow} T$ iff there exists a surmorphism $\theta : S \twoheadrightarrow T$ such that if $e_1, e_2 \in S$, $e_1 \neq e_2$, $e_1^2 = e_1$, $e_2^2 = e_2$ and $\theta(e_1) = \theta(e_2)$, then there exists $r_1, r_2 \in S$ such that $r_1 e_1 = e_2$, $r_2 e_2 = e_1$. Otherwise stated, $\theta : S \underset{\mathcal{L}'}{\twoheadrightarrow} T$ iff two idempotent[†] elements in S having the same image under θ generate the same principal left ideal of S.[3] For example, let G be a group and suppose (Y, G) and (X, T) are given. Let $S = (Y, G) \text{ w } (X, T)$ and let $(Y \times X, S) \twoheadrightarrow (X, T)$ be the natural surmorphism (given by restricting to the X-coordinate). Then it can be shown that this induces $S \underset{\mathcal{L}'}{\twoheadrightarrow} T$. See [M-L-S], Chapter 8. Also it can be shown that the restriction of an $\mathcal{L}'$-surmorphism to a subsemigroup is another $\mathcal{L}'$-surmorphism.

3.21. Axiom III. *Let* $S \underset{\mathcal{L}'}{\twoheadrightarrow} T$. *Then* $\theta(S) \leq \theta(T) + 1$.

Axiom III holds for $\#_G = \theta$. The proof is by standard techniques. See [M-L-S], Chapter 8, and [Survey].

The importance of the γ and $\mathcal{L}'$-surmorphism occurring in Axioms II and III is shown by the following

Proposition 3.22. *Let* $\varphi : S \twoheadrightarrow T$ *be given. Then there exists*

$$S \underset{\gamma}{\twoheadrightarrow} T_0 \underset{\mathcal{L}'}{\twoheadrightarrow} S_1 \underset{\gamma}{\twoheadrightarrow} \cdots \underset{\gamma}{\twoheadrightarrow} T_{n-1} \underset{\mathcal{L}'}{\twoheadrightarrow} S_n \underset{\gamma}{\twoheadrightarrow} T_n = T$$

such that the composite surmorphism is φ. *That is, every surmorphism can be decomposed into* γ *and* $\mathcal{L}'$ *surmorphisms.*

For a proof, see [M-L-S], Chapter 8. Now, using Proposition 3.22 we can introduce the following

---

*See [TilsonXII] for a streamlineed proof. -CLN

[†]Recall that $e \in S$ is said to be *idempotent* if $e^2 = e$. -CLN

[3]The prime in $\mathcal{L}'$ is used to denote that the condition must hold just for idempotent elements. When the restriction to idempotents is dropped the condition is denoted $\mathcal{L}$. However, unlike $\mathcal{L}'$, $\mathcal{L}$ is *not* preserved under restriction.

Definition 3.23. Let $\#_{\gamma-\mathcal{L}'}(S)$ be the smallest integer n such that there exists a series

$$S \twoheadrightarrow_{\gamma} T_0 \twoheadrightarrow_{\mathcal{L}'} S_1 \twoheadrightarrow_{\gamma} T_1 \twoheadrightarrow_{\mathcal{L}'} S_2 \twoheadrightarrow_{\gamma} \ldots \twoheadrightarrow_{\gamma} T_{n-1} \twoheadrightarrow_{\mathcal{L}'} S_n \twoheadrightarrow_{\gamma} T_n = \{1\} \ .$$

$\#_{\gamma-\mathcal{L}'}(S)$ is well defined since Proposition (3.22) can be applied to $S \twoheadrightarrow \{1\}$.

Clearly if θ satisfies Axioms II and III then $\theta(S) \leq \#_{\gamma-\mathcal{L}'}(S)$ for all S. But $\#_G$ satisfies Axioms II and III and thus we have

Theorem 3.24. $\#_G(S) \leq \#_{\gamma-\mathcal{L}'}(S)$ *for all* S.

Let $\theta_j : \mathcal{S}_{Fin} \to \mathbb{N}$ for $j = 1, 2$. We write $\theta_1 \leq \theta_2$ iff $\theta_1(S) \leq \theta_2(S)$ for all $S \in \mathcal{S}_{Fin}$. Now we introduce the final Axiom.

3.25. Axiom IV. θ *satisfies Axioms I, II and III, and if* θ_1 *also satisfies Axioms I, II and III, then* $\theta_1 \leq \theta$.

Then we have

Theorem 3.26. *There exists one and only one function satisfying Axioms I, II, III and IV. It is* $\#_G$ *and*

$$(3.27) \qquad \#_G(S) = \min\{\#_{\gamma-\mathcal{L}'}(T) : T \twoheadrightarrow S\}$$

For a proof, see [Axioms].

We have now presented the axioms for complexity in their entirety. Before leaving them we make the following comments. Axiom I states if $S \mid T$, then the complexity of T is not less than that of S. Also the complexity of $S \leq\leq S_1 \times \cdots \times S_n$ is the maximum of the complexity of the S_j's. This is as it should be for a complexity function, since intuitively complexity does not rise for a divisor nor become greater under direct or subdirect product. Direct products are bigger, costlier, etc. than the component pieces, but not more complex than the component pieces. Theorem (3.18) shows us that we must impose additional restraints on a complexity function since if two semigroups have only the same complexity we cannot expect their structure to be very similar. (Consider two mathematical theories or two computers of about the same complexity. How similar is their structure?) Proposition (3.22) gives us a method to impose reasonable additional restraints.

Axiom II states that complexity does not change under a surmorphism which is one-to-one on each subgroup. Also, since for groups such maps are isomorphisms and we are generalizing group theory, this Axiom seems natural and pleasing.

Now if G is a group $G \xrightarrow[\mathcal{L}']{} \{1\}$. Also if we developed some of the calculus of homomorphism of semigroups (see [M-L-S], Chapter 8) we could show that $S \xrightarrow[\mathcal{L}']{} T$ involves "punching out groups of S but leaving the global structure intact up to γ-maps". Thus since the complexity of non-trivial groups is one, we might expect that "punching out groups" lowers complexity at most one. This yields Axiom III.

Finally, Axiom IV says complexity is the 'most complex' function satisfying Axioms I, II and III. Now the function $\#_{\gamma-\mathcal{L}'}$ does not even satisfy Axiom I since there exist elementary examples such that $S \mid T$ but $\#_{\gamma-\mathcal{L}'}(T) < \#_{\gamma-\mathcal{L}'}(S)$. However, (3.27) shows if we "smooth" $\#_{\gamma-\mathcal{L}'}$ to θ so that it satisfies $S \mid T$ implies $\theta(S) \leq \theta(T)$ (which is just what the right-hand side of (3.27) does) then the smoothed $\#_{\gamma-\mathcal{L}'}$ becomes complexity. (Compare Definitions (3.13) and (3.23). Is equation (3.27) surprising?)

Axioms I, II and III are natural. The surprising thing is that some non-zero function satisfies them. (If the inequality of Axiom III is replaced by equality *no* function satisfies the old Axioms I and II and the changed Axiom III!)

We know complexity is important for semigroups by its strong and natural analogy with the important functions of group theory. This manifests itself in the following manner: even though of course the complexity of a semigroup will not determine its structure we will find that by developing the theory of complexity of finite semigroups, the *structure theory* of finite semigroups will most rapidly be developed (even for those caring nothing at all for complexity itself). This has already been shown to be the case. For example, Proposition (3.22) arose in pursuing the theory of complexity. There are numerous other examples. See [Synthesis]. The reason for this is that pursuing complexity asks difficult questions but questions which can be answered at our present state of knowledge. Also, the techniques which must be developed to further complexity will be of *general interest and applicability* to the structure of finite semigroups (for example [Synthesis]). Complexity has this 'magic' property since it generalizes group theory and the group theorists by long experience know what questions can be answered and what questions cannot be answered at the level of present techniques. (Solvability of groups of odd order is proved, but very little about p-groups is known. In general, questions about basic pieces, e.g. SNAGs, and groups built from given basic pieces with *no* restrictions on the extensions can be solved, or at least great progress has been made; on the other hand, questions about putting together even elementary pieces, e.g. $\mathbb{Z}_p$, with *restrictions* on the extensions *cannot* be solved, e.g. p-groups. Notice complexity asks only the first type of questions for semigroups and not the second type.) The theory of finite semigroups was stifled for years because it had no questions, no program. Since 1962 it has had both and is growing.

In another important technical sense complexity does not depend on 'extensions'. In fact, Axiom II (3.19), can be *equivalently* formulated as follows: If S is a semigroup with combinatorial ideal I, then $\#_G(S) = \#_G(S/I)$. Here I is an *ideal* iff $i \in I$, $s \in S$ implies $is, si \in I$. Clearly $I \leq S$. Further, $S/I = (S - I) \cup \{0\}$ with the multiplication defined so that $\eta : S \twoheadrightarrow S/I$ with $\eta(s) = s$ for $s \in S - I$ and $\eta(i) = 0$ for $i \in I$ is a surmorphism. (For a proof of the equivalence, see [M-L-S], Chapter 9, 9.34.) Thus the complexity does not change when a combinatorial semigroup I is extended (by an ideal extension) by a semigroup T. Thus complexity is defined 'as nearly as possible on the Grothendieck ring' where the 'Grothendieck ring identifies any two extensions of A by B' in the free Abelian group generated by all the (isomorphism classes of) objects.

Also we will see that complexity will have many applications to biology, physics, etc. See the later chapters of this book. We now turn to mathematics and list some corollaries resulting from the previous results.

Corollary 3.28 (Continuity of Complexity with Respect to Homomorphisms). *Let $\theta : S \twoheadrightarrow T$ be a surmorphism and let $\#_G(S) = n$ and $\#_G(T) = k$. Then there exist surmorphisms $S = S_n \twoheadrightarrow S_{n-1} \twoheadrightarrow \cdots \twoheadrightarrow S_k = T$ so that the composite surmorphism is θ, and $\#_G(S_j) = j$ for $j = k, \ldots, n$.*

Proof. For the proof use Axiom II (3.19), Axiom III (3.21), and Proposition (3.22). $\qquad\Box$

As a companion assertion to Corollary (3.28) we list the following *false conjecture* whose counterexample was found by Rhodes. In the following $T \leq S$ denotes that T is a subsemigroup of S.

False Conjecture 3.29 (Continuity of Complexity with Respect to Subsemigroups). *Let $T \leq S$ with $k = \#_G(T) \leq \#_G(S) = n$. Then there exists subsemigroups $T = S_k \leq S_{k-1} \leq \cdots \leq S_n = S$, so that $\#_G(S_j) = j$ for $j = k, \ldots, n$.*

An equivalent reformulation of (3.29) is the following. Say M is a *maximal subsemigroup* of S iff $M \leq S, M \neq S$ and $M \leq T \leq S$ implies $T = M$

or $T = S$.

False Conjecture 3.29. *Let M be a maximal subsemigroup of S. Then*
$$\#_G(M) \leq \#_G(S) \leq \#_G(M) + 1 .$$

However, the following appears to be true.

Theorem 3.29a. *(1) Let M be a maximal subsemigroup of S. Then*
$$\#_G(M) \leq \#_G(S) \leq 2(\#_G(M)) + 1 .$$
(2) For each positive integer n there exists T_n with maximal subsemigroup M_n so that
$$\#_G(M_n) = n \quad and \quad \#_G(T_n) = 2(\#_G(M_n)) + 1 .$$
In short, the bounds of (1) are sharp.

The proof of (1) follows easily from the classification of maximal sub-semigroups in [M-L-S], Chapter 7.3, and the Generalized Embedding Theorem (GET) in [M-L-S], Chapter 6.3, by taking V to be the ideal of J-classes below or equal to the J-class not in M and $T = M$.

There is an example of $M_1 < T_1$, i.e. $\#_G(M_1) = 1$, $\#_G(T_1) = 3$, M_1 a maximal subsemigroup of T_1. (See [KerSys] or [PL].) Using the ideas of [Synthesis], $T_1, T_2, T_3, \ldots, T_n, \ldots$ can be constructed. However not all the details have been rigorously checked.

See (6.92) for applications of (3.29a) to evolution.*

In a vein similar to (3.29) we have the following proposition. We recall I is an *ideal* of S iff for each $i \in I$ and $s \in S$, both $is \in I$ and $si \in I$. Note ideals are subsemigroups.

Proposition 3.30. *Let V be an ideal of S, with $\#_G(S) = n \geq k = \#_G(V)$. Then there exist ideals $V_n, \ldots, V_k$ of S with*
$$V = V_k \subseteq V_{k+1} \subseteq \cdots \subseteq V_n = S$$
and $\#_G(V_j) = j$ for $j = k, \ldots, n$.

For a not very elegant elementary proof, see [M-L-S], Theorem 6.3.7(b). There exist short elegant elementary proofs of (3.30). Also (3.30) is valid if 'ideal' is replaced by 'left ideal' or by 'right ideal', see, e.g. [PL]. Closely related to (3.30) is the following proposition. Let (X, S) be a given right mapping semigroup. Let $|Y|$ denote the cardinality of the set Y. Let $spectrum(X, S) = \{k > 1 : \text{there exists } s \in S \text{ such that } |X \cdot s| = |\{x \cdot s : x \in X\}| = k\}$.

---

*See also [BioCpx] and [AL].   -CLN

Proposition 3.31. *Let* (X, S) *be given with* $|X| = n$. *Then*

$$\#_G(S) \leq |spectrum(X, S)| \leq n - 1.$$

For a proof, use Proposition (3.30). For details see [M-L-S], Chapter 6.3.

Propositions (3.30) and (3.31) are elementary and do not depend on Axiom II (3.19). The following requires Axiom II together with the matrix representation theory of finite semigroups.

Corollary 3.32. *Let the finite semigroup* S *be given. Then there exists a finite-dimensional complex irreducible matrix representation* ρ *of* S *such that*

$$\#_G(S) = \#_G(\rho(S)) .$$

Proof. By [Character] or [CC] the direct sum of all the irreducible representations of S is a γ-surmorphism. Then the result follows from Axiom I (3.17) and Axiom II (3.19). $\qquad\square$

The previous results on complexity can be reworked for p-length. For brevity we will not enter into the details here. See [Survey], section 5 and [p-length]. However we will mention that Axioms I and II are valid for p-length as well as a modified Axiom III (see the above references). Hall and Higman's Theorem (2.0) gives an upper bound for the p-length of finite p-solvable groups and Theorem (3.27) gives an upper bound for the complexity of a finite semigroup. We can combine these results to give an upper bound to the p-length of a finite p-solvable semigroup, thus relating our generalization of finite group theory directly to group theory. First a definition,

Definition 3.33. $HH_p(S) = \max\{\ell_p(G) : G$ is a subgroup of $S\}$. If G is a group, the *exponent* of G is the smallest positive integer n such that $g^n = 1$ for all $g \in G$. If G is a group, then $p^{e_p(G)}$ is the exponent of a Sylow p-subgroup of G. By definition, $e_p(S) = \max\{e_p(G) : G$ is a subgroup of $S\}$.

Notice Theorem (2.0) restated is $HH_p(G) \leq 3e_p(G)$. Then the promised upper bound is the following.

Theorem 3.34. *Let* S *be a* p-solvable semigroup. *Then*

$$\ell_p(S) \leq \#_{\gamma-\mathcal{L}'}(S) \cdot HH_p(S) \leq \#_{\gamma-\mathcal{L}'}(S) \cdot 3e_p(S) .$$

Proof. For a proof, use the definition of $\#_{\gamma-\mathcal{L}'}$ (Definition 3.23), Axiom II for ℓ_p and modified Axiom III for ℓ_p (see [Survey], Theorem (5.8) and Proposition (5.14)) and Theorem (2.0). $\qquad\square$

From (3.27) we can obtain the following *lower bounds* to complexity.

Corollary 3.35 (Lower Bounds for Complexity). *Let $\theta : S_{Fin} \to \mathbb{N}$ satisfy*

(3.35a) $$S \twoheadrightarrow T \quad implies \quad \theta(T) \leq \theta(S)$$

and

(3.35b) $$\theta(S) \leq \#_{\gamma-\mathcal{L}'}(S) \quad for\ all\ S.$$

Then

$$\theta(S) \leq \#_G(S) \ for\ all\ S.$$

Proof. $T \twoheadrightarrow S$ implies $\theta(S) \leq \theta(T) \leq \#_{\gamma-\mathcal{L}'}(T)$, so $\theta(S) \leq \min\{\#_{\gamma-\mathcal{L}'}(T) : T \twoheadrightarrow S\} = \#_G(S)$ by (3.27). $\qquad\square$

In the papers [LB I],[LB II] two functions $\#_I$ and $\#_S$, respectively, are defined which satisfy (3.35a) and (3.35b). Thus $\#_I(S) \leq \#_G(S)$ and $\#_S(S) \leq \#_G(S)$ for all S by Corollary (3.35). (Actually $\#_I \leq \#_S \leq \#_G$.) Both $\#_I(S)$ and $\#_S(S)$ are defined in terms of the *subsemigroup structure* of S.

To a large extent what we know about complexity is summarized by

$$\#_S(S) \leq \#_G(S) \leq \#_{\gamma-\mathcal{L}'}(S),$$

but things are progressing rapidly.

This completes our generalization of finite group theory to finite semigroups. We note in closing the great importance in both finite group and semigroup theory of equations of the form

$$(X,S) \,|\, (X_n, S_n) \wr \cdots \wr (X_1, S_1)$$

where S is a semigroup (group) and the S_j's belong to a restricted class of semigroups (groups).

Guide to the Literature on Complexity. The best place to start is with the surveys: [Survey], [Nice], [TilsonXII], and the book [M-L-S] especially Chapters 1 and 5–9, and [p-length]. Then [LB I], [LB II], [Mapping], [Synthesis], [Bull], [AxII Weak], [AxII Strong], [Axioms], [Stiffler], and [2-J]. Then follow the references in these papers, and especially see Chapter 4 of [q-theory], the most complete book to date, for recent developments.

Bibliography

[2-J] B. R. Tilson. Complexity of two J-class semigroups. *Advances in Mathematics*, 11:215–237, 1973.

[AL] C. L. Nehaniv and J. Rhodes. The evolution and understanding of hierarchical complexity in biology from an algebraic perspective. *Artificial Life*, 6(1):45–67, 2000.

[AxII Strong] J. Rhodes. A proof of the fundamental lemma of complexity [strong version] for arbitrary finite semigroups. *Journal of Combinatorial Theory, Series A*, 16(2):209–214, 1974.

[AxII Weak] J. Rhodes. A proof of the fundamental lemma of complexity [weak version] for arbitrary finite semigroups. *Journal of Combinatorial Theory, Series A*, 10:22–73, 1971.

[Axioms] J. Rhodes. Axioms for complexity for all finite semigroups. *Advances in Mathematics*, 11:210–214, 1973.

[BioCpx] C. L. Nehaniv and J. Rhodes. On the manner in which biological complexity may grow. In *Mathematical and Computational Biology*. American Mathematical Society, Providence, RI, 1999. Lectures on Mathematics in the Life Sciences **26**, pp. 93–102.

[Bull] J. Rhodes. The fundamental lemma of complexity for arbitrary finite semigroups. *Bulletin of the American Mathematical Society*, 74:1104–1109, 1968.

[CC] J. Rhodes. Characters and complexity of finite semigroups. *Jour. Combinatorial Theory*, 6:67–85, 1969.

[Character] J. Rhodes and Y. Zalcstein. Elementary representation and character theory of finite semigroups and its applications. In John Rhodes, editor, *Monoids and Semigroups with Applications (Proceedings of the Berkeley Workshop in Monoids, Berkeley, 31 July - 5 August 1989)*, pages 334–367. World Scientific Press, Singapore, 1991.

[Cohn] P. M. Cohn. *Universal Algebra*. (first published 1965 by Harper & Row) revised edition, D. Reidel Publishing Company, Dordrecht, 1980.

[Gins] A. Ginsburg. *Algebraic Theory of Automata*. Academic Press, 1968.

[Glue] C. L. Nehaniv. Monoids and groups acting on trees: characterizations, gluing, and applications of the depth preserving actions. *International Journal of Algebra & Computation*, 5(2):137–172, 1995.

[Hall-Higman] P. Hall and G. Higman. The p-length of a p-soluble group, and reduction theorems for Burnside's problem. *Proceedings of the London Mathematical Society*, 3(7):1–42, 1956.

[KerSys] J. Rhodes. Kernel systems – a global study of homomorphisms. *Journal of Algebra*, 49:1–45, 1977.

[LB I] J. Rhodes and B. R. Tilson. Lower bounds for complexity of finite semigroups. *Journal of Pure & Applied Algebra*, 1(1):79–95, 1971.

[LB II] J. Rhodes and B. R. Tilson. Improved lower bounds for complexity of finite semigroups. *Journal of Pure & Applied Algebra*, 2:13–71, 1972.

[M-L-S] M. A. Arbib, editor. *Algebraic Theory of Machines, Languages and Semigroups*. Academic Press, 1968.

[Mapping] J. Rhodes. Mapping properties of wreath products of finite semigroups with some applications. manuscript.

[Nice] J. Rhodes. Algebraic theory of finite semigroups. *Séminaire Dubreil-Pisot: Algébre et Théorie des Nombres*, 23(Fascicule 2: Demi-groupes [1970. Nice], Exp. No. DG 10. (9 pp.)), 1969/70. http://www.numdam.org/numdam-bin/browse?id=SD_1969-1970__23_2.

[p-length] B. Tilson. *p*-length of *p*-solvable semigroups. In K. W. Folley, editor, *Semigroups: Proceedings of a Symposium on Semigroups Held at Wayne State University, Detroit, Michigan, June 27-29 1968*, New York, 1969. Academic Press.

[PL] B. Austin, K. Henckell, C. L. Nehaniv, and J. Rhodes. Subsemigroups and Complexity via the Presentation Lemma. *Journal of Pure & Applied Algebra*, 101(3):245–289, 1995.

[q-theory] John Rhodes and Benjamin Steinberg. *The q-theory of Finite Semigroups*. Springer Monographs in Mathematics, Springer Verlag, 2009. [The most complete treatment of group complexity to date].

[Results] J. Rhodes. Some results on finite semigroups. *Journal of Algebra*, 4:190–195, 1966.

[Stiffler] P. Stiffler, Jr. Extension of the fundamental theorem of finite semigroups. *Advances in Mathematics*, 11:159–209, 1973.

[Survey] J. Rhodes. Algebraic theory of finite semigroups, structure numbers and structure theorems for finite semigroups. In K. W. Folley, editor, *Semigroups: Proceedings of a Symposium on Semigroups Held at Wayne State University, Detroit, Michigan, June 27-29 1968*, pages 125–162, New York, 1969. Academic Press. (with an appendix by B. R. Tilson).

[Synthesis] J. Rhodes and D. Allen, Jr. Synthesis of the classical and modern theory of finite semigroups. *Advances in Mathematics*, 11:238–266, 1973.

[TilsonXII] B. Tilson. Complexity of semigroups and morphisms. In S. Eilenberg, editor, *Automata, Languages, and Machines*, volume B, chapter XII, pages 313–384. Academic Press, New York, NY, 1976.

[Trans] K. Krohn and J. Rhodes. Algebraic theory of machines. *Transactions of the American Mathematical Society*, 116:450–464, 1965.

Chapter 4

A Reformulation of Physics

In the following we introduce two principles which will allow us to reformulate physics. The first is well known and accepted (especially since Gibbs) while the second is new.

Experiments, Phase Spaces and Operators.
Let E be an experiment. Then Principle I states that there exists a phase space X and inputs A and an action $\lambda : X \times A \to X$ where for $x_0 \in X$, $\lambda(x_0, a) = x_1$ is the result of applying the input $a \in A$ to the experimental system E initially in the state x_0.

A standard example from classical physics is given by a system E of N bodies with the phase space $X = \mathbb{R}^{6N}$, which gives the position and momentum of each body, and $A = \mathbb{R}^1$ is time and $\lambda(x_0, t) = x_1$ gives the propagation of the system in time t from x_0 to x_1 via the appropriate classical differential equations (e.g. considering only gravitational forces between the N bodies, Newton's $F = m\ddot{x}$ and Newton's $F = \dfrac{m_1 m_2}{r^2}$ suffice to define the N-body problem of celestial mechanics).* Of course we may replace the cartesian coordinates by polar coordinates when this is desirable or by 'generalized coordinates' as in the formulations of mechanics associated with the names of Lagrange and Hamilton. These new coordinates necessitate a change in variable from the cartesian coordinates. However, in general the dynamical states of a system with f degrees of freedom (N bodies have $3N$ degrees of freedom) require a subset of $\mathbb{R}^{2f}$ for their phase space.

---

*According to this viewpoint, time appears, not only to index states at different instances, but also as a group of signed durations or displacements acting on states of the phase space. So if t is a duration, we may write $x_t = \lambda(x_0, t)$, and more generally $x_{t+t'} = \lambda(x_t, t')$ for a signed temporal displacement t'. See also the author's prologue for discussion of groups and semigroups as models of time . -CLN

The presently accepted foundations of statistical thermodynamics and the foundations of ergodic theory (begun by Gibbs and precisely formulated by modern mathematics) and even quantum mechanics amply demonstrate the exceedingly wide application of Principle I in physics. In thermodynamics and ergodic theory the inputs are time and the phase spaces are derived from the dynamical phase space mentioned previously. In quantum mechanics the phase space is a Hilbert space of wave functions defined over the classical dynamical coordinates. The wave function changes with time, according to Schrödinger's equation when the operator Hamiltonian is formulated in accordance with the physical situation.

In the previous examples all the phase spaces were subsets of Euclidean or Hilbert space, i.e. infinite collections of vectors over the real or complex numbers. The inputs were time. However, in biology we can find natural experiments which possess *finite* phase spaces. Following [K-L-R] we give one such example.

We consider intermediary *metabolism* in normal cells. We take the state space $X=$ the substrates (molecules usually of low molecular weight relative to the enzymes). We let the inputs $A=$ the coenzymes (or prosthetic groups). *The action λ is given by $\lambda(x_1, a) = x_2$ iff the reaction $x_1 \xrightarrow{a} x_2$ occurs* assuming all enzymes and inorganic ions are present in sufficient quantity. *If no reaction occurs then $\lambda(x_1, a) = x_1$.* Time does not occur explicitly in the model since we let the reaction $x_1 \xrightarrow{a} x_2$ go to completion.

We give a somewhat more detailed discussion by quoting from [K-L-R].

"We begin by stating our primary interest to be the investigation of the organization of normal cellular metabolism with an emphasis on the biochemical aspects. Thus in our model we will not attempt to describe such phenomena as membrane transport or diffusion, although such physical aspects are important, perhaps crucial, to the maintenance of many of the metabolic pathways which will be studied. Another aspect of our model is that it is not kinetic, i.e., it will not be concerned with rates or, strictly speaking, with equilibrium. The model will be built on the basis of biochemical reactions that are known to occur, but whose rates need not be known, especially when they occur as part of a complex system of reactions. Simplifications are usually necessary, sometimes even desirable, in any analysis, however distortions (and falsifications) are to be avoided as much as possible. Accordingly, the model has been conceived as being embeddable in a real — *in vitro* — situation.

Metabolism is viewed as a collection of biochemical reactions, and a metabolic pathway is a connected series of these. Since in living things there is very nearly a one-to-one correspondence between biochemical reactions and the specific enzymes that catalyze them, we will consider the terms metabolism, metabolic process, and multi-enzyme system to be synonymous, as we have done tacitly heretofore. We characterize a multi-enzyme system M as $M(E, S, I)$, where

E is a set of *enzymes* (high molecular weight catalytic proteins) denoted E_1, E_2, etc.;

S is the set of *substrates* or metabolites (molecules usually of low molecular weight relative to the enzymes) produced by reactions catalyzed by elements of E and denoted by $s_1, s_2, \ldots$; and

I is a set of *inorganic ions* or "minerals" (e.g., inorganic phosphate, ammonium, metallic activator cations, etc.) required for reactions to be catalyzed by elements of E.

Now we wish to define a "metabolic state-space" Q_M corresponding to $M = M(E, S, I)$:

$$Q_M = \{ q_i = E \cup I \cup \{s_i\} \mid s_i \in S \}, \qquad M = M(E, S, I).$$

Heuristically, we can regard each state $q_i \in Q_M$ as a suspension of all the enzymes of E in a solution with the inorganic ions of I to which some fixed amount has been added, say one mole (or one molecule), of substrate s_i. Elements of $E \cup I$ should be assumed to be in such high concentrations that they limit no reaction catalyzed by the system. Thus we can assume that any reaction catalyzed by elements of E for which the substrates are present goes to completion. We also assume that any state is an *open system* with respect to all substances produced by, or introduced into it except for elements of S. This is plausible since members of S may be thought of as being bound to specific members of E.[*]

The state space which we use is derived from Q_M by identifying with q_n all subsets $P_{q_n} = \{ q_i, q_j, \ldots, q_n \}$ of Q_M for which reaction sequences

$$s_i \to s_j \to \cdots \to s_n$$

exist that require only the elements of E or I (i.e., that require no cofactors).[†] Intuitively this makes sense since one mole of any element of P_{q_n} in

---

[*]More generally, with minor modifications one can similarly replace Q_M with all subsets of $X = E \cup I \cup S$, or include information about discretely graded concentration levels in addition to presence vs. absence of elements of X. -CLN

[†]Note: if the reactions involved require no cofactors, then this entails identification of the states which form a particular cycle into a single 'higher level state' (e.g. when $s_i = s_n$) including any states in transients to the cycle (e.g. when $s_k = s_n$ with $i < k \leq n$). In particular a derived state need not be an unchanging static equilibrium but could well contain a dynamical cycle. -CLN

the hypothetical solution yields state q_n (since time is not an explicit part of our model and all reactions go to completion). We will denote Q_M thus partitioned as Q_m and denote its elements by numbers. In general, Q_m will have fewer states than either substrates or enzymes, but not many fewer since most biochemical reactions do require some organic cofactor.

We can now study the activities of a multi-enzyme system $M(E, S, I)$ as experiments on its state space Q_m according to Principle I. To do this we need to define a set of basic inputs. In our present model this is A_m, where A_m is the set of all cofactors (usually coenzymes or prosthetic groups) required for reactions to be catalyzed by elements of E, and not included in I.

The stipulation that a true prosthetic group which is tightly bound to its enzyme, such as FAD, could be introduced by itself into a system, or that its reduced counterpart $FADH_2$ would then diffuse out in a reasonable period of time is, admittedly, quite artificial. There are technical ways around this, such as to let the entire flavoprotein, i.e. the FAD + apoenzyme complex, be the input. This is not necessarily more realistic in the event that the apoenzyme itself is part of an enzyme complex with other enzymes of the system. The safest route is perhaps to consider models in which the system is not entirely open. This viewpoint is not at all repugnant to our theory and has been considered by us. However, it induces complications not germane to the basic outline of the approach which we are attempting to present here."

See Chapter 6, Part I for a detailed example and further discussion.

The examples of the application of Principle I can be continued *ad infinitum*.

Often one has a dynamical system (e.g. classical physics or a biological cell) which under the natural input, time, arrives at some steady state. Then some external input a (e.g. external force or carbon source delivered by the bloodstream) or an external input b (e.g. another particle crashing into the system or a dilute solution of amino acids) disturbs the steady state of the system. This new disturbed state under the natural input, time, eventually arrives at a new steady state. Thus if a string of external inputs $a_1, a_2, \ldots, a_n$ is applied to the system the steady states of the system are bounced from q_0 to q_1 under a_1, then to q_2 under a_2 until q_n is finally obtained under the last input a_n. Modeled in this way, the inputs are the set of external inputs $\{a, b, \ldots\}$ and time does not occur explicitly but as the free action of the system.

For reference we formally state

4.1. Principle I. *Each experiment E yields a phase space $E(X) \equiv X$ and input set $E(A) \equiv A$ and an action $E(\lambda) \equiv \lambda : X \times A \to X$. $\lambda(x, a)$ is the experimentally observed 'effect' of a perturbation or stimulus $a \in A$ upon the system initially in state x.*

An experiment E gives rise to the phase space, inputs and action $(E(X) \equiv X, E(A) \equiv A, E(\lambda) \equiv \lambda)$. Principle II deals with the question of what is a *theory* explaining the results of E.

Theories and Coordinate Systems on Experiments.
A theory is to allow one to understand a phenomenon and to gain insight. A process may be working in a feedback manner and one might have a precise description of the feedback (e.g. in control theory, electrical engineering and analogue computing, differential equations can be represented as feedback or recursive systems) but still not understand the phenomena, not even know the qualitative properties of the solutions. Undoubtedly nature and man perform many, many processes via feedback or recursion, but the precise description of the recursion relations does *not* constitute *understanding* the process. If it did, writing down the differential equation (being a precise statement of the recursive relations) would settle the physics, but it usually just begins the analysis. More or less the phase space $E(X) \equiv X$, inputs $E(A) \equiv A$ and action $E(\lambda) \equiv \lambda$, yielding (X, A, λ), gives the possible results of the experiment E in *recursive form*. For example, if at time $t = 0$ the system (X, A, λ) is in state $x_0 \in X$ and stimuli $a_1, a_2, \ldots, a_n$ are applied at times $t = 1, t = 2, \ldots, t = n$, then the result y_t at time $t \leq n$ is given recursively by

$$y_0 = x_0,$$
(4.2) $$y_t = \lambda(y_{t-1}, a_t)$$

so λ encompasses the recursive relations. Thus recursion (via λ) completely describes the results of the experiment E but in a manner that usually (in fact almost always) does not allow one to immediately *understand* the process. But what do we mean by "understanding the experiment"? Informally (very precise definitions later), *understanding an experiment E means introducing coordinates into phase space of E which are in triangular form under the action of the inputs of E.* Why? First, triangular coordinates allow one to approximate to the process by ignoring all but the first few coordinates. For example, solve a differential equation for the power series solution $f(z) = \sum_{n=0}^{\omega} a_n z^n$ and then consider $g(z) = a_0 + a_1 z^1 + a_2 z^2$ as an approximate solution. Nature may not 'do' a process in triangular form, but only by introducing triangular coordinates can one readily understand the process.

Conservation Principles. Generalizing from the methods of science since Bacon, Galileo and Newton, we find that understanding an experiment or phenomenon very often consists of introducing some parameters or quantities (which can be computed for the experiment with some accuracy) which, while not completely describing the existing situation, do possess the following two important properties. First, the parameters or quantities determine some obviously important aspects of the system. Second, if the values of these parameters or quantities of the system are known at time t (denoted $Q(t)$) and it is also known what inputs are presented to the system from time t to time $t + \varepsilon$ (denoted $I[t, t + \varepsilon]$), then the new parameters or quantities of the system at time $t + \varepsilon$ are *uniquely determined* from only $Q(t)$ and $I[t, t + \varepsilon]$.

In short, if you want to understand a system (as opposed to building or simulating a system) then introduce some parameters which may not tell you exactly what the state the system is, but such that if you know these parameters and then what the system is being hit with, from this alone you can figure out the new parameters for the system. Examples from physics are many, e.g. the quantity energy.

Thus this is the way systems are understood. But this is triangular coordinate form since the parameters introduced give a first coordinate and the second coordinates are chosen so that together with the first parameters the system is uniquely determined. Now the second coordinate can be further divided by the introduction of additional parameters in an inductive way, etc.

To make this clearer, consider the standard quantities in physics: energy, momentum, angular momentum, etc. Notice if $e(x)$ denotes the energy of the state x, then (in most models) $e(x \cdot a)$ (the energy of the state resulting after applying the stimuli a to initial state x) depends only on $e(x)$ and a, but *not* on x. Thus $x \mapsto (x', e(x))$ is a triangularization of (X, A). Here x' is chosen to make the mapping $x \mapsto (x', e(x))$ one-to-one. (In the next section we will give 'good minimal' ways of choosing x' when the phase space is finite.) In general, *conservation principles give a first coordinate of a triangularization*. In the main a large part of physics can be viewed as discovering and introducing quantities (functions) e of the states q of the system such that under appropriate stimuli a, $e(q \cdot a)$ depends only on $e(q)$ and a, and not on q.

Symmetries. We next show that the (group of) *symmetries* of an experiment *give a last coordinate of a triangularization.* The intuitive notion of a symmetry of an experiment is well known and extensively used in the physical sciences. For example the 'mirror-image' experiment will yield (but only usually after 1956 by Lee, Yang and Madame Wu) the 'mirror-image' result, or if one 'rotates the experiment' θ degrees and the output rotates θ degrees, then 'rotation of θ degrees' is a symmetry of the experiment.

We first heuristically indicate the construction involved in going from the group of symmetries to the triangularization, and then precisely write it out in all pedantic detail. First we break the phase space up into symmetry classes, where two states are in the same class iff some symmetry carries the one onto the other. Next choose exactly one member from each class. Then (assuming that each state can be reached from any other by applying the appropriate sequence of stimuli) one can easily show that every member of the phase space can be written as a unique symmetry applied to the chosen member of its class. Thus x in the phase space can be coordinated by letting the symmetry class of x be the first coordinate, and letting that unique symmetry Π_x, which carries the chosen member of the class onto x, be the second coordinate. The precision follows.

The precise definition of symmetry is easy to come by via Principle I, (4.1).

Definition 4.3 (of Symmetry). Let an experiment E yield (X, A, λ) in accordance with Principle I, (4.1). Then, by definition, Π is a *symmetry* of E iff

a. Π is a permutation of X, i.e. $\Pi : X \to X$ and Π is one-to-one and onto. (For convenience we let Π act to the left of X.)
b. Π commutes with the action of each $a \in A$, that is, writing $\lambda(x, a)$ as $x \cdot a$,

$$(\Pi(x)) \cdot a = \Pi(x \cdot a)$$

for all $x \in X,\ a \in A$.

Let $x_0 \in X$ and $a_1, \ldots, a_n \in A$. In accordance with (4.2) we define

(4.4)
$$x_0 \cdot a_1 \ldots a_n \equiv \lambda(x_0, a_1 \ldots a_n) = y_n$$
$$\text{with } y_0 = x_0 \quad \text{and} \quad y_t = \lambda(y_{t-1}, a_t) \,.$$

Thus $x_0 \cdot a_1 \ldots a_n = y_n$ is the resulting state after $a_1, a_2, \ldots, a_n$ have been applied to the system, in that order, which was initially in state x_0. We say E or (X, A, λ) is *transitive* iff given $x_0, x_1 \in X,\ x_0 \neq x_1$, there exists

$a_1, \ldots, a_n \in A$ such that $x_0 \cdot a_1 \ldots a_n = x_1$. Thus (X, A, λ) is transitive iff 'each state can be reached from any other state under some application of stimuli'. We have the following trivial

Fact 4.5. *(a) The symmetries of E, $\mathrm{Sym}(E)$, form a group (a subgroup of the group of all permutations of X).*

(b) Assume E is transitive. Then each symmetry Π is a regular permutation, i.e. $\Pi(x_0) = x_0$ for some $x_0 \in X$, implies $\Pi(x) = x$ for all $x \in X$.

(c) Let $x \in X$. Then by definition the orbit $\mathcal{O}_x$ of x under $\mathrm{Sym}(E)$ equals $\{\Pi(x) : \Pi \in \mathrm{Sym}(E)\} \subseteq X$. X is the disjoint union of the orbits. Let $x_1, x_2, x_3, \ldots$ be chosen so that $\mathcal{O}_{x_1} + \mathcal{O}_{x_2} + \cdots = X$ (where $+$ denotes disjoint union). Then, assuming E is transitive, $\Pi(x_j) = x'_j$ has one and only one solution $\Pi \in \mathrm{Sym}(E)$, given $x'_j \in \mathcal{O}_{x_j}$. Thus $\Pi \mapsto x'_j = \Pi(x_j)$ is a one-to-one correspondence of $\mathrm{Sym}(E)$ onto $\mathcal{O}_{x_j}$, so all the orbits are in one-to-one correspondence.

(d) For each $\mathcal{O}_{x_j}$ and $a \in A$ there exists a unique k so that $\mathcal{O}_{x_j} \cdot a \subseteq \mathcal{O}_{x_k}$.

Proof. (a) is trivial. For (b), if $x \neq x_0$ there exists $a_1, \ldots, a_k$ so that $x_0 \cdot a_1 \ldots a_k = x$ by the transitivity of E. But then $\Pi(x) = \Pi(x_0 \cdot a_1 \ldots a_k) = \Pi(x_0) \cdot a_1 \ldots a_k = x_0 \cdot a_1 \ldots a_k = x$. Finally, (c) follows from (b) since $\Pi_1(x_j) = x'_j$ and $\Pi_2(x_j) = x'_j$ implies $\Pi_1^{-1}\Pi_2(x_j) = x_j$ which implies $\Pi_1^{-1}\Pi_2$ is the identity by (b) so $\Pi_1 = \Pi_2$. To prove (d) choose k such that $x_j \cdot a \in \mathcal{O}_{x_k}$. Then $\Pi(x_j) \cdot a = \Pi(x_j \cdot a) \in \mathcal{O}_{x_k}$. This proves Fact (4.5). $\square$

Now using the above we show how $\mathrm{Sym}(E)$ gives rise to a triangularization. We assume E is transitive. For $x \in X$, let $\overline{x}$ be that unique member of the list $x_1, x_2, \ldots$ such that $\mathcal{O}_x = \mathcal{O}_{\overline{x}}$. For $x \in X$ we define $c(x) = (\Pi_x, \mathcal{O}_x = \mathcal{O}_{\overline{x}})$ where Π_x is that unique member of $\mathrm{Sym}(E)$ such that $\Pi_x(\overline{x}) = x$. If $\overline{x} = x_j$, we write $\mathcal{O}_x = \mathcal{O}_{\overline{x}}$ as $\mathcal{O}_j$. Then $c : X \to \mathrm{Sym}(E) \times \{\mathcal{O}_k : k = 1, 2, 3, \ldots\}$ and c is a one-to-one onto map by Fact (4.5). Thus c provides the desired triangularization. As the reader might have already grasped, we can relate the above construction to the *wreath product* construction of the previous Chapters 2 and 3 as follows.

Definition 4.6. Let E be an experiment giving rise to (X, A, λ) via Principle I, (4.1). Let $E(S) \equiv S$, read "the *semigroup of the experiment E*", by definition equals $\{\cdot a_1 \ldots a_n : n \geq 1 \text{ and } a_j \in A\}$ where $\cdot a_1 \ldots a_n$ denotes the mapping of X into X given by $x \mapsto x \cdot a_1 \ldots a_n$ with $x \cdot a_1 \ldots a_n$ defined by (4.4).

Thus (X, S) is a right mapping semigroup with A generating S and the action given by λ and (4.4). Using the above notation, we have proved

Fact 4.7. *Let E be a transitive experiment giving rise to (X, A, λ) via Principle I, (4.1), with semigroup $E(S) \equiv S$. Let $G = \text{Sym}(E)$. Then the correspondence c gives rise to*

$$(X, S) \leq (G, G) \wr (\{\mathcal{O}_k : k = 1, 2, \dots\}, T)$$

where $T = \{*a_1 \dots a_n : n \geq 1 \text{ and } a_j \in A\}$ and $\mathcal{O}_k * a_1 \dots a_n$ is defined as that unique $\mathcal{O}_m$ so that $\mathcal{O}_k \cdot a_1 \dots a_n = \{x'_k \cdot a_1 \dots a_n : x'_k \in \mathcal{O}_k\} \subseteq \mathcal{O}_m$, which is well defined by Fact (4.5)(d). T is a homomorphic image of S.

Also we have

$$(\Pi, \mathcal{O}_k) \cdot a = (\Pi \cdot f_a(\mathcal{O}_k), \mathcal{O}_{k*a})$$

where $\mathcal{O}_{k*a}$ is uniquely defined by $\mathcal{O}_k \cdot a \subseteq \mathcal{O}_{k*a}$ and $f_a(\mathcal{O}_k) = \Pi' \in G$ is uniquely defined by

$$\Pi'(\overline{(x_k \cdot a)}) = x_k \cdot a ,$$

so $\cdot a = \text{w}(f_a, *a)$ using the notation of (2.3).

Coordinate Representations. Now let ρ be a $d \times d$ matrix representation of $G = \text{Sym}(E)$ with coefficients in the field F. Then just as in (2.8) we can define a $md \times md$ matrix representation of S with coefficients in F (where m is the number of distinct orbits) by

$$\cdot a_1 \dots a_n \mapsto \rho^*(M._{a_1}) \dots \rho^*(M._{a_n})$$

where $M._{a_j}$ is the $m \times m$ row monomial matrix with coefficients in G associated with $\text{w}(f_{a_j}, *a_j)$ as defined after (2.5), i.e., $M._{a_j}(\mathcal{O}_p, \mathcal{O}_q) = f_{a_j}(\mathcal{O}_p)$ when $\mathcal{O}_p * a_j = \mathcal{O}_q$ and zero otherwise, and ρ^* is defined after (2.8), i.e., $\rho^*(M._{a_j})$ is given by replacing the $(p, q)^{\text{th}}$ entry of $M._{a_j}$ by ρ of the $(p, q)^{\text{th}}$ entry.[*]

Recapitulating, given a transitive experiment E, with symmetries $\text{Sym}(E) = G$, and an irreducible (say) representation ρ of G, we obtain a representation ρ^* of $S \equiv E(S)$ which is not necessarily irreducible. Now suppose S is a *permutation group* on X which corresponds to the process E being *reversible*. Then ρ^* has kernel $N_\rho \trianglelefteq S$ or $N_\rho = \{s \in S : \rho^*(s) = 1\}$. But then using $N_\rho \trianglelefteq S$ and (2.4) we can introduce the Lagrange group coordinates of Chapter 2 and obtain

(4.8a) $$(S, S) \leq (N_\rho, N_\rho) \wr (S/N_\rho, S/N_\rho) .$$

---

[*]In the case the $(p, q)^{\text{th}}$ entry is zero, then $\rho(0)$ is interpreted as the $d \times d$ zero matrix. -CLN

Further, $(S, S) \twoheadrightarrow (X, S)$ so

(4.8b) $$(X, S) \,|\, (N_\rho, N_\rho) \wr (S/N_\rho, S/N_\rho) \;.$$

Also if $x \in X$ and $S_x = \{s \in S : x \cdot s = x\}$ and $S_x \leq N_\rho$ the above can be strengthened to

(4.8c) $$(X, S) \leq (N_\rho/S_x, N_\rho) \wr (S/N_\rho, S/N_\rho) \;.$$

Thus each irreducible (say) representation of ρ of $G = \mathrm{Sym}\ (E)$ lends to a triangularization (4.8b) or (4.8c) of the reversible transitive experiment E.

We believe a modified and more sophisticated version of the functor

$$\rho \mapsto \text{(the triangularization (4.8b) or (4.8c))}$$

is applicable to particle physics in explaining the 'eight-fold way' of Murray Gell-Mann and Yuval Ne'eman. Via our formal Principle II (soon to be introduced) the triangularizations (4.8) should correspond to physical situations (i.e. particles). Also we feel that certain state spaces of particle physics may be essentially finite (not vector or Hilbert spaces) and the basic tools of classification could also be the finite simple groups (SNAGs) and not just the simple Lie groups and unitary representations. We have similar opinions regarding biology. We cannot pursue our ideas of particle physics here but we do pursue our ideas of biology in this book. See Chapter 6.

Before leaving the present topic we would like to add some remarks in partial answer to the question "Why does the (Frobenius) theory of the direct sum decomposition of the tensor product of irreducible representations of groups via character theory come up so often in quantum mechanics and particle physics? For example, in Gell-Mann's work. Why do the characters and constituents have physical meaning?" Let S be a finite semigroup. Suppose S has a faithful irreducible matrix representation ρ with character $\chi(\rho) = \chi$. Then by a theorem χ determines ρ and

(4.9) $$(X, S) \leq (G^0, G^0) \wr (Y, T)$$

for naturally defined sets X, Y and group G and homomorphic image T of S. (See [Character] Proposition (3.17).) In fact X is the unique 0-minimal ideal of S, G is the unique non-zero maximal subgroup of X, and Y is the principal left ideals of X.) *Thus the irreducible characters code irreducible matrix representations of the semigroup of the experiment of E which in turn give rise to triangularizations of E whose last coordinate is a group.* This indicates the importance of characters. Essentially a converse statement is also true. Regarding tensor products and physical interpretations we postpone further discussion until the end of this Chapter 4.

We now formally state

4.10. Principle II. *Let E be an experiment giving rise to a phase space $E(X) \equiv X$, inputs $E(A) \equiv A$, action $E(\lambda) = \lambda$ and semigroup $E(S) \equiv S$ via Principle I, (4.1), and (4.6). Then a physical theory for E is a solution of*

$$(4.10a) \qquad (X, S) \mid (X_n, S_n) \wr \cdots \wr (X_1, S_1)$$

Also, to have theoretical and practical consequences, the terms S_j should be chosen from a restricted class of semigroups and (4.10a) should be a 'minimal' solution. (See the discussion following and Definition (4.13).)

When the phase space X is finite then the Prime Decomposition Theorem (3.4) can be applied to obtain solutions of (4.10a) where the S_j's are simple groups dividing S or equal the trivial U_3.

When E is a reversible experiment, corresponding to S being a permutation group, then $\mid$ can be replaced by $\leq$ in (4.10a) and Lagrange group coordinates can be introduced via (2.4) and induction. In somewhat more detail, suppose E is reversible and transitive. Let $S_x = \{s \in S : x \cdot s = x\}$. Then as is well known, $(X, S) \cong (S/S_x, S)$. Further, let $S_x = H_n < H_{n-1} < \cdots < H_1 < H_0 = S$ be a subgroup chain of S ending at S_x. Then using (2.4) and induction we obtain

$$(X, S) \cong (S/S_x, S) \leq (H_{n-1}/H_n, H_{n-1}) \wr \cdots \wr (H_0/H_1, H_0) \ .$$

Further, by using the results of [Maurer] or [Glue] one can show that essentially *all* solutions of (4.10a) for transitive reversible experiments E arise in the above manner as Lagrange group coordinates of subgroup series.

The situation for the general (non-reversible) process is quite different. In classical physics, since most of the famous equations are time reversible, the non-reversible case was not too important and could profitably be avoided. However in biology (as we shall see) it cannot be avoided. Thus we must deal with semigroups and not just groups. The first change this necessitates is that $\leq$ is not enough in (4.10a) and must be replaced by $\mid$. In fact if $(X_n, F_R(X_n))$ denotes the semigroup of all functions of $X_n = \{1, \ldots, n\}$ into itself considered as acting to the right of X_n, then it can be proved (see [M-L-S], Fact 5.3.20) that

$$(X_n, F_R(X_n)) \leq (Y_2, S_2) \wr (Y_1, S_1)$$

implies $F_R(X_n) \leq S_2$ or $F_R(X_n) \leq S_1$. Thus (4.10a) has no non-trivial solutions for $(X_n, F_R(X_n))$ if $\mid$ is replaced by $\leq$. Also for any semigroup S there exists an n (order of S plus one will do) so that $S \leq F_R(X_n)$. Finally,

$$(X_n, F_R(X_n)) \mid (X_2, S_2^*) \wr \cdots \wr (X_n, S_n^*)$$

where (X_k, S_k^*) denotes the group of all *permutations* on X_k together with the constant maps. Also

$$(X_k, S_k^*) \mid (U_3, U_3) \wr \cdots \wr (U_3, U_3) \wr (X_k, S_k)$$

with (X_k, S_k) the full permutation group (the symmetric group) on X_k.

Thus we see $\mid$ is necessary in (4.10a). Now the interpretation of $\leq$ is immediate. But in what sense does (4.10a) give a triangularization of (X, S)? When unwinding Definitions (2.2), (2.3) and (3.1) we come to the following result. Given (X, S), a solution of (4.10a) is constructed by choosing sets $X_1, \ldots, X_n$ and forming a set of coordinates $X_n \times \cdots \times X_1$, then for each $x \in X$ a non-empty subset $Q_x \subseteq X_n \times \cdots \times X_1$ is chosen so that $x \neq y$ implies $Q_x \cap Q_y = \emptyset$. If $Q_x = \{(x_n^{(p)}, \ldots, x_1^{(p)}) : 1 \leq p \leq |Q_x|\}$, then each member $(x_n^{(p)}, \ldots, x_1^{(p)})$ of Q_x is to be thought of as 'coordinates' for x. Thus distinct members of X never have the same 'coordinates' and each member of X has some 'coordinates', but two distinct 'coordinates' might describe the same point of X, and some points of $X_n \times \cdots \times X_1$ may be the 'coordinates' of no point in X. Further it is required that for each $s \in S$ there is a defined a function $\cdot \hat{s} : X_n \times \cdots \times X_1 \to X_n \times \cdots \times X_1$ so that $\cdot \hat{s}$ is in triangular form and has j^{th} component action in S_j (see Definition (2.2)–(2.5) and below) and if $(x_j^{(p)}, \ldots, x_2^{(p)}, x_1^{(p)})$ are the first j coordinates of some 'coordinates' of x, then $(x_j^{(p)}, \ldots, x_2^{(p)}, x_1^{(p)}) \cdot \hat{s} = (y_j^{(p)}, \ldots, y_1^{(p)})$ are the first j coordinates of some 'coordinates' of $x \cdot s$.

Thus while $\leq$ means that 'coordinates' and points of phase space uniquely correspond, $\mid$ means that 'coordinates' uniquely describe points, but that points do not uniquely describe 'coordinates' but describe a set of 'coordinates'. *A little thought shows this change does not much alter the physical meaning of coordinates but generalizes it in accordance with use.*

As is well known, any $n \times n$ matrix over the complex numbers can be put into triangular form. By inspecting the definition of the wreath product (Definition (2.2)–(2.3)) which consists of maps in triangular form with component action S_j in the jth coordinate, it seems natural to *consider* S_j the finite analogue of *'eigenvalues'* in the Euclidean and Hilbert space case since in both cases the component action and the 'eigenvalues' are the contribution of the new j^{th} coordinate to the old j^{th} coordinate, i.e., both are the *diagonal* terms of the triangular action. Thus when the phase space and action is linear (e.g. normal operators on Hilbert space), the eigenvalues are multiplication by a complex number via the Spectral Theorem, but when the phase space is finite the 'eigenvalues' are (by Theorem (3.4)) the irreducible semigroups, i.e. U_3, $\mathbb{Z}_p$ or SNAGs. Thus while building blocks of linear physics are multiplication by a complex number (or multiplication by a measurable function on L_2 of something) the *building blocks of finite phase space state physics are the simple groups, i.e. the SNAGs* (= finite simple non-abelian groups).* (We discuss this biological program in more detail later in this section and carry out the program in Chapter 6, Part I. Also, insofar as models can be made for particle physics with *finite* phase spaces (not just finite-dimensional), then SNAGs are the proper means of classification (not just Lie groups and unitary representations)). Thus since much of cellular metabolism can be studied as finite phase space physics, the building blocks of biological metabolism are the finite simple groups and thus the *biological reactions should be classified via the classification schemes of advanced group theory for the finite simple groups.*

We might mention here that examples of triangular coordinates in mathematics are very plentiful. Matrices in triangular form have already been noted. Writing real numbers in decimal notation we find $+$ and $\cdot$ (i.e., $(\mathbb{R}^1, \mathbb{R}^1, +)$ and $(\mathbb{R}^1, \mathbb{R}^1, \cdot)$) in triangular form and the usual algorithms taught to the young show how they lie in the wreath product. (What is the component action?) If $\ldots, a_n, a_{n-1}, \ldots, a_1, a_0$ are taken as coordinates for the power series $\sum_{n=0}^{\omega} a_n z^n$ then the power series under $+$ and $\cdot$ (i.e. $(PS, PS, +)$ and $(PS, PS, \cdot)$) are in triangular form. Triangular coordinates play a large role in many fields of mathematics; to take but one example, consider Probability Theory (see [Kol]). In considering sums of an independent random variables, the distribution functions are first determined. Now the sum of the random variables is the convolution of the distribution functions. Thus by Fourier transforming, the transform of the distribution function of the sum of the random variables go as the products of the transforms of the individual distribution functions. Then triangular coordinates are obtained by taking the log of the transform and expanding it into a power series and taking the coefficients $a_0, a_1, a_2, \ldots$ which are the *moments* of the process. Thus the moments introduced by Chebyshev are introducing triangular coordinates into probability theory. The examples

---

*From the Krohn-Rhodes Prime Decomposition Theorem (3.4), all finite simple groups can occur among the building blocks, but the author is excluding all the $\mathbb{Z}_p$'s (p prime) in the proposed classification of physical and biological systems. Basically, $\mathbb{Z}_p$'s are excluded on the view of this classification since "with just a little noise they approach random behavior, while the SNAGs are like Shannon error correcting codes surrounding the $\mathbb{Z}_p$ for p dividing the order of the SNAG". See the discussion on "Computing with SNAGs" at the end of Appendix A to Chapter 5, and also Chapter 6, Part II for more on the reasons that the $\mathbb{Z}_p$'s are less fundamental for the classification of physical and biological systems than SNAGs. -CLN

of triangular coordinates in mathematics could be continued *ad infinitum*.

We have now looked at physical experiments with finite phase spaces in a new way via our two principles and have seen that understanding the experiment consists of introducing triangular coordinates. However, we also saw in the previous sections of this book that in finite group theory and in finite semigroup theory introducing triangular coordinates was also very important and lead naturally to the important idea of complexity. But how does complexity fit in with our theory of finite phase space physics?

Well, from our viewpoint, all triangularizations of the experiment E (meaning solutions of (4.10a)) are different ways to understand E or are different theories of E. Now the most important parameters of E are those aspects which clearly expose themselves in *every* theory of E. Now by Theorem (3.4), if $P \in \mathbf{PRIMES}(E(S))$, then $P \mid S_k$ for some S_k of (4.10a), $k = 1, \ldots, n$. Thus we have that the functor

$$E \mapsto E(S) \mapsto \mathbf{PRIMES}(E(S))$$

is very important because in *every theory* of E, $\mathbf{PRIMES}(E(S))$ *must* appear (as multiples among the $S_1, \ldots, S_n$ of (4.10a)). In short (ignoring the trivial U_3), in each solution (4.10a), $\mathbf{PRIMES}(E(S))$ form the 'eigenvalues' (up to multiples) and a solution exists with exactly $\mathbf{PRIMES}(E(S))$ as 'eigenvalues' (and U_3). Thus *all* the theories of E have the same building blocks ($\mathbf{PRIMES}(E(S))$ plus U_3) but the order can be very different (analogous to the Jordan-Hölder Theorem). In the application section of this book, Chapter 6, Part I, we classify metabolic reactions by the method of $E \mapsto \mathbf{PRIMES}(E(S))$ using advanced group theory to classify the SNAGs. This gives a Mendeleev type table for metabolic reactions.

We could also count the number of repetitions of each $P \in \mathbf{PRIMES}(E(S))$ leading to Jordan-Hölder-type theorems, as was previously discussed in Chapter 3.

Getting back to the question of the relations of complexity and our theory of finite phase space physics, we see that *complexity* (of $E(S)$) *is the minimal number of physical coordinates needed to describe the experiment E* (with 'eigenvalues' the component action groups). Now this interpretation of the complexity of the experiment is an 'external' one — the internal workings need not be considered. However by using the lower bounds for complexity mentioned at the end of Chapter 3, we can define a lower bound to the complexity as the *longest chain of (essential) internal dependencies* (precise definitions next).

4.11. Essential Dependencies. If (X, S) is given and G is a subgroup of S, then (X, G) may not be a group of permutations (for example, consider a constant map on n points giving $(X_n, \{1\})$). However, it is easy to prove that each member of G has the same range $X(G)$ on X, i.e., for all $g \in G$, $X \cdot g = \{x \cdot g : x \in X\} = X \cdot e = \{x : x \cdot e = x\} \equiv X(G)$, where $e^2 = e$ is the identity of G. Also $(X(G), G)$ is a faithful permutation group.* Thus $G \mapsto X(G)$ maps subgroups of S into subsets of X. The transitive components of $X(G)$ (in $(X(G), G)$) can be considered as 'pools of feedback' or 'reaction chains' or 'natural subsystems', etc., depending on the appropriate interpretation stemming from the experiment E and its phase space X.

Let G_1, G_2 be two subgroups of S, with $e_k^2 = e_k$ the identity of G_k, for $k = 1, 2$. We define $e_1 > e_2$ iff $e_1 \neq e_2$ and $e_1 e_2 = e_2 e_1 = e_2$. It is easy to show that $>$ is a transitive relation on the collection of idempotents of S. Also denote $X(G_k)$ by X_k. Then it is not difficult to show that $e_1 > e_2$ implies $X_1 \supsetneq X_2$. Now a chain of *internal dependencies* is given by $e_1^2 = e_1 > e_2^2 = e_2 > \cdots > e_n^2 = e_n$ and non-trivial subgroups G_k with identities e_k so that

$$X_1 \supsetneq X_2 \supsetneq \cdots \supsetneq X_n \neq \emptyset .$$

But we must also require that each step $X_k \supsetneq X_{k+1}$ be 'essential' so we precisely define this next.

First, $\mathrm{Per}(X_k)$, read "the *permutator* of X_k", equals by definition $\{\cdot s : X_k \to X_k : s \in S$ and $x_k \mapsto x_k \cdot s$ is a permutation on $X_k\}$. Thus if $X_k = X(G_k)$, then $G_k \leq \mathrm{Per}(X_k)$.

Given subgroups G_k, G_{k+1} of S with identities $e_k^2 = e_k$ and $e_{k+1}^2 = e_{k+1}$, respectively, and $X_k = X(G_k)$, $X_{k+1} = X(G_{k+1})$ and $e_k > e_{k+1}$ so $X_k \supsetneq X_{k+1} \neq \emptyset$, define $\mathcal{X}_{k+1} = \{X_{k+1} \cdot g_k : g_k \in G_k\}$. Thus $\mathcal{X}_{k+1}$ is all those subsets of X_k given by letting (X_k, G_k) act or 'knock about' $X_{k+1} \subseteq X_k$. Now let $\mathcal{I}_{k+1} = \{f^2 = f \in S : X \cdot f = X_k \cdot f \in \mathcal{X}_{k+1}\}$. Thus $\mathcal{I}_{k+1}$ is all those idempotents of S whose range is some translation of X_{k+1} under some member of G_k. Let $\langle \mathcal{I}_{k+1} \rangle$ be the subsemigroup of S generated by $\mathcal{I}_{k+1}$. Then by definition $X_k \supsetneq X_{k+1}$ is *essential* iff

$$\langle \mathcal{I}_{k+1} \rangle \cap \mathrm{Per}\,(X_{k+1}) \supseteq G_{k+1} \neq \{e_{k+1}\} .$$

Then we have

---

*To see that $(X(G), G)$ is faithful, suppose that $g, g' \in G$ have the same action on all of $X(G)$. Let $z \in X$, then $z \cdot g = z \cdot eg = (z \cdot e) \cdot g = (z \cdot e) \cdot g' = z \cdot g'$, since $z \cdot e \in X(G)$. Thus g and g' have the same action on all $z \in X$, not just on $X(G)$, whence $g = g'$ since (X, S) is faithful. -CLN

Theorem 4.11a. *Let k be the length of the longest chain of essential dependencies of the experiment E. Then the complexity of E (i.e. the complexity of $E(S)$) is greater than or equal to k.*

For a proof of (4.11a) we can use Corollary (3.35). Actually, k, as defined above, is $\leq \#_I(S)$ as defined in [LB I]. See also [LB II].

See Chapter 6, Part I, for a detailed interpretation of the internal essential dependencies for models of intermediary cellular metabolism.

The definition of 'essential internal dependency' can be strengthened by using $\#_S$ of [LB II] or subsequent papers of Tilson and Rhodes.

Summary and Research Program: Minimal Coordinate Systems.
It is possible to summarize Chapters 2, 3 and 4 thus far by asserting that the central problem of finite group theory, finite semigroup theory, and finite phase state physics is the following: given (X, S), *to find all 'minimal' solutions of*

$$(4.12) \qquad (X, S) \mid (X_n, S_n) \wr \cdots \wr (X_1, S_1).$$

We will give a precise definition of 'minimal' below but for group theory this will boil down to finding the finite simple non-abelian groups (SNAGs). For finite semigroup theory it is the program of determining the *structure theory* of finite semigroups, and for finite phase space physics it is the problem of *determining all the theories* for experiment E. We assert that this mathematical program will be just as important as "physical intuition" in bringing forth understanding.

To make this program precise we must define 'minimal solution' (following some suggestions of Dennis Allen, Jr.).

Definition 4.13. Let (X, S) be given. Suppose

$$(4.14) \quad (X, S) \overset{(\theta,\rho)}{\longleftarrow} (Y, T) \leq (X_n \times \cdots \times X_1, (X_n, S_n) \text{ w } \cdots \text{ w } (X_1, S_1))$$

is given. Then we say (4.14) is *minimal* iff

(4.15a)   Each S_j is a non-trivial group or a non-trivial combinatorial semigroup.

(4.15b)   If $Y' \subseteq Y$ and $T' \leq T$ and $(X, S) \overset{(\theta,\rho)}{\longleftarrow} (Y', T')$, then $Y' = Y$ and $T' = T$.

(4.15c)   If $T \leq (X_n, S'_n) \text{ w } \cdots \text{ w } (X_1, S'_1)$ where $S'_j \leq S_j$, then $S'_j = S_j$ for $j = 1, \ldots, n$.

(4.15d) Let Π be a permutation in $(X_n, S_n) \wr \cdots \wr (X_1, S_1)$. Then we can consider

$$(4.16) \quad (X, S) \overset{(\theta', \rho')}{\leftarrow} (Y^\Pi, T^\Pi) \le (X_n \times \cdots \times X_1, (X_n, S_n)\mathrm{w} \cdots \mathrm{w}(X_n, S_1)),$$

where $Y^\Pi = \{y\Pi : y \in Y\}$ and $T^\Pi = \{\Pi^{-1}t\Pi : t \in T\}$ (so T is isomorphic with T^Π), $\theta'(y\Pi) = \theta(y)$ and $\rho'(\Pi^{-1}t\Pi) = \rho(t)$. We say (4.16) is (4.14) *conjugated by* Π. We demand (4.14) conjugated by Π (i.e. (4.16)) to satisfy (4.15c) for all Π.

(4.15e) Given (4.14), let stabilizer$(\theta) = \{f \in (X_n, S_n)\mathrm{w} \cdots \mathrm{w}(X_1, S_1) :$

$Y \cdot f \subseteq Y$ and $f\theta = \theta\} \le (X_n, S_n)\mathrm{w} \ldots \mathrm{w}(X_1, S_1)$.

Let $f^2 = f \in$ stabilizer(θ). Then we can consider

$$(4.17) \quad (X, S) \overset{(\theta', \rho')}{\leftarrow} (Y \cdot f, R) \le (X_n \times \cdots \times X_1, (X_n, S_n)\mathrm{w} \cdots \mathrm{w}(X_1, S_1))$$

with $Y \cdot f = \{y \cdot f : y \in Y\}$, R is the subsemigroup of (X_n, S_n) w $\cdots$ w (X_1, S_1) generated by $\{ftf : t \in T\}$, $\theta'(y \cdot f) = \theta(y \cdot f) = \theta(y)$ and $\rho'(ft_1f \cdots ft_nf) = \rho(t_1 \cdots t_n)$. We say that (4.17) is (4.14) *conjugated by* $f^2 = f$. We demand that (4.14) conjugated by $f^2 = f$ satisfy (4.15a)–(4.15d) for all $f^2 = f \in$ stabilizer(θ).

We can outline the *major program in finite semigroup theory* as follows. First, given any solution of (4.14), not necessarily minimal, we can define a series of *reductions* analogous to (4.15a)–(4.15e). For example, if there exists $Y' \subsetneqq Y, T' \le T$ and

$$(X, S) \overset{(\theta, \rho)}{\leftarrow} (Y', T')$$

then we replace (4.14) by

$$(X, S) \overset{(\theta, \rho)}{\leftarrow} (Y', T') \le (X_n, S_n) \wr \cdots \wr (X_1, S_1),$$

giving the reduction which is analogous to (4.15b), etc.

---

*Here "$f\theta = \theta$" means $(y \cdot f)\theta = (y)\theta$ for all $y \in Y$. -CLN

Then given a solution (α), not necessarily minimal, of (4.14), we can perform all reductions on α leading to a sequence of minimal solutions $\alpha_1, \ldots, \alpha_k$, where we count repetitions in a natural way. For example, the solution $(X, S) \mid (X, S)$ leads to all minimal solutions occurring at least once*, while if α is a minimal solution itself, $k = 1$ and $\alpha_1 = \alpha$.

Thus the major problems are

(4.18a)   Given (X, S), find all minimal solutions of (4.14).

(4.18b)   Given solution α of (4.14), determine $\alpha_1, \ldots, \alpha_k$, the minimal solutions (with repetitions) obtained by reducing α.

This program has been carried out in detail for finite groups [Maurer]. The theory of complexity is the beginning of the program in the theory of finite semigroups, since complexity is the minimal length of any solution, leading to some information on the minimal solutions.

Notice with two theories or solutions of (4.14) with coordinates $\ldots x_k, x_{k-1}, \ldots, x_1$ and $\ldots y_m, y_{m-1}, \ldots, y_1$ it might take all the coordinates of the x-theory to give the first coordinate of the y-theory. In fact, if X is a mechanical theory of E, the 'Fourier-transform' probability moment theory of E might be

$$y_1 = \sum x_i$$
$$y_2 = \sum x_i^2$$
$$\vdots$$
$$y_k = \sum x_i^k \,, \text{ etc.}$$

Thus two theories have the same building blocks $(\mathbf{PRIMES}(E(S)) + U_3)$ but can otherwise be very different.

Thus given two distinct theories of E (i.e. two distinct minimal solutions of α and β of (4.14)), how can we compare them? A usual procedure in mathematics (for example with differentiable structures on the 7-sphere) is to consider the direct product $\alpha \times \beta$ restricted to the diagonal. Thus, if

$$(X, S) \leftarrow (Y, T) \leq (X_n, S_n) \wr \cdots \wr (X_1, S_1) = W$$

is α, and

$$(X, S) \leftarrow (Y', T') \leq (X_n', S_n') \wr \cdots \wr (X_1', S_1') = W'$$

---

*This holds since if S is not a group or combinatorial, then (4.15a) is not satisfied, so it is necessary to perform a wreath decomposition reduction, and this can be then iterated in every possible way to yield all minimal solutions. -CLN

is β, then

$$(X \times X, S) \leftarrow (Y \times Y', T \times T') \leq W \times W \mid (X_n \times X_n', S_n \times S_n')\wr \cdots \wr(X_1 \times X_1', S_1 \times S_1')$$

is denoted $a \otimes \beta$ with $(x, x') \cdot s = (x \cdot s, x' \cdot s)$, $(y, y') \cdot (t, t') = (y \cdot t, y' \cdot t')$, etc.*,†

But then reducing $\alpha \otimes \beta$ to the minimal solutions (with repetitions) $\alpha_1, \ldots, \alpha_k$ gives a *measure of the difference* of the theories α and β.

We end this section with the suggestion of an approach to the problems (4.18(a)) and (4.18(b)). Since

$$(X, C) \text{ w } (Y, T) \overset{\gamma}{\twoheadrightarrow} (Y, T)$$

for any combinatorial semigroup C and

$$(X, G) \text{ w } (Y, T) \overset{\mathcal{L}'}{\twoheadrightarrow} (Y, T)$$

for any group G, the minimal solutions of (4.18(a)) (for (X, S)) will be in one-to-one correspondence with diagrams of the form

(4.19)
$$T \twoheadrightarrow T_n \underset{\mathcal{L}'}{\twoheadrightarrow} T_{n-1} \underset{\gamma}{\twoheadrightarrow} \gamma(T_{n-1}) \underset{\mathcal{L}'}{\twoheadrightarrow} T_{n-2} \cdots \twoheadrightarrow \{1\}$$
$$\downarrow$$
$$S$$

where $\#_G(S) = n$.

Now by using the character theory for finite semigroups (see [Character]), we can assume the surmorphisms $T \twoheadrightarrow T_{n-1}$, $T \twoheadrightarrow T_{n-2}$ are given by completely reducible matrix representations of T. Then if α and β are two minimal solutions of (4.14) for (X, S) and α is coded by (4.19) and β is coded by (4.19) with primes added, then consider the representations

$$\rho : T \twoheadrightarrow T_n$$
$$\rho' : T' \twoheadrightarrow T_n' \ .$$

Then taking the *tensor product* of ρ and ρ' yields

$$T \times T' \xrightarrow{\rho \otimes \rho'} T_n \otimes T_n' \ ,$$
$$\downarrow$$
$$S$$

---

*The 'restriction to the diagonal' in the tensor product is for the action of $\cdot s$, but is *not* a restriction to the diagonal on the states. -CLN

†If α and β are decompositions with different lengths n and m, respectively, with, say, $n > m$, then we may intersperse trivial coordinates, e.g. take $(X_i', S_i') = (\{1\}, \{1\})$, the one-element semigroup acting on itself, for each i with $n \geq i > m$. -CLN

which will code $\alpha \otimes \beta$ so the solution of (4.18(b)) can be reduced to computing tensor products of completely reducible matrix representations. Thus a *tensor product is a means of comparing different theories of E*.

Bibliography

On Physics

[Fermi] Enrico Fermi. *Thermodynamics*. Dover, new edition, 1956.

[Feynman] P. Feynman, R. B. Leighton, and M. Sands. *The Feynman Lectures on Physics*, volume 1-3. Addison-Wesley, 1964, 1966. Revised and reprinted as part of *The Feynman Lectures on Physics: The Definitive and Extended Edition*, 2005. (Excellent survey of physics.).

[Schrödinger] Erwin Schrödinger. *Statistical Thermodynamics*. Cambridge University Press, 1967.

For biology see the references at the end of Chapter 6, Part II.

Mathematical References

[Character] J. Rhodes and Y. Zalcstein. Elementary representation and character theory of finite semigroups and its applications. In John Rhodes, editor, *Monoids and Semigroups with Applications (Proceedings of the Berkeley Workshop in Monoids, Berkeley, 31 July - 5 August 1989)*, pages 334–367. World Scientific Press, Singapore, 1991.

[Glue] C. L. Nehaniv. Monoids and groups acting on trees: characterizations, gluing, and applications of the depth preserving actions. *International Journal of Algebra & Computation*, 5(2):137–172, 1995.

[K-L-R] K. Krohn, R. Langer, and J. Rhodes. Algebraic principles for the analysis of a biochemical system. *Journal of Computer and Systems Sciences*, 1(2):119–136, 1967.

[Kol] A. N. Kolmogorov and B. V. Gnedenko. *Limit Distributions for Sums of Independent Random Variables*. Addison-Wesley, 1968.

[LB I] J. Rhodes and B. R. Tilson. Lower bounds for complexity of finite semigroups. *Journal of Pure & Applied Algebra*, 1(1):79–95, 1971.

[LB II] J. Rhodes and B. R. Tilson. Improved lower bounds for complexity of finite semigroups. *Journal of Pure & Applied Algebra*, 2:13–71, 1972.

[Maurer] W. D. Maurer. *On Minimal Decompositions of Group Machines*. PhD thesis, Department of Mathematics, University of California at Berkeley, January 1965.

Chapter 5

Automata Models and the Complexity of Finite State Machines

Part I. The Prime Decomposition Theorem

What is a (sequential) machine? It's a 'black-box' which maps input strings to output strings such that the output b_k at time k depends only on the inputs $a_1, \ldots, a_k$ up to time k. Precisely,

Definition 5.1. Let A be a non-empty set. Then $A^+ = \{(a_1, \ldots, a_n) : n \geq 1 \text{ and } a_j \in A\}$. A *(sequential) machine* is by definition a function $f : A^+ \to B$ where A is the *basic input set*, B is the *basic output set* and $f(a_1, \ldots, a_n) = b_n$ is the output at time n if a_j is the input at time j for $1 \leq j \leq n$.

For example, the machine with inputs 0 and 1 which adds modulo p is $f : \{0,1\}^+ \to \{0,1,\ldots,p-1\}$ with $f(a_1,\ldots,a_n) = a_1 + \cdots + a_n \pmod{p}$.

The best example of a (sequential) machine is an organism which must respond in real time with its environment just to stay alive. Thus $f(a_1, \ldots, a_n) = b_n$ is f's response to stimuli $a_1, \ldots, a_n$ immediately at time n. Turing machine models (which we do *not* consider in this book) are 'off-line' computing. The answer need not be immediate, the machine can calculate over a period of time — an organism must respond in real time, immediately. Our machines are 'real-time' computation.

Given $a_1, \ldots, a_n$ we might like to know $(b_1, \ldots, b_n)$ and not just b_n, thus

Notation 5.2. Let $f : A^+ \to B$ be a machine. Then $f^+ : A^+ \to B^+$ is defined by $f^+(a_1, \ldots, a_n) = (f(a_1), \ldots, f(a_1, \ldots, a_n))$.

What could be inside the 'black-box', inside the machine?

Definition 5.3. $C = (A, B, Q, \lambda, \delta)$ is by definition a *circuit* with *basic inputs* A, *basic outputs* B, *states* Q, *next-state function* λ and *output function*

δ iff A and B are finite non-empty sets, Q is a non-empty set, $\lambda : Q \times A \to Q$ and $\delta : Q \times A \to B$. *

How do we 'run' or 'use' a circuit C to get a machine f?

Definition 5.4. Let $C = (A, B, Q, \lambda, \delta)$ be a circuit. Let $q \in Q$. Then $C_q : A^+ \to B$, read "*the machine given by starting C in state q*", is defined inductively by

$$C_q(a_1) = \delta(q, a_1)$$

and

$$C_q(a_1, \ldots, a_n) = C_{\lambda(q, a_1)}(a_2, \ldots, a_n)$$

for $n \geq 2$. We say C *realizes* the machine $f : A^+ \to B$ iff there exists a $q \in Q$ such that $C_q = f$.

Suppose we write $\lambda(q, a_1)$ as $q \cdot a_1$ and define $q \cdot a_1 \ldots a_n = (q \cdot a_1 \ldots a_{n-1}) \cdot a_n$. Then $C_q(a_1, \ldots, a_n) = \delta(q \cdot a_1 \ldots a_{n-1}, a_n)$. Thus under the input $a_1, \ldots, a_n$ the state at time 0 is $q_0 = q$, the state at time 1 is $q_1 = q \cdot a_1, \ldots$, the state at time n is $q_n = q \cdot a_1 \ldots a_n$. The output at time 1 is $\delta(q_0, a_1)$, the output at time 2 is $\delta(q_1, a_2), \ldots$, the output at time n is $\delta(q_{n-1}, a_n)$. Writing $q \xrightarrow{a} q'$ if $q \cdot a = q'$ and $q \xrightarrow{a} b$ if $\delta(q, a) = b$ we have

$$q = q_0 \xrightarrow{a_1} q_1 \xrightarrow{a_2} q_2 \xrightarrow{a_3} \cdots \xrightarrow{a_{n-1}} q_{n-1} \xrightarrow{a_n} q_n$$

$$a_1 \downarrow \qquad a_2 \downarrow \qquad a_3 \downarrow \qquad\qquad a_n \downarrow$$

$$b_1 \qquad\quad b_2 \qquad\quad b_3 \qquad\qquad b_n$$

with $C_q(a_1, \ldots, a_k) = b_k$ for $k = 1, \ldots, n$, i.e.

$$C_q^+(a_1, \ldots, a_n) = (b_1, \ldots, b_n) .$$

It is natural to ask if every machine can be realized by some circuit. The answer will be "yes"! Toward this end we introduce the following

Notation 5.5. *Let A be a non-empty set. Then $A^+ = \{(a_1, \ldots, a_n) : n \geq 1\}$ forms an infinite semigroup under concatenation, i.e.*

$$(a_1, \ldots, a_n) \cdot (b_1, \ldots, b_m) = (a_1, \ldots, a_n, b_1, \ldots, b_m).$$

(A^+ is the free non-commutative semigroup with generators A.)

For each $t \in A^+$ we can consider the left translation $L_t : A^+ \to A^+$ with $L_t(r) = t \cdot r$. By convention, $L_1 : A^+ \to A^+$ with $L_1(r) = 1 \cdot r = r$ for all $r \in A^+$. Notice $L_{t_1} L_{t_2} = L_{t_1 \cdot t_2}$ since $L_{t_1} L_{t_2}(r) = L_{t_1}(t_2 r) = t_1 t_2 r = L_{t_1 t_2}(r)$.

---

*A *circuit* (with finitely many states) is often referred to in the literature also as a *finite-state transducer* or as a *Mealy automaton*. -CLN

Proposition 5.6. *Let the machine* $f : A^+ \to B$ *be given. Define* $C(f)$, *read "the circuit of* f *", by*

$$C(f) = (A, B, \{fL_t : t \in A^+ \cup \{1\}\}, \lambda, \delta)$$

with $\lambda(fL_t, a) = fL_tL_a = fL_{ta}$ *and*

$$\delta(fL_t, a) = (fL_t)(a) = f(ta) .$$

Then

$$C(f)_{fL_t} = fL_t$$

and so in particular $C(f)_f = f$. *Thus* f *can be realized as the machine given by* $C(f)$ *when started in state* f.

Proof. Very easy! The reader should supply the proof. $\qquad\square$

From the above proposition we have seen that any machine f can be realized by some circuit, in particular by $C(f)$. We next show that $C(f)$ is the *unique minimal* circuit which realizes f.

Toward this end we introduce

Notation 5.7. *Let* $C = (A, B, Q, \lambda, \delta)$ *and* $C' = (A, B, Q', \lambda', \delta')$.

(a) C' *is a subcircuit of* C *iff* $Q' \subseteq Q$, *and for each* $a \in A$, $q' \in Q'$ *we find* $\lambda(q', a) \in Q'$ *and* $\lambda'(q', a) = \lambda(q, a)$ *and* $\delta'(q', a) = \delta(q', a)$.

(b) C *and* C' *differ only up to a renaming of states iff there exists* $j : Q \to Q'$ *so that* j^{-1} *exists and* $j(\lambda(q, a)) = \lambda'(j(q), a)$ *and* $\delta(q, a) = \delta(j(q), a)$ *for all* $q \in Q$, $a \in A$.

Proposition 5.8. *(a) Let* $C = (A, B, Q, \lambda, \delta)$ *be a given circuit. Then* $\mathcal{M}(C) \equiv \{C_q : q \in Q\}$, *read "the machines realized by* C *", is closed under left translation, i.e.,* $g \in \mathcal{M}(C)$ *implies* $gL_t \in \mathcal{M}(C)$ *for all* $t \in A^+$.

(b) Let $\mathcal{M}$ *be a collection of machines all mapping from* A^+ *into* B *which is closed under left translation. Define* $C(\mathcal{M})$, *read "the circuit of* $\mathcal{M}$ *", by* $C(\mathcal{M}) = (A, B, \mathcal{M}, \lambda, \delta)$ *with* $\lambda(g, a) = gL_a$, $\delta(g, a) = g(a)$. *Then* $C(\mathcal{M})_g = g$ *for all* $g \in \mathcal{M}$.

(c) Let $C = (A, B, Q, \lambda, \delta)$ *be a circuit. Then the circuit* $C(\mathcal{M}(C)) \equiv C^*$ *is, by definition, the reduction of* C. *If* $C' = (A, B, Q', \lambda', \delta')$ *is any circuit such that* $\mathcal{M}(C) = \mathcal{M}(C')$ *and* $q' \mapsto C'_{q'}$ *is a one-to-one map, then up to a renaming of states* C' *is the reduction of* C. C *is said to be reduced iff* $C = C^*$ *up to renaming.*

(d) Let $f : A^+ \to B$ *be a given machine. Let* $C = (A, B, Q, \lambda, \delta)$ *be any circuit realizing* f. *Then* C^* *contains* $C(f)$ *as a subcircuit.*

Proof. (a) It is easy to verify that $C_q L_a = C_{\lambda(q,a)}$ for all $a \in A$. Then since $L_{a_1 \ldots a_n} = L_{a_1} \cdots L_{a_n}$, (a) follows.

(b) Easy by direct verification.

(c) Let C be a circuit such that $q \mapsto C_q$ is one-to-one. Then since $C_{\lambda(q,a)} = C_q L_a$ it is easy to verify that C and C^* differ only up to the renaming of the states by $q \mapsto C_q$. Thus (c) follows.

(d) Notice $C(f) = C(\{f L_t : t \in A^+ \cup \{1\}\})$. Now clearly $\mathcal{N} \subseteq \mathcal{M}$ implies $C(\mathcal{N})$ is a subcircuit of $C(\mathcal{M})$. Also $f \in \mathcal{M}(C)$ implies $\{f L_t : t \in A^+ \cup \{1\}\} \subseteq \mathcal{M}(C)$ by (a). Thus $C(f)$ is a subcircuit of $C(\mathcal{M}(C)) \equiv C^*$, proving (d).

This proves Proposition (5.8).   □

Thus if we have a 'black-box' or machine which from the outside does f (i.e., we put in $a_1, \ldots, a_n$ and get out $(f(a_1), f(a_1, a_2), \ldots, f(a_1, \ldots, a_n))$) we 'know' $C(f)$ is inside the 'black-box' doing f. In short there is a *unique minimal circuit which realizes f*.

The situation is entirely different for Turing machines or for computable functions. In general there is no best program to compute a given recursive function. We could mention M. Blum's Speed-up Theorem [Blum] which states (roughly) that *there exists* a computable function f such that for *any* program P which computes f (on the universal Turing machine T) there exists another program P' which is exponentially faster than P. Thus for this particular f of Blum's, there is certainly no best or minimal or unique way of 'doing' f.

We strongly hold the opinion that for the class of computable functions there does *not exist* unique minimal programs to compute the function. Consider generating the primes. Some programs are favored now, but as hardware changes the fashion will change.

The reason why Turing machine programs to realize a computable f are not unique and the circuit which realizes the (sequential) machine f (namely $C(f)$) is unique is not hard to fathom. In the sequential machine model we are given much more information. It is 'on-line' computing; we are told what is to happen at each unit of time. The Turing machine program is 'off-line' computing; it just has to get the correct answer — there is no time restraint, no space restraint, etc.

When there is no canonical construction (as from computable f to program P to realize f) the theory by necessity is *combinatorial*. When there is a canonical construction (as from machine f to $C(f)$) the theory becomes *algebraic.* Constructions and invariants are derived from canonical $C(f)$, etc. We will see that this is the case here. (Of course this is nearly a tautology since algebra = those combinatoric problems having well developed and powerful methods.)

However general (sequential) machines and circuits are not too important since if the state set Q is infinite (say, $Q = \mathbb{Z}$, the integers) then the output function $\delta : Q \times A \to B$ could be a badly non-computable function, with all the interesting things taking place in the output map δ, and we are back to recursive function theory. See Proposition (5.74).

Thus we will generally assume from now on that the state sets are finite as well as the input and output sets. As an immediate consequence of Propositions (5.6) and (5.8) we have

Proposition 5.9. *The machine $f : A^+ \to B$ is realizable by a finite state circuit iff $C(f)$ is a finite state circuit iff $\{fL_t : t \in A^+ \cup \{1\}\}$ is a finite set. Further, if $n = |\{fL_t : t \in A^+ \cup \{1\}\}|$ is finite, then any circuit realizing f has at least n states and any circuit realizing f with exactly n states is $C(f)$ (up to a renaming of states).*

Let $X \subseteq A^+$. Then it is usual to say that X is *recognizable* by a finite state circuit (or machine) iff its characteristic function ψ_X ($\psi_X : A^+ \to \{0,1\}$ with $\psi_X(t) = 1$ if $t \in X$ and 0 otherwise) is realizable by a finite state circuit. Thus X is recognizable by a finite state circuit iff $\{\psi_X L_t : t \in A^+ \cup \{1\}\}$ is a finite set. For example, if $A = \{0,1\}$ and $X_1 = \{0^n 10^n : n = 1, 2, 3, \dots\}$ and 0^n denotes $0 \dots 0$ (n times) and $X_2 = \{t = a_1 \dots a_n \in \{0,1\}^+ :$ there is an odd number of 1's among $a_1, \dots, a_n\}$. Then by the computing $\psi_{X_k} L_t$ it is easy to see that X_1 is not recognizable and X_2 is recognizable by a finite state circuit.

We next want to introduce the idea of one machine being 'more capable' than another. Roughly, g is 'more capable' than f if $f = Bg\alpha$ where B and α are 'trivial maps' or 'trivial codes'. Precisely,

Definition 5.10. (a) $H : A^+ \to B^+$ is a *trivial code* iff H is a homomorphism. Thus $H(a) \in B^+$ for each $a \in A$ and $H(a_1, \dots, a_n) = H(a_1) \dots H(a_n)$. We say H is a *length preserving* (ℓp) *trivial code* iff H is a trivial code with $|H(t)| = |t|$ for all $t \in A^+$ where $|x_1 \dots x_p| = p$. Clearly, $H : A^+ \to B^+$ is an ℓp trivial code iff there exists $h : A \to B$ so that $H(a_1, \dots, a_n) = (h(a_1), \dots, h(a_n))$. In this case we denote H by $\overline{h}$. Notice for all trivial codes $|H(t)| \geq 1$ for all t.

(b) Let $f : A^+ \to B$ and $g : C^+ \to D$ be two machines. Then f is *less than or equal in capability to* g, written $f \mid g$, iff there exists a trivial code $H : A^+ \to C^+$ and a function $j : D \to B$ so that $f = jgH$. That is, the following diagram commutes.

$$
\begin{array}{ccc}
C^+ & \xrightarrow{\ g\ } & D \\[4pt]
H \big\uparrow & & \big\downarrow j \\[4pt]
A^+ & \xrightarrow{\ f\ } & B
\end{array}
$$

We then say that f *divides* g, or say there is a *division* by f of g. We write $f \mid g$ (ℓp) iff $f = jgH$ with H ℓp iff $f = jg\overline{h}$ for some $h : A \to B$, $j : D \to B$.

Clearly "$\mid$" (division) and "$\mid$ (ℓp)" (length-preserving division) are transitive relations on the collection of all machines.

(Definition (5.10) is continued below.)

Notice from an engineering viewpoint, an ℓp trivial code $h : B \to C$ is just a 'set of wires' from the basic outputs B to the basic inputs C realized by 'running a wire' from $b \in B$ to $h(b) \in C$. They are truly trivial. Trivial codes (not necessarily ℓp) are somewhat more complicated to implement since $H(a)$ can be long. However no 'computational depth' is necessary. We will mainly use ℓp trivial codes.

We next wish to combine machines in a natural way.

Definition 5.10 (continued). Let $f : A^+ \to B$ and $g : C^+ \to D$ be two machines.

(c) The *parallel combination* of f and g, denoted $f \times g$, is by definition $f \times g : (A \times B)^+ \to C \times D$ with $f \times g((a_1, b_1), \ldots, (a_n, b_n)) = (f(a_1, \ldots, a_n), g(b_1, \ldots, b_n))$. The definition is extended to $f_1 \times \cdots \times f_n$ in the obvious manner, i.e., $f_1 \times \cdots \times f_n = (f_1 \times \cdots \times f_{n-1}) \times f_n$.

(d) Let $h : B \to C$ be given with associated ℓp trivial code $\overline{h}$. Then the *series combination* of f by g with connection ℓp trivial code h is by definition $g\overline{h}f^+ = k$. So $k(a_1, \ldots, a_n) = g(h(f(a_1)), h(f(a_1 a_2)), \ldots, h(f(a_1, \ldots, a_n)))$.

Let $\mathcal{M}$ be a collection of machines. We would like to define the collection of all machines which can be built by using members of $\mathcal{M}$ (any finite number of times) and the operations of series combination, parallel combination, and ℓp division. Precisely,

Definition 5.11. Let $\mathcal{M}$ denote a collection of machines. Then $SP(\mathcal{M})$, read "*series-parallel* of $\mathcal{M}$", is the collection of all machines $g : A^+ \to B$

such that

(5.11a)
$$g = h_n g_n \overline{h}_{n-1} \ \cdots \ g_2^+ \overline{h}_1 g_1^+ \overline{h}_0$$

where $g_k = f_{k1} \times \cdots \times f_{k\pi(k)}$ with $f_{kj} \in \mathcal{M}$ and $g_k : A_{k-1}^+ \to A_k'$ and $h_0 : A \to A_0, \ h_1 : A_1' \to A_1, \dots, h_n : A_n' \to A_n \equiv B$.

The question naturally arises: Given $\mathcal{M}$, what can be built by using series and parallel combination and ℓp division machines from $\mathcal{M}$, i.e., what is $SP(\mathcal{M})$? Also, what machines *cannot* be constructed from $\mathcal{M}$ without 'essentially' being themselves in $\mathcal{M}$? We want a 'Prime Decomposition Theorem' for finite state machines. Toward this end we introduce some special machines.

Definition 5.12. (a) D_1, read "the *delay-one machine*" over $\{a, b\}$, by definition is $D_1 : \{a, b\}^+ \to \{a, b, *\}$ with $D_1(a_1) = *$ and $D_1(a_1, \dots, a_n) = a_{n-1}$.

(b) Let S be a semigroup. Then S^M, read the *machine of S*, is by definition the machine $S^M : S^+ \to S$ defined by $S^M(s_1, \dots, s_n) = s_1 \cdots \cdots s_n$ with $\cdot$ the product of S.

(c) A^r denotes the semigroup with elements A and multiplication $a_1 a_2 = a_2$. Thus $A^{rM} : A^+ \to A$ with $A^{rM}(a_1, \dots, a_n) = a_n$ so $A^{rM+} : A^+ \to A^+$ with $A^{rM+}(a_1, \dots, a_n) = (a_1, \dots, a_n)$, i.e., A^{rM+} is the identity map on A^+.

By (5.12b) we see that each semigroup gives rise naturally to a machine. This process is reversible via the following definition. (See also the Appendix at the end of this Chapter, especially Proposition (5.44).)

Definition 5.13. Let $f : A^+ \to B$ be a machine. Then f^S, read "the *semigroup of f*", is given by the congruence $\equiv_f$ on A^+ where for $t, r \in A^+$, $t \equiv_f r$ iff $f(\alpha t \beta) = f(\alpha r \beta)$ for all $\alpha, \beta \in A^+ \cup \{1\}$. Then if $[t]_f$ denotes the equivalence class of the equivalence relation $\equiv_f$ containing t, we have $f^S = \{[t] : t \in A^+\}$ and $[t]_f \cdot [r]_f = [tr]_f$ which is well defined (where tr denotes the product in A^+ and $\cdot$ denotes the product in f^S) since $[t]_f [r]_f \subseteq [tr]_f$.

In the following when we write f^{SM} we mean $(f^S)^M$, etc.

The importance of f^S is shown by the following elementary but very important

Proposition 5.14. *Let $f : A^+ \to B$ be a machine. Then $f \mid f^{SM} (\ell p)$. In fact we have the following* fundamental expansion

(5.15)
$$f = j_f f^{SM} \overline{h}_f$$

where $h_f : A \to f^S$ with $h_f(a) = [a]_f$ and $j_f([t]_f) = f(t)$.

That is, for any machine f, the following diagram commutes.

$$
\begin{array}{ccc}
(f^S)^+ & \xrightarrow{\;(f^S)^M\;} & f^S \\
\overline{h_f}\big\uparrow & & \big\downarrow j_f \\
A^+ & \xrightarrow{\quad f \quad} & B
\end{array}
$$

Proof. Easy, just calculate. □

Thus every machine has less than or equal capability with the machine of its semigroup. See the Appendix at the end of this Chapter for a definition of f^S in terms of circuits (Proposition (5.44)) and a conceptual proof of (5.15).

We also have that the operations of series, parallel combination, and division preserve machines realizable by finite state circuits. Precisely,

Proposition 5.16. *Let $\mathcal{M}$ be a collection of machines each realizable by a finite state circuit and each with finite basic input and finite basic output sets. Then*

(a) $f \mid g$ and $g \in \mathcal{M}$ implies f is realizable by a finite state circuit.

(b) $f \in SP(\mathcal{M})$ implies f is realizable by a finite state circuit.

(c) Let $f : A^+ \to B$ with A and B sets of finite order. Then f is realizable by a finite state machine iff f^S is a semigroup of finite order.

Proof. For a proof see (5.46) in the Appendix at the end of this Chapter.

In the following, $\mathcal{S}$ denotes a collection of finite semigroups, $\mathcal{S}^M = \{S^M : S \in \mathcal{S}\}$, and **PRIMES, UNITS** and U_3 are defined as in Chapter 3. Then the promised 'Prime Decomposition Theorem for Finite State Machines' first proved by Krohn and Rhodes is the following

Theorem 5.17 (Prime Decomposition for Finite State Machines). *(In the following, all machines f considered have finite basic input and output sets and are realizable by finite state circuits, so f^S is a finite semigroup.)*

(a) Let f be a machine. Then
$$f \in SP(\mathcal{S}^M \cup \{D_1, U_3^M\}) \text{ iff } \mathbf{PRIMES}(f^S) \subseteq \mathbf{PRIMES}(\mathcal{S}).$$
In particular,

(5.18) $f \in SP((\mathbf{PRIMES}(f^S))^M \cup \{D_1, U_3^M\})$.

(b) The following two statements are equivalent:
   (i) $P \in$ **PRIMES** $\cup$ **UNITS**
   (ii) If $P \mid f^S$ *and* $f \in SP(\mathcal{M})$, *then* $P \mid g^S$ *for some* $g \in \mathcal{M}$.

Proof. See the Appendix at the end of this Chapter, or [Trans] or [M-L-S].

Corollary 5.19. *The following statements are equivalent:*
(a) f^S *is a (finite) combinatorial semigroup.*
(b) $f \in SP(\{D_1, U_3^M\})$.

We state another corollary after the following

Definition 5.20. (a) Let $\mathcal{M}_k = \{f : A^+ \to B : f = C_q$ for some circuit $C = (A, B, Q, \lambda, \delta), q \in Q$, where Q has less than or equal to k states$\}$.
   (b) Define **size**$(f) = \min\{k : f \in SP(\mathcal{M}_k)\}$.

Thus **size**(f) is the smallest integer n so that f can be built series-parallel using n state machines but f cannot be built series-parallel from $n - 1$ state machines.

Corollary 5.21. *Let* $f : A^+ \to B$ *with* **size**$(f) \geq 2$. *Then* **size**(f) *equals the maximum of* $\{$**size**$(P^M) : P \in$ **PRIMES**$(f^S)\} \cup \{2\}$.

Also **size**$(\mathbb{Z}_p^M) = p$ if p is a prime integer ≥ 2, and **size**(P^M), if P is a SNAG, is the smallest number of letters on which P has a faithful transitive permutation representation. Notice there are machines of such size n, except size 4, since **size**$(\{1\}^M) = 1$, **size**$(\mathbb{Z}_2^M) = 2$, **size**$(\mathbb{Z}_3^M) = 3$, **size**$(A_n^M) =$ **size**$(S_n^M) = n$, $n \geq 5$. *There are no machines of size 4* since **PRIMES**$(S_4) = \{\mathbb{Z}_2, \mathbb{Z}_3\}$. Here of course A_n denotes the alternating group on n letters and S_n denotes the symmetric group on n letters.

Let $\mathcal{C}$ denote $\{\mathbb{Z}_p^M : p \geq 2$ is a prime integer$\}$. $\mathcal{C}$ is the *set of prime counter machines.* A natural question is, what is $SP(\mathcal{C} \cup \{D_1, U_3^M\})$? From the Prime Decomposition Theorem we know $f \in SP(\mathcal{C} \cup \{D_1, U_3^M\})$ iff the maximal subgroups of f^S are *solvable.* Thus $P \in$ **PRIMES**$\setminus \mathcal{C}$ implies P^M is *not* in $SP(\mathcal{C} \cup \{D_1, U_3^M\})$. For example, $A_n^M \notin SP(\mathcal{C} \cup \{D_1, U_3^M\})$ for $n \geq 5$.

We can use the powerful solvability criteria of Burnside and Feit-Thompson to yield necessary conditions that f is constructable from counter machines, 'flip-flops' and delays. For example,

Corollary 5.22 (Feit-Thompson). *Let* $f = C_q$ *where* $C = (A, B, Q, \lambda, \delta)$ *and no input sequence* $t \in A^+$ *reverses two states (i.e. there exists no*

$t \in A^+$ and $q_1 \neq q_2$, $q_1, q_2 \in Q$ such that $q_1 \cdot t = q_2$ and $q_2 \cdot t = q_1$). Then $f \in SP(\{\mathbb{Z}_p^M : p \geq 3 \text{ is prime integer }\} \cup \{D_1, U_3^M\})$.

Proof. It is not difficult to show (see Proposition (5.48c)) that the hypothesis implies the maximal subgroups of f^S are of odd order. Thus by Feit-Thompson the maximal subgroups are solvable so $\mathbf{PRIMES}(f^S) \subseteq \{\mathbb{Z}_p : p \geq 3 \text{ is a prime integer}\}$. Then the result follows from (5.18). $\square$

We can strengthen the previous corollary via the following definition and corollary.

Definition 5.23a. Let $C = (A, B, Q, \lambda, \delta)$ be a circuit. Then **prime-loop**$(C) = \{p : p \geq 2$ is a prime integer and there exists an input *sequence* $t \in A^+$ and states $q_1, q_2, \ldots, q_p$ of Q such that $q_1 \cdot t = q_2, \ldots, q_{p-1} \cdot t = q_p, q_p \cdot t = q_1\}$.

Corollary 5.23b (Constructability from Counters). *Let $f : A^+ \to B$ be a machine, then*
(i) $f \in SP(\{D_1, U_3^M\})$ iff **prime-loop**$(C(f))$ is empty.
(ii) [Sylow] $f \in SP(\{\mathbb{Z}_p^M, U_3^M, D_1\})$, where p is a prime integer, iff **prime-loop**$(C(f)) \subseteq \{p\}$.
(iii) [Burnside] $f \in SP(\{\mathbb{Z}_p^M, \mathbb{Z}_q^M, U_3^M, D_1\})$, where p and q are two distinct prime integers, iff **prime-loop**$(C(f)) \subseteq \{p, q\}$.
(iv) [Feit-Thompson] $f \in SP(\{\{C_p^M : p \in \Pi\}, D_1, U_3^M\})$, where Π is a collection of odd prime integers, iff **prime-loop**$(C(f)) \subseteq \Pi$.

Proof. It is not difficult to show (see Proposition (5.48c)) that **prime-loop**$(C(f)) = \{p : p \geq 2$ is a prime integer which divides the order of some *subgroup* of $f^S\}$. Now, by Sylow, p-groups are solvable; by Burnside, groups of order $p^\alpha q^\beta$ are solvable; and by Feit-Thompson, groups of odd order are solvable. Now the corollary follows from (5.18). $\square$

Notice **prime-loop**$(A_5^M) = \{2, 3, 5\}$ but A_5^M does *not* lie in $SP(\{\mathbb{Z}_2^M, \mathbb{Z}_3^M, \mathbb{Z}_5^M, D_1, U_3^M\})$.

We notice that in some ways the Prime Decomposition Theorem for machines is analogous with the Fundamental Theorem of Arithmetic. We have the notion of divides (namely, $\mid$) for machines, the notions of products (namely, series combination and parallel combination), we have found the primes ($\mathbf{PRIMES}^M$) and proved each machine is a product of the prime machines which divide f^{SM} (namely, $(\mathbf{PRIMES}(f^S))^M$) together with the unit and delay machines (namely, $\textbf{UNITS}^M \cup \{D_1\}$). Thus $\textbf{PRIMES}$ for machines play the role of the prime integers in arithmetic and $\textbf{UNITS}^M \cup \{D_1\}$ play the role of $\{\pm 1\}$, the units in arithmetic. A limited form of uniqueness of the decomposition was also proved, namely (5.17b).

In any 'prime decomposition theorem' in algebra there are three main questions: What are the primes? Can every element be decomposed into primes? How different are the various decompositions of an element into primes, i.e., what about uniqueness? For machines the first two questions are completely answered by Theorem (5.17), namely, the primes are the $\textbf{PRIMES}$ of group theory (i.e. $\textbf{PRIMES}^M$) plus $\textbf{UNITS}^M$ and D_1, i.e. simple groups plus a 'flip-flop' and a 'delay'. The remaining question is to find all the 'minimal' decompositions of a given machine f into primes, units and delays. Current research is directed toward answering this question. See Part II of this Chapter following and the Appendix at the end of this Chapter.

$SP(\mathcal{M})$ intuitively connects members of $\mathcal{M}$ together in a loop-less manner or in a manner allowing *no feedback* from the nth machine g_n of (5.11a) to the ith-machine g_i, $i < n$. (Similarly, no feedback is allowed from the jth-machine g_j to any machine g_i with $i < j$.) Why do we allow no feedback? The problem lies in the proper definition of feedback. Suppose we are given $C = (A, B, Q, \lambda, \delta)$ with $f = C_{q_0}$. Suppose we allow feedback of the form α (output b_t at time t, input a_{t+1} at time $t + 1$) yields a new input $a'_{t+1} = \alpha(b_t, a_{t+1})$ which is actually entered into the machine at time $t+1$. Then in the Appendix at the end of this Chapter (Proposition (5.77)) we make the above precise and easily prove all machines can be realized by *one-state* machines allowing feedback of type α. We simply put the next-state function λ into the feedback α.

The reduced circuit $C(f) = (A, B, Q_f, \lambda_f, \delta_f)$ is essentially the *minimal way of writing f in feedback or recursive form* since if $f^+(a_1, \ldots, a_n) = (b_1, \ldots, b_n)$, then

$$C(f)_{q_0} = f$$
$$q_1 = \lambda(q_0, a_1)$$
$$q_{t+1} = \lambda(q_t, a_{t+1})$$
$$b_1 = \delta(q_0, a_1)$$
$$b_{t+1} = \delta(q_t, a_{t+1}),$$

which is recursive form. The SP decompositions are at the other end of the scale — they write f as a collection of machines in loop-less or *no feedback* form, each machine of which cannot be further decomposed into simpler machines in this manner. The two ends of the scale are mathematically natural. Things in between are not natural mathematically. If something in between is desired it must come up from some application like the Jacob and Monod model in biology [J-M], or shift registers in electrical engineering [Shift]. Then when these 'in-between models' are precisely defined we believe techniques like those used to prove the Prime Decomposition Theorem will also be very useful.

5.24. To show again that $C(f)$ is the minimal recursive or feedback form of f we notice that usual design procedures for constructing f by transistors (say) proceed as follows:

a. Given f, compute $C(f)$.
b. If $C(f) = (A, B, Q, \lambda, \delta)$, let n be the smallest integer not less than than $\log_2 |Q|$ and choose $j : Q \to \{0,1\}^n$ so that j is one-to-one. The map j is called a *state assignment*.
c. For ease of exposition, assume $A = B = \{0,1\}$. For each $a \in A$ define $B_a : \{0,1\}^{n+1} \to \{0,1\}^n$ by $B_a(j(q),a) = j(\lambda(q,a))$ and otherwise B_a is defined in an arbitrary but fixed way. Also define $B : \{0,1\}^{n+1} \to \{0,1\}$ by $B(j(q),a) = \delta(q,a)$ and otherwise B is defined in an arbitrary but fixed way.
d. By standard techniques realize the Boolean functions B_a and B by transistors (say by the Shannon-Quine method).

For further comments on state-assignments, see (5.80) in the Appendix at the end of this Chapter.

In the main, the importance of series-parallel decompositions does not stem from their aid in *constructing* devices. The best modes of construction are usually feedback or recursive methods. However, as we belabored in Chapter 4, a series-parallel decomposition allows one to *understand* what the machine f 'is doing'. Writing f as

$$f = h_n g_n \overline{h_{n-1}} g_{n-1}^+ \ldots \overline{h_1} g_1^+ \overline{h_0}$$

writes f as *approximated by*

$$h_1 g_1 \overline{h_0}, \; h_2 g_2 \overline{h_1} g_1^+ \overline{h_0}, \; \ldots, \; h_n g_n \overline{h_{n-1}} g_{n-1}^+ \overline{h_{n-2}} \ldots \overline{h_1} g_1^+ \overline{h_0}.$$

Notice if feedback were allowed the functions $h_1 g_1 \overline{h_0}$, etc., would *not* be defined. Thus series-parallel is an approximation theory of f and in this way it allows one to *understand* f.

Now, roughly, the *complexity of f equals how hard it is to understand it*. We formalize this in the next part of this Chapter immediately following.

Part II. Complexity of Finite State Machines

In this part we want to develop, as intuitively as possible, the theory of complexity of finite state machines. All machines considered will have finite basic input and output sets and will be realizable by a finite state circuit.

What properties should the complexity of finite state machines satisfy? First, the complexity will take values in the non-negative integers $\mathbb{N} = \{0, 1, 2, \ldots\}$. Then the next natural question is, how does complexity relate to the operations of division, parallel combination, and series combinations?

In the following, $\mathcal{M}_{Fin}$ denotes the collection of all machines (with finite basic input and output sets and realizable by a finite state circuit). Also, Θ denotes a function $\Theta : \mathcal{M}_{Fin} \to \mathbb{N}$.

The first condition is

5.25. Axiom A. *(a)* $f \mid g$ *implies* $\Theta(f) \leq \Theta(g)$; $\Theta(fL_t) \leq \Theta(f)$.
(b) $\Theta(f_1 \times \cdots \times f_n) = \max\{\Theta(f_i) : i = 1, \ldots, n\}$.

The first part of Axiom A(a) demands that if f has less than or equal the capability of g, then the complexity of f is less than or equal to the complexity of g, a very reasonable condition. Further, since fL_t is simply f after the input string t has been presented to f, the second part of Axiom A(a) demands that fL_t is not more complicated than f, another very reasonable condition. Axiom A(b) requires that if two machines are placed in parallel so that the individual operations are *completely* independent of each other then the resulting complexity is the maximum of the two individual machines. The machine $f = f_1 \times f_2$ may be costlier and larger than f_1 and f_2 but not more complex. The machine f is no harder to understand than the most complex one of f_1 and f_2. By pondering the matter, the reader should convince himself or herself that Axiom A(b) is a very natural property of complexity and any function not possessing this property does not deserve the name complexity.

Using the well known theorem of G. Birkhoff [Cohn], one can show $\Theta(f) = \Theta(g)$ for *all* functions satisfying Axiom A iff $f \mid \prod g$ and $g \mid \prod f$ where $\prod f$ denotes $f \times \cdots \times f$ (n times) for some finite n.

In the following, given machines f_1 and f_2, we will write "f_2 series f_1" for any series combination of f_1 by f_2 (see Definition (5.10d)). The next condition is

5.26. Axiom B. *For all machines f_1 and f_2,*

$$\Theta(f_2 \ series \ f_1) \leq \Theta(f_2) + \Theta(f_1) \ .$$

Notice we have inequality and not equality. Axiom B is very intuitive since one way to understand f_2 series f_1 is by understanding f_1 and then understanding f_2 since there is *no feedback from f_2 to f_1*. If there was feedback no such condition on the complexity would hold since the feedback condition might be very complicated and need to be accounted for, giving $\Theta(f_2 \text{ general } f_1) \leq \Theta(f_2) + \Theta(f_1) + \Theta(\text{feedback of } f_2 \text{ to } f_1)$, where in the f_2 series f_1 case, $\Theta(\text{feedback of } f_2 \text{ to } f_1)=0$. However inequality and not equality holds since it might be possible to understand $f = f_2$ series f_1 in a quicker or better way than by composing a theory for f_1 with a theory for f_2, especially if the connecting ℓp trivial codes are badly non-one-to-one. Notice also that a little reflection shows that $+$ is the natural function in Axiom B. Of course if one prefers multiplication instead of addition, one could replace the complexity by its exponential.

For *any* theory of complexity (whether for finite state machines, finite semigroups, Boolean functions, etc.) we expect analogues of division, parallel combination and series combination to be defined (in semigroups: division, direct product, semidirect product; in Boolean functions: dropping or combining variables, direct product, composition) and Axioms A and B to hold (or something very close).

Our philosophical position is the following: for *finite* algebraic or combinatorial systems, devices, etc. (e.g. finite state machines, Boolean functions, finite semigroups, finite strings of zeros and ones) *there is a natural theory of complexity* (in Plato's heaven if you like) but there is *no* natural theory (equals there is *no* theory) of complexity for wider classes off finite, as for example, recursive functions. (The existence of the universal Turing machines convinces me of the non-existence of a theory of complexity for recursive functions.) There is probably not a theory of complexity, even for context-free languages.

Intuitively, every universal Turing machine has complexity $+\infty$, every language (which is used) has complexity $+\infty$ since 'anything' can be written in it. Their 'efficiency' may be different, their 'entropy' or 'compression' may be different, but they are all as complicated as possible since each can do everything.

Further, the 'correct' axioms for the complexity functions on the finite-type objects are Axioms A and B (or something very close). Winograd, Papert and Minsky, Kolmogorov, and Rhodes have all initiated work on the complexity of systems of finite type (see [Wino], [P-M] and [Kol1, Kol2, Kol3]). We will briefly return to discussing these various 'complexities' and Axioms A and B later.

In the following we say a machine f is *nice* iff $C(f) = (A, B, Q, \lambda, \delta)$ with $\lambda = \delta$.

Now not only do we desire Axiom B (5.26), but also the *continuity of Axiom B* whose following precise statement stems from some ideas of Joel Karnofsky.

5.27. Continuity of Axiom B. Let f be a machine with $\Theta(f) = n$. Suppose $n = a + b$ with a, b non-negative integers. Then there exists nice machines f_1 and f_2 such that

$$f \quad \text{divides} \quad f_2 \text{ series } f_1$$

and $\Theta(f_1) = a, \quad \Theta(f_2) = b$.

A precise formulation of the continuity of Axiom A is the following: Given a machine f and non-negative integers n and k with $\Theta(f) = n \geq k$, there exists a nice machine g such that $g \mid f \times \cdots \times f$ (r times) for some r with $\Theta(g) = k$.

Now we will not need to postulate continuity of Axiom A since it will automatically follow from other axioms. However, the general principle is that we require not only an axiom but also continuity of that axiom. Intuitively, complexity should vary as 'smoothly' as possible.

Now, granting that Axioms A and B are necessary for any complexity function, we find however that they are not sufficient to uniquely determine the complexity function. *One must decide what machines will have complexity zero.* Many different choices are possible. It will depend on the applications to determine what one wants to consider trivial. However for most of the natural choices similar mathematical techniques will apply. Here we will make the 'biggest natural' choice and declare

5.28. Axiom C (Choice of Complexity Zero).

$$\Theta(U_3^M) = 0 \ ,$$
$$\Theta(D_1) = 0 \ ,$$

or, in words, the flip-flop machine and the delay-one machine have complexity zero.

We observe that Axioms A, B and C together with Corollary (5.19) imply $\Theta(f) = 0$ if f^S is a combinatorial semigroup.[*]

The last axiom, Axiom D, will simply state that Axioms A, B, Continuity of B, and C are the 'only properties true for Θ'.

---

[*]'only if' will follow for the complete set of Axioms. -CLN

5.29a. Axiom D (First version, denoted D_1). *Axioms A, B, Continuity of B, and C are a complete set of conditions for Θ in the sense that an equation E holds among the values of Θ iff E holds among the values of every function satisfying Axioms A, B, Continuity of B, and C.*[1]

5.29b. Axiom D (Second version, denoted D_2). Θ *satisfies Axioms A, B, Continuity of B, and C. If $\Theta' : \mathcal{M}_{Fin} \to \mathbb{N}$ satisfies Axioms A, B, Continuity of B, and C, then $\Theta' \leq \Theta$ meaning $\Theta'(f) \leq \Theta(f)$ for all $f \in \mathcal{M}_{Fin}$.*

It will turn out that the first and second versions of Axiom D are equivalent. See (5.83) of the Appendix at the end of this Chapter.

We have come to the end of the Axioms of complexity for finite state machines. Axiom A(a) and Axiom B are critical. Continuity of Axiom B is pleasant, but the mathematical theory will go through minor changes if Continuity of Axiom B is not assumed. However in this case we must make a choice of units of complexity (like the CGS System in mechanics using centimeters, grams and seconds). We will discuss this later in this Chapter. Axiom C amounts to a choice of what is complexity zero. Thus Continuity of Axiom B and Axiom C are essentially a choice of units of complexity. The choices made here are very natural, but the mathematical theory adapts to any arbitrary choice in a fairly straightforward manner. Axiom D simply says the previous axioms are the 'only properties' of complexity. Axioms A, B, Continuity of B, C, and D should seem very natural to the reader.

We now have

Theorem 5.30. *There exists exactly one function Θ satisfying Axioms A, B, Continuity of B, C, and D, and it is*

$$\Theta(f) = \#_G(f^S) \ .$$

Proof. See (5.81) of the Appendix at the end of this Chapter.

Reader, were you surprised by Theorem (5.30)?

Via (5.30) all the results regarding $\#_G$ of Chapter 3 apply. For example, by analogy with Remark (3.16) we have

Remark 5.31. Let Θ be the unique function satisfying Axioms A, B, Continuity of B, C, and D.

---

[1]See (5.82(9)) of the Appendix, where 'an equation E holds among the values of Θ' is made precise using equational theory of universal algebra.

(a) $\Theta(f) = 0$ iff f^S is combinatorial iff $f \in SP(\{U_3^M, D_1\})$.

(b) If U_3 does not divide f^S, then $\Theta(f) \leq 1$. In particular, if $f(a_1, \ldots, a_n) = f(a_{\Pi(1)}, \ldots, a_{\Pi(n)})$ for all permutations Π (i.e., f is an Abelian machine), then $\Theta(f) \leq 1$.

(c) Let f be realizable by the circuit $C = (A, B, Q, \lambda, \delta)$ with $|Q| = n$. Then $\Theta(f) \leq n - 1$.

Proof. See (5.83) of the Appendix at the end of this Chapter.

(d) Let $C = (\{a, b, c\}, \{0, 1\}, \{1, \ldots, n\}, \lambda, \delta)$ be defined for $n \geq 2$ by $\lambda(i, a) = i + 1 \pmod{n}$; $\lambda(1, b) = 2$, $\lambda(2, b) = 1$, $\lambda(x, b) = x$ for $x \neq 1, 2$; $\lambda(1, c) = 2, \lambda(y, c) = y$ for $y \neq 1$; and $\delta(q, d) = q \pmod 2$ for $q \in \{1, \ldots, n\}$, $d \in \{a, b, c\}$. Then $\Theta(C_q) = n - 1$ for all $q \in Q$. Thus machines of each complexity exist!

Proof. See (5.83) of the Appendix at the end of this Chapter.

We can strengthen Theorem (5.30) by introducing two definitions.

Definition 5.32 (of α and β). (All machines considered are in $\mathcal{M}_{Fin}$.)[*]
(a) We recall from (5.12c) that $A^{rM+} : A^+ \to A^+$ denotes the identity map on A^+, i.e., $A^{rM+}(a_1, \ldots, a_n) = (a_1, \ldots, a_n)$. Then $\alpha(f)$ is the smallest non-negative integer n so that
(b) f divides f_n series g_n series $\cdots$ series f_1 series g_1 series f_0 with $f_j \in SP(\{U_3^M, D_1\})$ for $j = 0, \ldots, n$ (or equivalently by (5.19)) f_j^S is a combinatorial semigroup) and $g_k = g_{k1} \times A_k^{rM}$ with A_k a finite set and $g_{k1} \in SP(\mathbf{PRIMES}^M)$ (or equivalently g_{k1}^S is a group). Thus f_j are combinatorial machines and g_k^+ are group machines in parallel with identity machines (0-time delays). $\alpha(f)$ is well defined by (5.17).
(c) (We assume the reader is familiar with the definition and notation of *cascade* combination, (5.50).) $\beta(f)$ is the smallest integer k so that
(d) f is realized by the cascade
$$\mathbf{C} = (A, B, C_k, G_k, C_{k-1}, G_{k-1}, \ldots, C_1, G_1, C_0,$$
$$N_k, X_k, N_{k-1}, X_{k-1}, \ldots, N_1, X_1, N_0, j)$$
with **prime-loop**(C_j) empty for $j = 0, \ldots, k$ (or, equivalently, using the notation of (5.40), C_j^S is a combinatorial semigroup by (5.45)), and each

---

[*]For a finite state sequential machines f, $\alpha(f)$ can be consider the "series-parallel complexity of f" and $\beta(f)$ can be considered the "cascade complexity of f". -CLN

input of G_m permutes the states of G_m (i.e., $q \mapsto \lambda(q, a)$ is one-to-one onto for each input a of G_m) for $m = 1, \ldots, k$. Thus G_m^S is a group.

Then the strengthened Theorem (5.30) is

Theorem 5.32 (e). *There exists exactly one function* Θ *satisfying Axioms A, B, Continuity of B, C, and D, and it is*

$$\Theta(f) = \#_G(f^S) = \alpha(f) = \beta(f) \ .$$

Proof. See (5.81) of the Appendix at the end of this Chapter.

As is usual in any system of physical measurement, *we must make a choice of 'units of complexity'* (like centimeters, grams, and seconds in mechanics, etc.). This is often equivalent to choosing the machines of complexity 1. Again the specific choice will be dictated by the application one has in mind. Again, however, for most of the choices similar mathematical techniques will apply. Now Continuity of Axiom B (5.27) amounts to the following (for a proof see Lemma (5.82)).

Choice of the 'Units of Complexity'. Let f be a machine with f^S a nontrivial group G. Then $\Theta(f) = 1$.

Other 'natural choices' of the 'units of complexity' occur to one. For example, $\Theta(f) = 1$ if $f^S \in$ **PRIMES**; or perhaps $\Theta(f) = \text{size}(f)$ when $f^S \in$ **PRIMES**, etc. However, the 'natural choices' of the 'units of complexity' all lead to a similar mathematical theory. If a successful theory with the present Axioms can be developed then it can be modified to yield a successful theory with the other choices. However, for the other choices, Continuity of Axiom B will *not* hold. See Lemma (5.82).

The present axioms say that if f can be realized with a circuit $C = (A, B, Q, \lambda, \delta)$ such that each input *permutes* the states (i.e., for each $a \in A$, $q \mapsto \lambda(q, a)$ is a one-to-one map of Q onto Q), then the complexity of f is at most one (and, in fact, exactly one unless each such permutation is the identity map). (See Proposition (5.75) and Lemma (5.82b) in the Appendix at the end of this Chapter for a proof.) The intuition is that *generalized arithmetic operations have complexity one*. To see this, suppose $f^S = G$ is a *solvable* group. Then it follows that $f = j\mathbb{Z}_{p_n}^M$ series $\cdots$ series $\mathbb{Z}_{p_1}^M \bar{h}$, i.e., $f \in SP(\{\mathbb{Z}_p^M : p \geq 2 \text{ is a prime integer}\})$. In this case f is clearly an arithmetic operation, being the iteration of adding modulo p for various prime integers p. Now assuming SNAGs are a non-commutative generalization of $\mathbb{Z}_p$ (i.e. passing from simple abelian groups to simple groups) we find f^S is a group iff $f = jS_n^M$ series $\cdots$ series $S_1^M\overline{h}$ where $S_j \in$ **PRIMES**, so *f is a generalized arithmetic operation iff f^S is a group*. See (5.94) and (5.95) and the discussion following for a proof of a 'Stone-Weierstrass' theorem for SNAGs and an *interpretation of* SNAGs *as mod p counters with error correction*. A possibility of using SNAGs as 'Shannon codes' is also suggested.

Thus Axioms C and Continuity of B boil down to saying: *combinatorial operations have complexity zero and generalized arithmetic operations have complexity one.* But we emphasize again that Axioms A and B are critical while Axioms C and Continuity of B are just a choice of 'the units of complexity' and may be varied to suit the situation.

Thus by Theorem (5.32e), *the complexity of f equals the minimal number of times non-arithmetic and arithmetic operations* (machines) *must be alternated* (in series) *to yield f* (when by convention only the number of arithmetic operators is counted).

With reference to our previous long discussion between Axiom B (5.26) and Continuity of Axiom B (5.27), we would like to say that investigating the relations between the various theories of 'complexity' for finite-type objects of Winograd, Papert-Minsky, Kolmogorov, and the one just presented here, and their relations to Axioms A (5.25) and B (5.26) is a very important and promising area of research. We plan to present a paper on this topic. It would take us too far astray to go into the details here. However we will present one result.

We consider the models of Winograd [Wino] to define the complexity of a function defined on a finite set. Essentially, quoting the definitions from [Spira], we consider logical circuits composed to elements each having at most r input lines and one splittable output line, and unit delay in computing their output. Each line carries values from the set $\{0, 1, \ldots, d-1\}$. The input lines are partitioned into n sets with $I_{c,j}$ the set of possible configurations on the jth $(j = 1, \ldots, n)$. O_c is the set of possible output configurations. Such a circuit is called a (d, r)-*circuit*.

Definition 5.34. Let $\phi : X_1 \times \cdots \times X_n \to Y$ be a function on finite sets. A circuit c is said to *compute* ϕ *in time* τ iff there exists maps $g_j : X_j \to I_{c,j}$ $(j = 1, \ldots, n)$ and a one-to-one function $h : Y \to O_c$ such that if c receives constant input $(g_1(x_1), \ldots, g_n(x_n))$ from time 0 through time $\tau - 1$, then the output at time τ will be $h(\phi(x_1, \ldots, x_n))$. We define

$W_{(d,r)}(\phi)$ to be the smallest time τ such that a (d,r)-circuit can compute' ϕ in time τ.

Notice if S is a finite semigroup, then the multiplication is a map $\cdot : S \times S \to S$ and $W_{(d,r)}(\cdot)$ is defined. We denote $W_{(d,r)}(\cdot)$ by $W_{(d,r)}(S)$. Then our result, which follows from the results of Winograd-Spira and the axioms for complexity, is

Proposition 5.35. *Let* $f_1, f_2, \ldots, f_n, \ldots$ *be a sequence of machines whose complexity tends to* ∞ *with* n. *Then for each fixed* (d,r), $W_{(d,r)}(f_n^S)$ *also tends to* ∞ *with* n. *In fact, for each finite semigroup* S,

$$(5.36) \qquad \log_r \log_d(\#_G(S)) \leq W_{(d,r)}(S) .$$

Proof. See (5.84) in the Appendix at the end of this Chapter.

In this Chapter we have introduced intuitive axioms for the complexity of finite state machines and found that the definition agrees with that for finite semigroups (via (5.32)). Previously in Chapters 2 and 3 we have seen that the complexity of finite semigroups was a natural generalization of the important concepts and techniques of finite group theory. In Chapter 4 we found that complexity was the minimal number of physical coordinates necessary to understand a physical system having a finite state space. In all cases 'complexity' is the maximum chain of 'dependencies'.

Thus our theory of complexity arises naturally from three diverse viewpoints, namely, finite groups and semigroups, finite state machines, and finite state physics. It is also given by a natural set of axioms. Thus it must be natural!

Now instead of stating some main theorems of complexity in the abstract, we will turn to applications (presented in the next Chapter) and state and use the theorems in specific cases.

For those who are already familiar with the 'machine techniques' of the "Algebraic Theory of Machines" [Trans] or for those who plan to read the Appendix following Chapter 5, we make a few more comments on complexity.

In attempting to write the machine f as a member of $SP(\mathcal{M})$ we find U_3^M machines and delays (both D_1 and A^{rM}) come up as 'nuisances' all over the place. If from the same 'idea' two different individuals write out f as a member of the $SP(\mathcal{M})$ giving (5.11a), then in the main the decompositions will be the same but differ in small details usually involving U_3^M, D_1, A^{rM} machines, and the $\overline{h_j}$'s. Thus it is natural to desire to obtain properties of machines which do *not* depend on such 'nuisances' but are more fundamental. Also if two circuits differ slightly, only in the output map δ, we would hope that the properties under consideration would not change. One way to guarantee this is to have the property depend not on the machine but only on the semigroup of the machine. In fact, for reasons similar to those given in the long discussion (3.2.8) of $\#_G$ above, (3.28), if the properties one is interested in depend on the output maps, the 'nuisances' U_3^M, A^{rM}, D_1, or on exactly how the various machines are 'hooked' together (i.e., on the h_j's), they probably cannot be solved successfully. The $\overline{h}_j$'s correspond in algebra to extensions (see (2.4), (2.5) and (5.53b) with $C' = C'(G^M)$) and we know from previous discussion that only properties defined 'on the Grothendieck ring' are generally successfully solved.

The complexity of machines has all the desirable properties. It only depends on f^S and not on f via (5.32). It does not depend on the 'nuisances' via Axiom C, (5.28). *Complexity does not depend on the explicit form of the h_j's, but only on the pieces they 'hook' together.* In a very philosophic way, Axioms A, B, Continuity of B, C, and D amount to asserting that Θ is 'defined on the Grothendieck ring as nearly as possible'.

However, in closing this Chapter we note that there are three main sources of intuition for the theory of finite state machines, namely:

5.37. (a) electrical engineering models of electronic circuitry; see, for example [SM], [P-M], [Haring], [Shift].

(b) finite semigroup theory.

(c) biology and cybernetics via considering an organism or chemical reaction, etc., as modeled by a finite state machine. See, for example, the following Chapter 6.

Our opinion is that (5.37a) has been overworked! Sequential machines do not occur very naturally in electronic circuitry. (They do though in linear system theory via the idea of Proposition (5.6). See Kalman's [Kal].)

The approach (5.37b) has been worked very successfully, mainly by the author, and has greatly increased our knowledge both of finite state machines and finite semigroups.

The approach of (5.37c) had not yet been pushed technically to date. Our next Chapter 6 pushes this viewpoint strongly. We feel major advances will come by pushing (5.37c) strongly, coupled with strong algebraic mathematical technique.

The reader can judge for himself or herself.

Bibliography

[Blum] M. Blum. A machine-independent theory of the complexity of recursive functions. *Journal of the Association for Computing Machinery*, 14(2):322–336, 1967.

[Cohn] P. M. Cohn. *Universal Algebra*. (first published 1965 by Harper & Row) revised edition, D. Reidel Publishing Company, Dordrecht, 1980.

[Haring] D. R. Haring. *Sequential-Circuit Synthesis*. Research Monograph No. 31. Massachusetts Institute of Technology Press, Cambridge, Massachusetts, 1966.

[J-M] F. Jacob and J. Monod. Genetics regulatory mechanisms in the synthesis of proteins. *Journal of Molecular Biology*, 3:318–356, June 1961.

[Kal] R. E. Kalman. *Introduction to the Algebraic Theory of Linear Dynamical Systems*, volume 11 of *Lecture Notes in Operation Researchs and Mathematical Economics*. Springer-Verlag, Berlin, 1969.

[Kol1] A.N. Kolmogorov. On tables of random numbers. *Sankhya: The Indian Journal of Statistics*, Series A 25:369–376, 1963.

[Kol2] A. N. Kolmogorov. Three approaches to the quantitative definition of information. *International Journal of Computer Mathematics*, 2:157–168, 1968.

[Kol3] A. N. Kolmogorov. Logical basis for information theory and probability theory. *IEEE Transactions on Information Theory*, IT-14(5):662–664, September 1968.

[M-L-S] M. A. Arbib, editor. *Algebraic Theory of Machines, Languages and Semigroups*. Academic Press, 1968.

[P-M] M. L. Minsky and S. A. Papert. *Perceptrons*. Massachusetts Institute of Technology Press, Cambridge, Massachusetts, 1969. Expanded edition published 1987.

[q-theory] John Rhodes and Benjamin Steinberg. *The q-theory of Finite Semigroups*. Springer Monographs in Mathematics, Springer Verlag, 2009. [The most complete treatment of group complexity to date].

[Shift] Robert L. Martin. *Studies in Feedback-Shift-Register Synthesis of Sequential Machines*. Research Monograph No. 50. Massachusetts Institute of Technology Press, Cambridge, Massachusetts, 1969.

[SM] E. F. Moore. *Sequential Machines*. Addison-Wesley, 1964.

[Spira] P. M. Spira. The time required for group multiplication. *Journal of the Association for Computing Machinery*, 16(2):235–243, 1969.

[Trans] K. Krohn and J. Rhodes. Algebraic theory of machines. *Transactions of the American Mathematical Society*, 116:450–464, 1965.

[Wino] S. Winograd. On the time required to perform multiplication. *Journal of the Association for Computing Machinery*, 14(4):793–802, 1967.

Appendix to Chapter 5

In this Appendix we gather together some additional elementary material on finite state machines and provide some proofs for some of the unproved assertions of Chapter 5. For a basic reference, see [M-L-S], especially Chapter 5 and also [Auto Th.].

We first wish to clarify when $f \mid g$ holds for machines.

Definition 5.38. (a) Let $f : A^+ \to B$ be a machine. Then $NF(f)$, read the *normal form of* f, by definition equals (f^S, f^P) where f^S was defined in Definition (5.13) and f^P is the partition (or equivalence relation) on f^S given by $[t]_r \equiv [r]_f \pmod{f^P}$ iff $f(t) = f(r)$. Let $j_f : f^S \to B$ be defined by $j_f([t]_f) = f(t)$, so j_f induces the partition f^P.

(b) Let (S_1, P_1) and (S_2, P_2) be two semigroups with partitions. Then $(S_1, P_1) \mid (S_2, P_2)$, read (S_1, P_1) *divides* (S_2, P_2) iff there exists a subsemigroup $S \leq S_2$ and a surmorphism $\theta : S \twoheadrightarrow S_1$ such that $s \equiv s' \pmod{P_2}$ implies $\theta(s) \equiv \theta(s') \pmod{P_1}$.

Proposition 5.39. *Let $f : A^+ \to B$ and $g : C^+ \to D$ be two machines.*

(a) $f \mid j_f f^{SM}$ (ℓp) and $j_f f^{SM} \mid f$ (where the latter division cannot in general be taken to be ℓp).

(b) $f \mid g$ iff $NF(f) \mid NF(g)$.

Proof. To prove (a) we define $h_f : A \to f^S$ by $h_f(a) = [a]_f$ yielding $f = j_f f^{SM} \overline{h_f}$ (the fundamental expansion (5.15)), where the equality is proved by an easy direct computation. This shows $f \mid j_f f^{SM}$ (ℓp). Next we define the homomorphism (or trivial code) $H_f : (f^S)^+ \to A^+$ by $H_f(s) = (a_1, \ldots, a_n)$ where $(a_1, \ldots, a_n)$ is any fixed sequence of A^+ such that $[(a_1, \ldots, a_n)]_f = s$. Then $j_f f^{SM} = f H_f$ is established by an easy direct computation which shows $f \mid j_f f^{SM}$. This proves (a).

We next show $f \mid g$ implies $NF(f) \mid NF(g)$. Suppose $hgH = f$ with H a homomorphism (or trivial code). Consider $S_g = \{ [H(t)]_g \in g^S : t \in A^+ \}$. S_g is a subsemigroup of g^S since H is a homomorphism. Then $[H(t)]_g \mapsto [t]_f$ is a well defined surmorphism of S_g onto f^S satisfying the conditions of Definition (5.38b) as is proved by an easy direct verification.

We now show $NF(f) \mid NF(g)$ implies $f \mid g$. Let $S_g \leq g^S$ and $\theta : S_g \twoheadrightarrow f^S$ a surmorphism satisfying the conditions of Definition (5.38b). Now by (a) it suffices to show $j_f f^{SM} \mid j_g g^{SM}$. Let j'_g be j_g restricted to S_g. Let $H : (f^S)^+ \twoheadrightarrow (S_g)^+$ be a homomorphism such that for each $s \in f^S$, $H(s) = \overline{s}$ with $\theta(\overline{s}) = s$. Also, since θ 'carries' the partitions as is required by Definition (5.38b), there exists a function h so that $j_f \theta = h j_g'$. Then $j_f f^{SM} = j_f(\theta S_g^M H) = h j_g' S_g^M H = h j_g g^{SM} H$ so $j_f f^{SM} \mid j_g' S_g^M \mid j_g g^{SM}$. This proves (b) and Proposition (5.39). $\qquad\square$

The above proposition reduces the machine division $f \mid g$ to an algebraic condition $NF(f) = (f^S, f^P) \mid (g^S, g^P) = NF(g)$. Notice

(5.39c) $\qquad f \mid g$ implies $f^S \mid g^S$.

We next wish to define the semigroup of a machine, f^S, in terms of circuits canonically associated with f and then give a conceptual proof of the fundamental expansion (5.15). We begin by defining the semigroup of a circuit.

Definition 5.40. Let $\lambda : Q \times A \to Q$. Then for $t = (a_1, \ldots, a_n) \in A^+$, $\cdot t : Q \to Q$ is defined by $q \cdot a_1 = \lambda(q, a_1)$ and $q \cdot a_1 \ldots a_n = (q \cdot a_1 \ldots a_{n-1}) \cdot a_n$ for $q \in Q$. Then $\{\cdot t : t \in A^+\}$ forms a semigroup under composition of maps of Q into Q where we write the variable on the left, so letting $\circ$ denote composition, $\cdot t_1 \circ \cdot t_2 = \cdot t_1 t_2$, i.e., $(q)(\cdot t_1 \circ \cdot t_2) = ((q) \cdot t_1) \cdot t_2 = (q) \cdot t_1 t_2$ for $q \in Q$.

We denote the semigroup $\{\cdot t : t \in A^+\}$ by λ^S, read "the *semigroup of* λ". If $C = (A, B, Q, \lambda, \delta)$ is a circuit, then λ^S is also occasionally denoted by C^S, read "the *semigroup* of C". Notice C^S is the collection of all transformations of the states of Q given by input *sequences* $t \in A^+$ to C.

Unfortunately, f^S and $C(f)^S$ are not quite isomorphic (see Proposition (5.44)). To remedy this slight defect we must change our definition of circuit a little and let the output depend only on the state and not on both the state and the input. Then all will be well. We did not do this initially since this new model is somewhat less clear in details and in ease of exposition.

Definition 5.41. $C' = (A, B, Q, \lambda', \delta')$ is a *state-output-dependent circuit** iff A, B and Q are non-empty sets, $\lambda' : Q \times A \to Q$ and $\delta' : X \to B$ for some set X with $\lambda'(Q \times A) \subseteq X \subseteq Q$.

For $q \in Q$, $C_q' : A^+ \to B$ is defined by $C_q'(a_1) = \delta'(q \cdot a_1)$ and $C_q'(a_1, \ldots, a_n) = C_{q \cdot a_1}'(a_2, \ldots, a_n)$ where, as before, $q \cdot a_1 = \lambda'(q, a_1)$ and $q \cdot a_1 \ldots a_n = ((q \cdot a_1 \ldots a_{n-1}) \cdot a_n)$. Thus $C_q'(a_1, \ldots, a_n) = \delta'(q \cdot a_1 \ldots a_n)$.

---

*Such a circuit is a sometimes called a *Moore automaton* in the literature. -CLN

Writing $q \xrightarrow{a} q'$ iff $\lambda'(q, a) \equiv q \cdot a = q'$ and writing $q \to b$ iff $\delta'(q)$ is defined and equals b we have

$$q = q_0 \xrightarrow{a_1} q_1 \xrightarrow{a_2} q_2 \xrightarrow{a_3} \cdots \xrightarrow{a_n} q_n$$

$$\downarrow \qquad\qquad \downarrow \qquad\qquad\qquad\qquad \downarrow$$

$$b_1 \qquad\qquad b_2 \qquad\qquad\qquad\qquad b_n$$

with $C'_q(a_1, \ldots, a_k) = b_k$ for $k = 1, \ldots, n$.

In the following, *notation with primes (like C') will denote state-output-dependent circuits.*

We next wish to define the analogue of $C(f)$ (see Definition (5.6)) for state-output-dependent circuits.

Definition 5.42. Let $f : A^+ \to B$ be a machine. Then $C'(f)$, read "the *reduced state-output-dependent circuit of f*", is defined as follows: Let $1 \notin A^+ \cup B$ and let $(A^+)^1 = A^+ \cup \{1\}$ and $B^1 = B \cup \{1\}$. Let 1 be the identity of $(A^+)^1$ so $1 \cdot t = t \cdot 1 = t$ for all $t \in (A^+)^1$. Define $\hat{f} : (A^+)^1 \to B^1$ by $\hat{f}(1) = 1$ and $\hat{f}(t) = f(t)$ for $t \in A^+$. For $t \in (A^+)^1$ define $\hat{L}(t) :$ $(A^+)^1 \to (A^+)^1$ by $\hat{L}(t)(r) = tr$ for all $r \in (A^+)^1$ so $\hat{L}(t_1 t_2) = \hat{L}(t_1) \cdot \hat{L}(t_2)$. Let $\{\hat{f}\hat{L}(t) : t \in (A^+)^1\}^*$ denote $\{\hat{f}\hat{L}(t) : t \in (A^+)^1\}$ if $f \neq fL_t$ for all $t \in A^+$ and otherwise it denotes $\{\hat{f}\hat{L}(t) : t \in A^+\}$. Finally let $C'(f)$ be the state-output-dependent circuit defined by

$$C'(f) = (A, B, \{\hat{f}\hat{L}(t) : t \in (A^+)^1\}^*, \lambda', \delta')$$

with $\lambda'(\hat{f}\hat{L}(t), a) = \hat{f}\hat{L}(t)\hat{L}(a) = \hat{f}\hat{L}(ta)$ and for $t \neq 1$, $\delta'(\hat{f}\hat{L}(t)) = \hat{f}\hat{L}(t)(1) = f(t)$.

Proposition 5.43. $C'(f)_{\hat{f}\hat{L}(t)} = fL_t$. *Thus there exists a state q_0 of $C'(f)$ such that $C'(f)_{q_0} = f$.*

Proof. Let $q = \hat{f}\hat{L}(t)$. Then $C'(f)_q(a_1, \ldots, a_n) = \delta'(q \cdot a_1 \ldots a_n) = \delta'(\hat{f}\hat{L}(t) \cdot a_1 \ldots a_n) = \delta'(\hat{f}\hat{L}(t)\hat{L}(a_1) \cdots \hat{L}(a_n)) = \delta'(\hat{f}\hat{L}(ta_1 \ldots a_n)) = f(ta_1 \ldots a_n) = fL_t(a_1 \ldots a_n)$. This proves Proposition (5.43).* $\qquad\square$

---

*Observe that both $L_t : A^+ \to A^+$ and $\hat{L}(t) : (A^+)^1 \to (A^+)^1$ occur in the above definition and proposition. The last assertion of the proposition is clear from the first, since by the definition of the states of the reduced state-output-dependent circuit C', either there exists a $t \in A^+$ with $f = fL_t$, and so $\hat{f}\hat{L}(t)$ is a state of C', or otherwise $\hat{f}\hat{L}(1)$ is a state of C'. -CLN

Proposition 5.44. *Let $f : A^+ \to B$ be a machine and let $C(f)$ and $C'(f)$ be as defined in Definition (5.6) and (5.42), respectively. Then*

(a) The semigroups f^S and $(C'(f))^S$ are isomorphic.

(b) The semigroup of the left action of f^S on itself is isomorphic with $(C(f))^S$, so in particular

$$f^S \underset{\gamma}{\twoheadrightarrow} C(f)^S .$$

Proof. Let $C'(f) = (A, B, Q', \lambda', \delta')$. Then with respect to λ', $\cdot a_1 \ldots a_n = \cdot a'_1 \ldots a'_m$ iff $\hat{f}\hat{L}(t)a_1 \ldots a_n = \hat{f}\hat{L}(t)a'_1 \ldots a'_m$ for all $t \in (A^+)^1$ iff $\hat{f}\hat{L}(t)\hat{L}(a_1) \cdots \hat{L}(a_n) = \hat{f}\hat{L}(ta_1 \ldots a_n) = \hat{f}\hat{L}(ta'_1 \ldots a'_m) = \hat{f}\hat{L}(t)\hat{L}(a'_1) \cdots \hat{L}(a'_m)$ for all $t \in (A^+)^1$ iff $\hat{f}\hat{L}(ta_1 \ldots a_n)(r) = \hat{f}\hat{L}(ta'_1 \ldots a'_m)(r)$ for all $t, r \in (A^+)^1$ iff $f(ta_1 \ldots a_n r) = f(ta'_1 \ldots a'_m r)$ for all $t, r \in (A^+)^1$ iff $[a_1 \ldots a_n]_f = [a'_1 \ldots a'_m]_f$, proving (a).*

Let $C(f) = (A, B, Q, \lambda, \delta)$. Then with respect to λ, a calculation similar to the above shows that $\cdot a_1 \ldots a_n = \cdot a'_1 \ldots a'_m$ iff $f(\alpha a_1 \ldots a_n \beta) = f(\alpha a'_1 \ldots a'_m \beta)$ for all $\alpha \in (A^+)^1$ and $\beta \in A^+$ (but $\beta \neq 1$). But $[a_1 \ldots a_n]_f [\beta]_f = [a'_1 \ldots a'_m]_f [\beta]_f$ for all $\beta \in A^+$ (i.e., $[a_1, \ldots, a_n]_f$ and $[a'_1, \ldots, a'_m]_f$ induce the same mapping in the left action of f^S on itself) iff $f(\alpha a_1 \ldots a_n \beta \gamma) = f(\alpha a'_1 \ldots a'_m \beta \gamma)$ for all $\alpha, \gamma \in (A^+)^1$, $\beta \in A^+$ iff $f(\alpha a_1 \ldots a_n \beta) = f(\alpha a'_1 \ldots a'_m \beta)$ for all $\alpha \in (A^+)^1$, $\beta \in A^+$ (but $\beta \neq 1$). This proves (b) and Proposition (5.44).† $\square$

Using Propositions (5.43), (5.44a), and Definition (5.41) it is trivial to give a conceptual proof of the fundamental expansion (5.15). In fact, from (5.41), $f(a_1, \ldots, a_n) = \delta'(q_0 \cdot a_1 \ldots a_n)$, so defining $j_f : C'(f)^S \to B$ by $j_f(\cdot a_1 \ldots a_n) = \delta'(q_0 \cdot a_1 \ldots a_n)$ and defining $h_f : A \to C'(f)^S$ by $h_f(a) = \cdot a$ we prove $f = j_f C'(f)^{SM} \overline{h_f}$.

In the proof of Corollaries (5.21) and (5.22) we require the following

---

*The proof here explicitly covers the case when $\hat{f}\hat{L}(1) \in Q'$, i.e. each $\hat{f}\hat{L}(t) \in Q'$ where t can take the value 1. But if $\hat{f}\hat{L}(1) \notin Q'$, then by definition of the reduced state-dependent-output circuit, there is a $t_0 \in A^+$, $f = fL_{t_0}$, in which case $f(a_1 \ldots a_n r) = fL_{t_0}(a_1 \ldots a_n r) = f(t_0 a_1 \ldots a_n r)$, whence the result also follows since, in this case, $f(ta_1 \ldots a_n r) = f(ta'_1 \ldots a'_m r)$ holds for all $t, r \in (A^+)^1$ iff $f(ta_1 \ldots a_n r) = f(ta'_1 \ldots a'_m r)$ holds for all $t \in A^+, r \in (A^+)^1$. -CLN

†To see easily that the mapping $f^S \twoheadrightarrow C(f)^S$ is a γ-homomorphism, observe that if $g, g' \in f^S$ are members of a subgroup G of f^S and are identified under the map, then they have the same left action on all elements of f^S, whence $g' = g'e = ge = g$, where e denotes the identity of $G \subseteq f^S$. In other words, the homomorphism is one-to-one when restricted to any subgroup. -CLN

Proposition 5.45. *Let* $f : A^+ \to B$ *be a machine realizable by a finite state circuit with A and B finite sets. Then $C'(f)$ is a finite state-output-dependent circuit and* **prime-loop**$(C(f)) =$ **prime-loop**$(C'(f)) = \{p : p \geq 2$ *is a prime integer and* $\mathbb{Z}_p$ *divides* $f^S\}$.

Proof. First, $C'(f)$ is finite state since clearly

$$|Q'| = |\{\hat{f}\hat{L}(t) : t \in (A^+)^1\}^*| \leq |\{fL_t : t \in (A^+)^1\}| \cdot |B| = |Q| \cdot |B|$$

and Q is a finite set by Proposition (5.9). Now $C'(f)^S$ is isomorphic with f^S by Proposition (5.44a), so (Q', f^S) is a right mapping semigroup. Now by using the well known fact that a prime p divides the order of a group G iff G has an element of order p and canonical facts concerning mapping representations of $\mathbb{Z}_p$ (see [Hall]), we see **prime-loop**$(Q', f^S) = \{\mathbb{Z}_p : p \geq 2$ is a prime integer and $\mathbb{Z}_p$ divides $f^S\} \equiv X_1$. Similarly, **prime-loop**$(Q, C(f)^S) = \{\mathbb{Z}_q : q \geq 2$ is a prime integer and $\mathbb{Z}_q$ divides $C(f)^S\} \equiv X_2$. But $f^S \underset{\gamma}{\twoheadrightarrow} C(f)^S$ by Proposition (5.44b), so clearly $X_1 = X_2$. This proves Proposition (5.45). $\qquad\square$

We next prove Proposition (5.16).

5.46. Proof of Proposition (5.16). The proof will follow from the following assertions and constructions.

(a) Let $f : A^+ \to B$ be a machine with A and B finite sets. Then f is realizable by a finite state circuit iff $C'(f)$ is a *finite* state-output-dependent circuit.

Proof of (a). By Proposition (5.9), f is realizable by a finite state circuit iff $\{fL_t : t \in (A^+)^1\} = Q$ is a finite set. But, clearly, $|Q| \leq |\{\hat{f}\hat{L}(t) : t \in (A^+)^1\}^*| \leq |Q| \cdot |B|$ proving (5.46a).

(b) Let f be a machine and let $C' = (A, B, Q', \lambda', \delta')$ be a state-output-dependent circuit. Suppose $f = C'_{q_0}$. Let $g : C^+ \to D$ and suppose $g = jfH$ for some homomorphism (trivial code) $H : C^+ \to A^+$ and some map $j : B \to D$. Let $C^\#$ be the state-output-dependent circuit $C^\# = (C, D, Q', \lambda^\#, \delta^\#)$ with $\lambda^\#(q', c) = q' \cdot H(c)$ (with respect to λ') and $\delta^\# = j\delta'$. Then $g = C^\#_{q_0}$.

Proof of (b). Let $\lambda'(q', a)$ be denoted by $q' \cdot a$ and let $\lambda^\#(q', c)$ be denoted by $q' * c$. Let $(c_1, \ldots, c_m) \in C^+$. Then $g(c_1, \ldots, c_m) = jf(H(c_1) \ldots H(c_m)) =$

$j\delta'(q_0 \cdot H(c_1) \ldots H(c_m)) = \delta^{\#}(q_0 * c_1 \ldots c_m) = C'_{q_0}(c_1, \ldots, c_m)$. This proves (5.43b).

(c) Let $C'_k = (A_k, B_k, Q'_k, \lambda'_k, \delta'_k)$ for $k = 1, 2$ be two state-output-dependent circuits. Suppose $f_k = (C'_k)_{q^*_k}$. Let $C'_1 \times C'_2 = (A_1 \times A_2, B_1 \times B_2, Q'_1 \times Q'_2, \lambda', \delta')$ with $\lambda'((q_1, q_2), (a_1, a_2)) = (\lambda'_1(q_1, a_1), \lambda'_2(q_2, a_2))$ and $\delta'(q_1, q_2) = (\delta'_1(q_1), \delta'_2(q_2))$. Then $(C_1 \times C_2)_{(q^*_1, q^*_2)} = f_1 \times f_2$.

Proof of (c). Trivial.

(d) Let $C'_k = (A_k, B_k, Q'_k, \lambda'_k, \delta'_k)$ for $k = 1, 2$ be two state-output-dependent circuits. Suppose $f_k = (C'_k)_{q^*_k}$. Let $h : B_1 \to A_2$. Define $C'_2 \overline{h} C'_1$ as the state-output-dependent circuit equaling $(A_1, B_2, Q'_2 \times Q'_1, \lambda', \delta')$ with $\lambda'((q_2, q_1), a_1) = (\lambda'_2(q_2, h\delta'_1(\lambda'_1(q_1, a_1))), \lambda'_1(q_1, a_1))$ or equivalently, in the obvious notation, $(q_2, q_1) \circ a = (q_2 * h\delta'_1(q_1 \cdot a), q_1 \cdot a)$. (Reader, notice this is a wreath product action!) Also, $\delta'(q_2, q_1) = \delta'_2(q_2)$. Then $(C'_2 \overline{h} C'_1)_{(q^*_2, q^*_1)} = f_2 \overline{h} f_1^+$.

Proof of (d). Let $(a_1, \ldots a_m) \in A_1^+$. Then

$f_2 \overline{h} f_1^+(a_1, \ldots, a_m) = f_2(h(f_1(a_1)), h(f_1(a_1, a_2)), \ldots, h(f_1(a_1, \ldots, a_m)))$.

On the other hand,

$(C'_2 \overline{h} C'_1)_{(q^*_2, q^*_1)}(a_1, \ldots, a_m)$

$= \delta'((q^*_2, q^*_1) \circ a_1 \ldots a_m)$

$= \delta'(q^*_2 * h(f_1(a_1))h(f_1(a_1, a_2)) \ldots h(f_1(a_1 \ldots a_m)), q^*_1 \cdot a_1 \ldots a_m)$

$= \delta'_2(q^*_2 * h(f_1(a_1)) \ldots h(f_1(a_1 \ldots a_m)))$

$= (C'_2)_{q^*_2}(h(f_1(a_1)), \ldots, h(f_1(a_1, \ldots, a_m)))$

$= f_2(h(f_1(a_1)), \ldots, h(f_1(a_1, \ldots, a_m)))$.

This proves (5.43d).

Now Proposition (5.16a) follows immediately from (5.46a) and (5.46b). Proposition (5.16b) follows from (5.43a–d) and a very easy induction.

We next prove Proposition (5.16c). If f^S is a finite semigroup, then defining $C(S) = (S, S, S^1, \cdot, \cdot)$, where 1 is a 2-sided identity adjoined to S if S has no identity, we find $C(S)_1 = f^S$ and $C(S)$ is a finite state circuit. But then (5.15) and Proposition (5.16a) imply f is realizable by a finite state circuit. On the other hand, if f is realizable by a finite state circuit, then by (5.46a), $C'(f)$ is a finite state-output-dependent circuit and thus $C'(f)^S$ is a finite semigroup (since $(Q', C'(f)^S)$ is a right mapping semigroup and Q' is a finite set) and $C'(f)^S$ is isomorphic to f^S by Proposition (5.44a). This proves Proposition (5.16).   □

We remark that it is *not* the case that $g \in SP(\mathcal{M})$ and $f \mid g$ implies $f \in SP(\mathcal{M})$. For example, $D_1 \notin SP(U_3^M)$ but $D_1 \mid g \in SP(U_3^M)$. See [MATMI].*

Techniques centered around the proof of the Prime Decomposition Theorem for Machines, Theorem (5.17).

Notation 5.47. *Let $f : A^+ \to B$ be a machine. Then $C'(f) = (A, B, Q', \lambda', \delta')$ is defined in Definition (5.42) and $(Q', C'(f)^S)$ is a right mapping semigroup as defined in Definition (5.40). Also by Proposition (5.44a), $C'(f)^S$ is isomorphic to f^S. Thus we will write Q' as f^Q and $(Q', C'(f)^S)$ as (f^Q, f^S).*

We first show how machine equations give rise to semigroup conditions of the form $S \mid (X_n, S_n) \mathrm{w} \cdots \mathrm{w}(X_1, S_1)$, etc.

Proposition 5.48. *(a) $f \mid g$ implies $f^S \mid g^S$. Also $S_1 \mid S_2$ iff $S_1^M \mid S_2^M$. Finally, $S^{MS} = S$.*

(b) $(f \times g)^S \leq\leq f^S \times g^S$

(c) Let $C' = (A, B, Q', \lambda', \delta')$ be a state-output-dependent circuit. Let $f = C'_{q_0}$. Then $f \mid (C')^{SM}$ (ℓp) and $(f^Q, f^S) \mid (Q', (C')^S)$ (see Definition (3.1b)).

(d) $(f_2 \overline{h} f_1^+)^S \mid (f_2^Q, f_2^S) \mathrm{w} (f_1^Q, f_1^S)$ and $((f_2 \overline{h} f_1^+)^Q, (f_2 \overline{h} f_1^+)^S)$ divides $(f_2^Q, f_2^S) \wr (f_1^Q, f_1^S) = (f_2^Q \times f_1^Q, (f_2^Q, f_2^S) \mathrm{w} (f_1^Q, f_1^S))$ (see Definition (3.1b)).

Proof. The first assertion of (a) is just (5.39c). Further, if $S_1 \xleftarrow{\theta} S_2' \leq S_2$, then $S_1^M = \theta \, S_2'^M \, \overline{h}$ where $h : S_1 \to S_2$ is any map such that $\theta(h(s_1)) = s_1$. Then letting θ^* be any extension[†] of θ to S_2 yields $S_1^M = \theta \, S_2'^M \, \overline{h} = \theta^* \, S_2^M \, \overline{h}$ proving $S_1 \mid S_2$ iff $S_1^M \mid S_2^M$.[‡]

---

[*] Alternatively, the reader may refer to Fact 2.32 and the exercises associated to Remark 2.33 in Chapter 5, "The Prime Decomposition Theorem of the Algebraic Theory of Machines" by K. Krohn, J. L. Rhodes, and B. R. Tilson, of the book [M-L-S]. Although D_1 divides a series-parallel construction of U_3^M's, no such division is length-preserving.
   -CLN

[†] Note that $\theta^* : S_2 \to S_1$ is just a function and will not in general be a homomorphism.
-CLN

[‡] For the converse part, if $S_1^M = j \, S_2^M \, H$ is a division of machines as in Definition (5.10),

Now since $S^M(s_1, \ldots, s_n) = s_1 \ldots s_n$ we find $[s_1, \ldots, s_n]_{SM} = [s'_1, \ldots, s'_m]_{SM}$ iff $s_1 \cdots s_n = s'_1 \cdots s'_m$ showing $S^{MS} = S$. This proves (a).

To prove (b) let $f : A^+ \to B$ and let $g : C^+ \to D$. Clearly $[(a_1, c_1), \ldots, (a_m, c_m)]_{f \times g} \mapsto [(a_1, \ldots, a_m)]_f$ is a surmorphism θ_f of $(f \times g)^S$ onto f^S. Similarly θ_g is a surmoprhism of $(f \times g)^S$ onto g^S. But also clearly $\theta_f(s) = \theta_f(s')$ and $\theta_g(s) = \theta_g(s')$ for $s, s' \in (f \times g)^S$ implies $s = s'$. This proves (b).

Now if $f = C'_{q_0}$ then by (5.41), $f(a_1, \ldots, a_n) = \delta'(q_0 \cdot a_1 \ldots a_n)$ so defining $h : A \to (C')^S$ by $h(a) = \cdot a$ and defining $j : (C')^S \to B$ by $j(\cdot a'_1 \ldots a'_m) = \delta'(q_0 \cdot a'_1 \ldots a'_m)$ we have $f = j (C')^{SM} \overline{h}$, proving $f \mid (C')^{SM}$ (ℓp). To show $(f^Q, f^S) \mid (Q', (C')^S)$ define* $\theta : \{q_0 \cdot t : t \in (A^+)^1\} \twoheadrightarrow f^Q$ by $\theta(q_0 \cdot a_1 \ldots a_n) = \hat{f}\hat{L}(a_1 \ldots a_n)$ for $(a_1, \ldots, a_n) \in A^+$ and define $\rho : (C')^S \twoheadrightarrow f^S$ by $\rho(\cdot a_1 \ldots a_n) = [a_1 \ldots a_n]_f$. The map θ is well defined since if $q_0 \cdot a_1 \ldots a_n = q_0 \cdot a'_1 \ldots a'_m$, then for $r \geq 0$, $q_0 \cdot a_1 \ldots a_n x_1 \ldots x_r = q_0 \cdot a'_1 \ldots a'_m x_1 \ldots x_r$ so $\hat{f}\hat{L}(a_1 \ldots a_n)(x_1 \ldots x_r) = \delta'(q_0 \cdot a_1 \ldots a_n x_1 \ldots x_r) = \delta'(q_0 \cdot a'_1 \ldots a'_m x_1 \ldots x_r) = \hat{f}\hat{L}(a'_1 \ldots a'_m)(x_1 \ldots x_r)$ so $\hat{f}\hat{L}(a_1 \ldots a_n) = \hat{f}\hat{L}(a'_1 \ldots a'_m)$. Similarly ρ is a well defined surmorphism. Now $\theta(q_0 \cdot a_1 \ldots a_n) \cdot [a'_1 \ldots a'_m]_f = \theta(q_0 \cdot a_1 \ldots a_n a'_1 \ldots a'_m)$ so $(f^Q, f^S) \mid (Q', (C')^S)$. This proves (c).

To prove (d) let $C'_k = (A_k, B_k, Q'_k, \lambda'_k, \delta'_k)$ for $k = 1, 2$ be two state-dependent-output circuits. Suppose $C'_k = C'(f_k)$ and $f_k = (C'_k)_{q^*_k}$ and suppose $h : B_1 \to A_2$. Now let $C'_2 \overline{h} C'_1$ be defined as in the proof of (5.46d) and writing f for $f_2 \overline{h} f_1^+ = (C'_2 \overline{h} C'_1)_{(q^*_2, q^*_1)}$, we have by (c) that $f \mid (C'_2 \overline{h} C'_1)^{SM}$ and thus by (a), $f^S \mid (C'_2 \overline{h} C'_1)^{SMS} = (C'_2 \overline{h} C'_1)^S$. But by examining the definition of λ' where $C'_2 \overline{h} C'_1 = (A_1, B_2, Q'_2 \times Q'_1, \lambda', \delta')$ we see immediately that $(C'_2 \overline{h} C'_1)^S \leq (Q'_2, (C'_2)^S)$ w $(Q'_1, (C'_1)^S) = (f^Q_2, f^S_2)$ w (f^Q_1, f^S_1). The other assertion of (d) follows in the same manner, thus proving (d) and Proposition (5.48). □

Proposition (5.48) above yields one of the main applications of machine theory to semigroup theory. If $g \in SP(\mathcal{M})$ with g given by (5.11a), then

---

simply let the semigroup S'_2 be the image of S_1^+ in S_2 under the trivial code $H : S_1^+ \to S_2^+$ composed with S_2^M, i.e. $S'_2 = \langle H(S_1) \rangle$. The equation for the division guarantees that j restricts to a well-defined surmorphism $S'_2 \twoheadrightarrow S_1$, showing $S_1 \mid S_2$. -CLN

*Additionally, $\theta(q_0)$ is defined as $\hat{f}\hat{L}(1)$ if and only if there is no $t = a_1 \ldots a_n \in A^+$ with $f = f L_t$. In this case, there is clearly no $t = a_1 \ldots a_n \in A^+$ with $q_0 = q_0 \cdot a_1 \ldots a_n$. -CLN

by the above

$$(g^Q, g^S) \mid (f^Q_{n1} \times \cdots \times f^Q_{n\pi(n)}, f^S_{n1} \times \cdots \times f^S_{n\pi(n)}) \wr \cdots \wr (f^Q_{11} \times \cdots \times f^Q_{1\pi(1)}, f^S_{11} \times \cdots \times f^S_{1\pi(1)}) \, .$$

so in particular

(5.49)

$$g^S \mid (f^Q_{n1} \times \cdots \times f^Q_{n\pi(n)}, f^S_{n1} \times \cdots \times f^S_{n\pi(n)}) \mathrm{w} \cdots \mathrm{w} (f^Q_{11} \times \cdots \times f^Q_{1\pi(1)}, f^S_{11} \times \cdots \times f^S_{1\pi(1)}) \, .$$

Now very often it is easier to write down an equation like (5.11a) than find a division like (5.49). In many fundamental papers in finite semigroups, machine equations are used to arrive at results similar to (5.49). See [Trans] and [AxII Weak].

But in machine theory we want to go in the other direction — from semigroup conditions like (5.49) to machine equations like (5.11a), so as to reduce the proofs to semigroup theory. We next make this precise.

There are two possible approaches. One is to define a 'stronger' form of combination of machines than series-parallel, namely, *cascade* combination. The definition of cascade was "cooked-up" to make the procedure converse to Proposition (5.48) valid. However, cascade combination is fairly close to series-parallel combination. The other approach is to stay with series-parallel, but then it will be necessary to introduce many 'extra' delay machines in the constructions to obtain a converse. We will present both approaches.

The history of these concepts is amusing. Initially Krohn and Rhodes used series-parallel and reduced the machine problems to obtaining results similar to (5.49) which, of course, used the wreath product, and then used semigroup theory to complete the solution; see [Trans]. Initially, automata theorists (notably, Arbib and Zeiger) in expositing the Prime Decomposition Theorem, attempted to do away with the wreath product. Of course the concept was essential, so finding this out they simply reversed themselves and *renamed* the wreath product the cascade combination. Series-parallel is the more intuitive concept; also in use it is easier to write down series-parallel decomposition than cascade decompositions. However, cascade is more elegant and so after everything is known it is much easier to present things using cascade combinations. However, all in all, the two concepts are trivial variations of each other.

Definition 5.50. Let $C'_k = (A_k, B_k, Q'_k, \lambda'_k, \delta'_k)$ be state-output-dependent circuits for $k = 1, \ldots, n$ with A_k, B_k and Q'_k finite non-empty sets and

$\delta'_k : Q'_k \to B_k$. Then a *cascade* $\mathbf{C}$ of C'_1 by C'_2 by ... by C'_n with *connections* (or *wires* or *connecting maps*) $N_1, N_2, \ldots, N_n$ and *basic inputs* A and *basic outputs* B and final *output function* j is a $(2n + 3)$-tuple: $\mathbf{C} = (A, B, C'_n, \ldots, C'_1, N_n, \ldots, N_1, j)$ with A and B finite non-empty sets with

$$N_1 : A \to A_1$$
$$N_2 : B_1 \times A \to A_2$$
$$\vdots$$
$$N_n : B_{n-1} \times \cdots \times B_1 \times A \to A_n$$

and

$$j : B_n \times \cdots \times B_1 \to B .$$

The *associated* (state-output-dependent) *circuit* of $\mathbf{C}$ equals, by definition,

$$C'(\mathbf{C}) = (A, B, Q'_n \times \cdots \times Q'_1, \lambda', j\delta')$$

where for $a \in A$ and $(q'_n, \ldots, q'_1) \in Q'_n \times \cdots \times Q'_1$,

(5.51a)
$$\lambda'((q'_n, \ldots, q'_1), a) \equiv (q'_n, \ldots, q'_1) \cdot a$$
$$= (q'_n \cdot N_n(\delta'_{n-1}(q'_{n-1}), \ldots, \delta'_1(q'_1), a), \ldots, q'_2 \cdot N_2(\delta'_1(q'_1), a), q'_1 \cdot N_1(a))$$

(where $q'_1 \cdot N_1(a)$ denotes $\lambda'_1(q'_1, N_1(a))$, etc.) and

(5.51b)
$$\delta' : \lambda'(Q'_n \times \cdots \times Q'_1 \times A) \to B_n \times \cdots \times B_1$$
$$\text{with } \delta'(q_n, \ldots, q_1) = (\delta'_n(q_n), \ldots, \delta'_1(q_1)) .$$

By definition $\mathbf{C}$ *realizes* the machine f iff $C'(\mathbf{C})_{\bar{q}} = f$ for some $\bar{q} \in Q'_n \times \cdots \times Q'_1$. This completes Definition (5.50).*

Remark 5.52. The idea of how the cascade $\mathbf{C}$ 'works' is as follows: suppose at time t, C'_k is in state $(q'_k)_t$ and suppose $(a)_{t+1}$ is the basic input of A put in at time $t + 1$. Then the state of C'_1 at time $t + 1$, $(q'_1)_{t+1}$, equals $(q'_1)_t \cdot N_1((a)_{t+1})$. Now let $(b_1)_t$ be the output of C'_1 at time t, so $(b_1)_t = \delta'_1((q'_1)_t)$. Then $N_2((b_1)_t, (a)_{t+1}) = (a_2)_{t+1}$ determines the input $(a_2)_{t+1}$ to C'_2 at time $t + 1$. Then $(q'_2)_{t+1}$, the state of C'_2 at time $t + 1$, equals $(q'_2)_t \cdot (a_2)_{t+1}$ and $(b_2)_{t+1}$, the output of C'_2 at time $t + 1$, equals $\delta'_2((q'_2)_t \cdot (a_2)_{t+1})$, etc. The final output at time $t + 1$ is $j(\delta'_n((q'_n)_{t+1}), \ldots, \delta'_1((q'_1)_{t+1}))$. Sometimes a heuristic diagram like the following is drawn:

---

*The definition here gives the notion of cascade for state-output-dependent circuits. Similarly, a definition of cascade for machines $f_1, \ldots, f_n$ is given simply by letting the C'_i's be the state-output-dependent circuits $C'(f_i)$ realizing the f_i's. -CLN

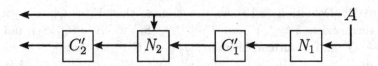

Now the reader may not intuitively like the 'time-delay' assumptions in cascade combination, namely that at time $t + 1$ the circuit C'_k receives an input $(a_k)_{t+1}$ which is a function of $(a)_{t+1}$ and $(b_{k-1})_t, \ldots, (b_1)_t$. One might think that instead C'_k should receive a function of $(a)_{t+1}$ and $(b_{k-1})_{t+1}$ at time $t + 1$. If one makes the latter choice, in place of the one made in cascade combination, one is back to series-parallel. Also for the special case of those circuits for which δ_k is the identity map, so $(b_k)_x = (q_k)_x$ at all times x, then $(a)_{t+1}$ and $(b_{k-1})_t = (q'_{k-1})_t, \ldots, (b_1)_t = (q'_1)_t$ determine $(a)_{t+1}$ and $(b_{k-1})_{t+1} = (q'_{k-1})_{t+1}, \ldots, (b_1)_{t+1} = (q'_1)_{t+1}$. Thus in this case cascade is 'stronger than' series-parallel. However, by putting in a delay to transform $(b_x)_j = (q'_x)_j$ into $(b_x)_{j-1} = (q'_x)_{j-1}$, series-parallel plus delays yields cascade. Thus in the special case (but usually general enough for all applications) cascade 'equals' series-parallel plus a few extra delays. We now proceed to make all this precise.

First the 'converse' to Proposition (5.48).

Proposition 5.53. *(a) Let* $C'_k = (A_k, B_k, Q'_k, \lambda'_k, \delta'_k)$ *be state-output-dependent circuits for* $k = 1, \ldots, n$. *Suppose* $f : A^+ \to B$ *is a machine which is realized by the cascade* $\mathbf{C} = (A, B, C'_n, \ldots, C'_1, N_n, \ldots, N_1, j)$. *Then we have*

$$(5.54) \qquad (f^Q, f^S) \mid (Q'_n, (C'_n)^S) \wr \cdots \wr (Q'_1, (C'_1)^S).$$

(b) Let $f : A^+ \to B$ *be a machine and suppose* $C' = (A, B, Q', \lambda', \delta')$ *is a state-output-dependent circuit realizing* f. *Further suppose there exists (right) mapping semigroups* $(Q'_n, S_n), \ldots, (Q'_1, S_1)$ *so that*

$$(5.55) \qquad (Q', (C')^S) \mid (Q'_n, S_n) \wr \cdots \wr (Q'_1, S_1).$$

Then there exists $C'_k = (A_k, B_k, Q'_k, \lambda'_k, \delta'_k)$ *with* C'_k *a state-output-dependent circuit and* $(Q'_k, (C'_k)^S)$ *isomorphic with* (Q'_k, S_k) *for* $k = 1, \ldots, n$ *and there exists a cascade* $\mathbf{C} = (A, B, C'_n, \ldots, C'_1, N_n, \ldots, N_1, j')$ *realizing* f.

Proof. (a) By Proposition (5.48c), $(f^Q, f^S) \mid (Q'_n \times \cdots \times Q'_1, C'(\mathbf{C})^S)$. But clearly $(Q'_n \times \cdots \times Q'_1, C'(\mathbf{C})^S) \leq (Q'_n, (C'_n)^S) \wr \cdots \wr (Q'_1, (C'_1)^S)$, proving (a).

(b) By Definition (3.1b) we can assume that $Y = Q'_n \times \cdots \times Q'_1$ and there exists $Y' \subseteq Y$ and $T' \le (Q'_n, S_n)$ w $\cdots$ w (Q'_1, S_1) and $\theta : Y' \twoheadrightarrow Q'$ and a surmorphism $\rho : T' \twoheadrightarrow (C')^S$ so that $Y' \cdot T' \subseteq Y'$ and $\theta(y' \cdot t') = \theta(y') * \rho(t')$ for all $y' \in Y'$, $t' \in T'$.* Also by Proposition (5.48c) and its proof there exists $h : A \to (C')^S$ and $j : (C')^S \to B$ so that $f = j(C')^{SM}\overline{h}$ and there exists $q_0 \in Q'$ so that $j(s) = \delta'(q_0 * s)$ so $f(a_1, \ldots, a_m) = \delta'(q_0 * h(a_1) \ldots h(a_m))$.

Now set $A_k = S_k$ and $B_k = Q'_k$ for $k = 1, \ldots, n$ and define $\lambda'_k : Q'_k \times S_k \to Q'_k$ by $\lambda'_k(q'_k, s_k) = q'_k \cdot s_k$. Also define $\delta'_k : Q'_k \to Q'_k$ by $\delta'_k(q'_k) = q'_k$. This defines C'_k for $k = 1, \ldots, n$. Clearly, $(Q'_k, (C'_k)^S)$ is isomorphic with (Q'_k, S_k).

Now for each $a \in A$ choose $w(a) \in T'$ so that $\rho(w(a)) = h(a)$. Then there exists $(w_n(a), \ldots, w_1(a))$ with $w_1(a) \in S_1$, $w_2(a) : Q'_1 \to S_2$, $\ldots$, $w_n(a) : Q'_{n-1} \times \cdots \times Q'_1 \to S_n$ so that

$$(5.56) \quad (q'_n, \ldots, q'_1) \cdot w(a) = (q'_n \cdot (w_n(a)(q'_{n-1}, \ldots, q'_1)), \ldots, q'_1 \cdot w_1(a)).$$

Now define $N_1 : A \to S_1$ by $N_1(a) = w_1(a)$ and define $N_2 : Q'_1 \times A \to S_2$ by $N_2(q'_1, a) = w_2(a)(q'_1)$, $\ldots$, and define $N_n : Q'_{n-1} \times \cdots \times Q'_1 \times A \to S_n$ by $N_n(q'_{n-1}, \ldots q'_1, a) = w_n(a)(q'_{n-1}, \ldots, q'_1)$. Finally define $j' : Q'_n \times \cdots \times Q'_1 \to B$ by $j'(y') = \delta'(\theta(y'))$ for $y \in Y'$ and otherwise j' is defined in a fixed but arbitrary manner. This defines **C**.

Let $C'(\mathbf{C}) = (A, B, Q'_n \times \cdots \times Q'_1, \lambda^*, \delta^*)$ be the associated circuit of **C**. Then clearly for $t = (a_1, \ldots, a_m) \in A^+$ and $(q'_1, \ldots, q'_n) \in Q'_n \times \cdots \times Q'_1$ we have

$$(5.57) \qquad (q'_n, \ldots, q'_1) \cdot t = (q'_n, \ldots, q'_1) \cdot w(a_1) \ldots w(a_m)$$

where $\cdot t$ is with respect to λ^* and $\cdot w(a_1) \cdots w(a_m)$ is the action in the wreath product $(Q'_n, S_n) \wr \cdots \wr (Q'_1, S_1)$. The definitions of $N_1, \ldots, N_n$ were chosen just to make (5.57) true. Now choose $(\overline{q}_n, \ldots, \overline{q}_1) \in Y'$ so that $\theta(\overline{q}_n, \ldots, \overline{q}_1) = q_0$. Then for $t = (a_1, \ldots, a_m) \in A^+$ we have

$$
\begin{aligned}
\delta'(\theta((\overline{q}_n, \ldots, \overline{q}_1) \cdot t)) &= \delta'(\theta((\overline{q}_n, \ldots, \overline{q}_1) \cdot w(a_1) \cdots w(a_m))) \\
&= \delta'(\theta(\overline{q}_n, \ldots, \overline{q}_1) * \rho(w(a_1) \cdots w(a_m))) \\
(5.58) \qquad &= \delta'(\theta(\overline{q}_n, \ldots, \overline{q}_1) * \rho(w(a_1)) \cdots \rho(w(a_m))) \\
&= \delta'(q_0 * h(a_1) \ldots h(a_m)) \\
&= f(a_1, \ldots, a_m).
\end{aligned}
$$

---

*Here $\cdot$ denotes the action in the wreath product while $*$ denotes the action in $(Q', (C')^S)$. -CLN

But if $\bar{q} = (\bar{q}_n, \ldots, \bar{q}_1)$ and $\cdot$ is taken with respect to λ^*, we have

(5.59) $C'(\mathbf{C})_{\bar{q}}(a_1, \ldots, a_m) = j'(\bar{q} \cdot t) = \delta'(\theta(\bar{q} \cdot t)).$

Now (5.58) and (5.59) show $C'(\mathbf{C})_{\bar{q}} = f$. This proves (b) and proves Proposition (5.53). $\square$

Essentially, Proposition (5.53) says solving f = a cascade of C'_1 by C'_2 by $\ldots$ by C'_n and solving (f^Q, f^S) divides $(Q'_n, (C'_n)^S) \wr \cdots \wr (Q'_1, (C'_1)^S)$ is the same problem. Now the first is a *machine problem* and the second is an *algebraic problem*. We can solve machine problems by using algebraic methods on the corresponding algebraic problem or vice versa.

Analogously with $SP(\mathcal{M})$ we can define cascade($\mathcal{M}$) as follows:

Definition 5.60. Let $\mathcal{M}$ be a collection of machines with finite basic input and output sets and realizable by a finite state circuit. Then cascade($\mathcal{M}$) $= \{g : g = C'(\mathbf{C})_{\bar{q}}$ where $\mathbf{C}$ is a cascade equaling $(A, B, C'_n, \ldots, C'_1, N_n, \ldots, N_1, j)$ where $C'_i = C'(f_i)$ for some $f_i \in \mathcal{M}$ and $f_i = C'(f_i)_{\bar{q}_i}$ and $\bar{q} = (\bar{q}_n, \ldots, \bar{q}_1)$ and $n, A, B, N_n, \ldots, N_1$ and j are arbitrary$\}$.

Informally, cascade($\mathcal{M}$) is all cascade combinations of $C'(f)$ such that $f \in \mathcal{M}$ (but $C'(f)$ may occur in the cascade any finite number of times) and $C'(f)$ is always started for the state $\bar{q}$ such that $C'(f)_{\bar{q}} = f$.

A natural question which we discussed heuristically before is — what is the relation between $SP(\mathcal{M})$ and cascade($\mathcal{M}$)? We proceed to answer this by the following two constructions

Construction 5.61 (Passing from Series-Parallel to Cascade Combination). Let $g : A^+ \to B$ and $g \in SP(\mathcal{M})$ and suppose

(5.62) $g = h_{n+1} g_n \overline{h_n} g^+_{n-1} \overline{h_{n-1}} \ldots g^+_2 \overline{h_2} g^+_1 \overline{h_1}$

with $g_k = f_{k1} \times \cdots \times f_{k\pi(k)}$ and $f_{kj} \in \mathcal{M}$. Let $C'(f_{kj}) = (A_{kj}, B_{kj}, Q'_{kj}, \lambda'_{kj}, \delta'_{kj})$ and $C'(f_{kj})_{\bar{q}_{kj}} = f_{kj}$. Now consider (5.46)(b),(c) and (d). In (5.46b) the construction $C' \to C^\#$ was defined; in (5.46c) the construction $C'_1, C'_2 \to C'_1 \times C'_2$ was defined; in (5.46d) the construction $C'_2, h, C'_1 \to C'_2 \overline{h} C'_1$ was defined.

Now we define the *state-output-dependent circuit M associated with the series-parallel combination* (5.62) as follows: Let the state-output-dependent circuit $M(g_k) = C'(f_{k1}) \times \cdots \times C'(f_{k\pi(k)})$ for $k = 1, \ldots, n$ where the construction of (5.46c) is used. Then define another state-output-dependent circuit $M^\#(g_1)$ using $M(g_1), \overline{h_1}$ and the construction of (5.46b).

Similarly define the state-output-dependent circuit $M^{\#}(g_n)$ using h_{n+1}, $M(g_n)$ and the construction of (5.46b).* Then define the state-output-dependent $M(g_2)\overline{h_2}M^{\#}(g_1)$ using the construction of (5.46d). Continue by induction to define the state-output-dependent circuit

$$M^{\#}(g_n)\overline{h_n}M(g_{n-1})\ldots\overline{h_3}M(g_2)\overline{h_2}M^{\#}(g_1) = M.$$

Now the following properties of M are easily verified:

5.63a. M is a state-output-dependent circuit with
$$M = (A, B, Q^* \equiv Q'_{n\pi(n)} \times \cdots \times Q'_{n1} \times \cdots \times Q'_{1\pi(1)} \times \cdots \times Q'_{11}, \lambda', \delta').$$

5.63b. $\delta' : Q^* \to B$ with
$$\delta'(q'_{n\pi(n)}, \ldots, q'_{n1}, \ldots, q'_{1\pi(1)}, \ldots, q'_{11}) = h_{n+1}(\delta'_{n\pi(n)}(q'_{n\pi(n)}), \ldots, \delta'_{n1}(q'_{n1})).$$

5.63c. Let $Q^*_k = Q'_{k\pi(k)} \times \cdots \times Q'_{k1}$. Then there exists functions
$$N_{1k} : A \to A_{1k} \quad \text{for} \quad 1 \le k \le \pi(k)$$

and

$$N_{2k} : Q^*_1 \times A \to A_{2k} \quad \text{for} \quad 1 \le k \le \pi(2)$$

and ... and

$$N_{nk} : Q^*_{k-1} \times \cdots \times Q^*_1 \times A \to A_{nk} \quad \text{for} \quad 1 \le k \le \pi(n)$$

such that (letting $q^*_k = (q'_{k\pi(k)}, \ldots, q'_{k1})$),

$$
\begin{aligned}
& \lambda'((q'_{n\pi(n)}, \ldots, q'_{n1}, \ldots, q'_{1\pi(1)}, \ldots q'_{11}), a) \\
(5.64) \quad & = (q_{n\pi(n)} \cdot N_{n\pi(n)}(q^*_{n-1}, \ldots, q^*_1, a), \ldots, q'_{n1} \cdot N_{n1}(q^*_{n-1}, \ldots, q^*_1, a), \\
& \quad \ldots, q'_{1\pi(1)} \cdot N_{1\pi(1)}(a), \ldots, q'_{11} \cdot N_{11}(a)),
\end{aligned}
$$

where $q'_{11} \cdot N_{11}(a)$ is taken with respect to λ'_{11}, etc.

5.63d. Let $\overline{q} = (\overline{q}_{n\pi(n)}, \ldots, \overline{q}_{n1}, \ldots, \overline{q}_{1\pi(1)}, \ldots, \overline{q}_{11})$. Then $M_{\overline{q}} = g$.

We leave the easy proofs of (5.63a–d)) to the reader (who should explicitly compute the exact formula for the N_{ij} when $n = 2$). However, we notice in (5.63c) two peculiarities of the functions N_{ij}. First, N_{ij} depends on $(q'_{i-1\pi(i-1)}, \ldots, q'_{11})$ and *not* on $(b_{i-1\pi(i-1)}, \ldots, b_{11})$, i.e., N_{ij} *depends*

---

*In applying (5.46b), for $M^{\#}(g_1)$ take $j =$ the identity map, $f = g_1$ and $H = \overline{h_1}$, for $M^{\#}(g_n)$ take $j = h_{n+1}$, $f = g_n$ and $H =$ the identity map. Note this $C \to C^{\#}$ construction is only used for these two circuits at the 'ends', but *not* for the $M(g_i)$ with $1 < i < n$. -CLN

on the previous states and not just on the previous outputs. Second, N_{ij} depends just on $(q^*_{i-1}, \ldots, q^*_1)$ and not on $(q'_{ij-1}, \ldots, q'_{i1}, q^*_{i-1}, \ldots, q^*_1)$.

Now let $C^*(f_{kj}) = (A_{kj}, Q'_{kj}, Q'_{kj}, \lambda'_{kj}, \text{id})$ where id: $Q'_{kj} \to Q'_{kj}$ with $\text{id}(q'_{kj}) = q'_{kj}$, i.e., id is the identity map. Then, by definition, *the cascade* $\mathbf{C}$ *associated with the series-parallel combination (5.62) equals*

$$\mathbf{C} = (A, B, C^*(f_{n\pi(n)}), \ldots, C^*(f_{n1}), \ldots, C^*(f_{1\pi(1)}), \ldots,$$
$$C^*(f_{11}), N_{n\pi(n)}, \ldots, N_{n1}, \ldots, N_{1\pi(1)}, \ldots, N_{11}, j')$$

with

$$j'(q'_{n\pi(n)}, \ldots, q'_{n1}, \ldots, q'_{1\pi(1)}, \ldots, q'_{11}) = h_{n+1}(\delta'_{n\pi(n)}(q'_{n\pi(n)}), \ldots, \delta'_{n1}(q'_{n1}))$$

and the N_{ij} are defined in (5.63c).

Then we have

Proposition 5.65. *The state-output-dependent circuit associated with the cascade of the series-parallel combination (5.62) equals the state-output-dependent circuit associated with the series-parallel combination (5.62).*

Proof. Use (5.64). $\qquad\square$

We note that $C^*(f_{kj})$ appears in the definition of the cascade associated with the series-parallel combination and *not* $C'(f_{jk})$. Thus the passage from series-parallel to cascade is not completely successful in the sense that the $C'(f_{kj})$ appears in the series-parallel while $C^*(f_{kj})$ occurs in the cascade. However for most applications this construction suffices. For example, since $C'(S^M) = (S, S, S^1, \cdot, \text{id})$ (where S^1 is S with a two-sided identity added if S has no right identity and otherwise S, $\cdot$ is the multiplication in S^1 and id: $S \to S$ with $\text{id}(s) = s$) we have $C'(S^M) = C^*(S^M)$. Thus we have

Proposition 5.66. *Let $\mathcal{S}$ be a collection of finite semigroups. Then* $cascade(\mathcal{S}^M) \supseteq SP(\mathcal{S}^M)$.

Proof. Use $C'(S^M) = C^*(S^M)$ proved above, construction (5.61) and Proposition (5.65). $\qquad\square$

Construction 5.67 (Passing from Cascade to Series-Parallel Combination). Let $\mathbf{C} = (A, B, C'_n, \ldots, C'_1, N_n, \ldots, N_1, j)$ be a cascade combination of C'_1 by C'_2 by $\ldots$ by C'_n where C'_j is a state-output-dependent circuit with $C'_j = (A_j, B_j, Q'_j, \lambda'_j, \delta'_j)$. Let $\bar{q} = (\bar{q}_n, \ldots, \bar{q}_1) \in Q'_n \times \cdots \times Q'_1$

and let $f = C(\mathbf{C})_{\bar{q}}$. Then the *series-parallel combination* $SP(\mathbf{C})$ associated with $\mathbf{C}$ and $\bar{q}$ is defined as follows: Let $f_k = (C'_k)_{\bar{q}_k} : A_k^+ \to B_k$ for $k = 1, \ldots, n$. If X is any set X^r is the semigroup with elements X and multiplication $x_1 \cdot x_2 = x_2$ for all $x_1, x_2 \in X$. Then $X^{rM+} : X^+ \to X^+$ is the identity map, i.e., $X^{rM+}(x_1, \ldots, x_n) = (x_1, \ldots, x_n)$. Also, by definition, $D_X : X^+ \to X \cup \{*\} \equiv X^*$ (where $*$ is an element not in X) and $D_X(x_1) = *$, $D_X(x_1, \ldots, x_n) = x_{n-1}$ for $n > 1$. Notice $D_1 = D_{\{a,b\}}$. If X is any set, X^* denotes $X \cup \{*\}$. In the following for $b_j \in B_j$, $b'_j = b_j$ and $*' = \delta'_j(\bar{q}_j)$ for $* \in B_j^*$.

Then define

$$g_{11} = \quad f_1 \times A^{rM}$$

$$g_{12} = D_{B_1} \times A^{rM} \times B_1^{rM}$$

$$g_{21} = \quad f_2 \times B_1^{*rM} \times A^{rM} \times B_1^{rM}$$

$$g_{22} = D_{B_2} \times B_1^{*rM} \times A^{rM} \times B_2^{rM} \times B_1^{rM}$$

$$\vdots$$

$$g_{n-1,1} = \quad f_{n-1} \times B_{n-2}^{*rM} \times \cdots \times B_1^{*rM} \times A^{rM} \times B_{n-2}^{rM} \times \cdots \times B_1^{rM}$$

$$g_{n-1,2} = D_{B_{n-1}} \times B_{n-2}^{*rM} \times \cdots \times B_1^{*rM} \times A^{rM} \times B_{n-1}^{rM} \times \cdots \times B_1^{rM}$$

and $g_n = f_n \times B_{n-1}^{rM} \times \cdots \times B_1^{rM}$.

Then $SP(\mathbf{C})$ equals by definition

(5.68) $f = h_{n+1} g_n \overline{h_n} g_{n-1,2}^+ g_{n-1,1}^+ \overline{h_{n-1}} \ldots \overline{h_3} g_{2,2}^+ g_{2,1}^+ \overline{h_2} g_{1,2}^+ g_{1,1}^+ \overline{h_1}$

with $h_1 : A \to A_1 \times A$

defined by $h_1(a) = (N_1(a), a)$,

$h_2 : B_1^* \times A \times B_1 \to A_2 \times B_1^* \times A \times B_1$

defined by $h_2(b_{11}, a, b_{12}) = (N_2(b'_{11}, a), b_{11}, a, b_{12})$,

$$\vdots$$

$h_{n-1} : B_{n-2}^* \times B_{n-3}^* \times \cdots \times B_1^* \times A \times B_{n-2} \times \cdots \times B_1 \to$

$\qquad A_{n-1} \times B_{n-2}^* \times \cdots \times B_1^* \times A \times B_{n-2} \times \cdots \times B_1$

defined by $h_{n-1}(b_{n-2,1}, b_{n-3,1}, \ldots, b_{11}, a, b_{n-2,2}, \ldots, b_{12}) =$

$(N_{n-1}(b'_{n-2,1}, \ldots, b'_{11}, a), b_{n-2,1}, \ldots, b_{11}, a, b_{n-2,2}, \ldots, b_{12})$,

and

$h_n : B_{n-1}^* \times B_{n-2}^* \times \cdots \times B_1^* \times A \times B_{n-1} \times \cdots \times B_1 \to A_n \times B_{n-1} \times \cdots \times B_1$

defined by

$h_n(b_{n-1,1}, \ldots, b_{11}, a, b_{n-1,2}, \ldots, b_{12}) = (N_n(b'_{n-1,1}, \ldots, b'_{11}, a), b_{n-1,2}, \ldots, b_{12})$

and finally $h_{n+1} : B_n \times B_{n-1} \times \cdots \times B_1 \to B$ with $h_{n+1} = j$.

First observe that if $\mathbf{C}$ is replaced by the new cascade $\mathbf{C}_1$, which is exactly the same as $\mathbf{C}$ except that C'_k is replaced by $C'(f_k)$, then $C(\mathbf{C})_{\bar{q}} = C(\mathbf{C}_1)_{\bar{q}} = f$. Thus we may assume, with no loss of generality, that $C'_k = C'(f_k)$ and $C'(f_k)_{\bar{q}_k} = f_k$. In this case $Q'_k = \{\hat{f}\hat{L}(t) : t \in (A^+)^1\}^*$ (see Definition (5.42)).

Proof of (5.68). Let $t = (a_1, \ldots, a_m) \in A^+$ so $m \geq 1$. With respect to λ^*, the next state function of the state-output-dependent circuit associated with $\mathbf{C}$ (see Definition (5.50)), define $(\bar{q}_n, \ldots, \bar{q}_1) \cdot t = (q_n^t, \ldots, q_1^t)$ and $(\bar{q}_n, \ldots, \bar{q}_1) \cdot a_1 \ldots a_{m-1} = (q_n^{t-1}, \ldots, q_1^{t-1})$. Then define $(b_n^t, \ldots, b_1^t)$ and $(b_n^{t-1}, \ldots, b_1^{t-1})$ by $b_j^x = \delta'_y(q_j^x)$.

Then we prove if p_1 denotes projection onto the first coordinate and

$x_k = p_1\, h_k\, g_{k-1,2}\, g_{k-1,1}^+\, \overline{h_{k-1}} \ldots \overline{h_3}\, g_{22}^+\, g_{21}^+ \overline{h_2}\, g_{12}^+\, g_{11}^+\, \overline{h_1}$ (for $1 \leq k \leq n$) and

$y_k = p_1\, g_{k2}\, g_{k1}^+\, \overline{h_k}\, g_{k-1,2}^+\, g_{k-1,1}^+\, \overline{h_{k-1}} \ldots \overline{h_2}\, g_{12}^+\, g_{11}^+ \overline{h_1}$ (for $1 \leq k < n$)

(5.69a) $(q_n^t, \ldots, q_1^t) = (\hat{f}_n\hat{L}(x_n^+(t)), \ldots, \hat{f}_1\hat{L}(x_1^+(t)))$,

(5.69b)
$$(b_{n-1}^{t-1}, \ldots, b_1^{t-1}) = (\delta'_{n-1}(q_{n-1}^{t-1}), \ldots, \delta'_1(q_1^{t-1}))$$
$$= (y_{n-1}(t), \ldots, y_1(t)) .$$

To prove (5.69a) we proceed by induction on $|t|$. For $|t| = 0$, $(q_n^1, \ldots, q_1^1) = (\bar{q}_n, \ldots, \bar{q}_1)$ and (5.69a) is valid. Thus assume (5.69a) holds for $t \in (A^+)^1$ and let $a \in A$. Then it is easily verified that

$$x_1(ta) = N_1(a)$$
$$x_2(ta) = N_2(\delta'_1(q_1^t), a)$$
$$\vdots$$
$$x_n(ta) = N_n(\delta'_{n-1}(q_{n-1}^t), \ldots, \delta'_1(q_1^t), a) .$$

Thus $q_k^{t \cdot a} = q_k^t \cdot x_k(ta)$ (with $\cdot$ taken with respect to λ'_k of $C'(f_k)$) which proves (5.69a) for ta. Finally*, (5.69b) follows from (5.69a) since $y_k(ta) = b_k^t$. This proves (5.69a) and (5.69b).

---

*To see $y_k(ta) = b_k^t$ observe that the first coordinate of g_{k2} is a one-step delay applied to the output of the first coordinate of g_k^+, i.e. to f_k^+, and so $y_k(ta) = D_{B_k}(f_k^+(x_k^+(ta))) = f_k(x_k^+(t)) = \delta'_k(\bar{q}_k \cdot x_k^+(t)) = \delta'_k(\hat{f}_k\hat{L}(x_k^+(t))) = b_k^t$. -CLN

Now (5.68) follows immediately from (5.69a) since if f' denotes the right-hand side of (5.68) then $f'(t) = h_{n+1}(f_n(x_n^+(t)), b_{n-1}^t, \ldots, b_1^t) = h_{n+1}(\delta_n'(q_n^t), \delta_{n-1}'(q_{n-1}^t), \ldots, \delta_1'(q_1^t)) = j(\delta_n'(q_n^t), \ldots \delta_1'(q_1^t)) = C(\mathbf{C})_{\bar{q}}(t) = f(t)$. This proves (5.68). $\qquad\square$

We can now complete the statement of the relations between series-parallel and cascade combination. We recall that $U_1 = \{a, b\}^r$ so U_1 has elements $\{a, b\}$ and product $x \cdot y = y$ for all $x, y \in \{a, b\}$. Thus $\{a, b\}^{rM+}$ is the identity map on $\{a, b\}^+$. Also we recall that $U_3 = U_1^1$.

Proposition 5.71. *(a) Let $\mathcal{M}$ be a collection of machines:*

$$cascade(\mathcal{M}) \subseteq SP(\mathcal{M} \cup \{D_1, U_1^M\}).$$

(b) $D_1 \in cascade(U_1^M)$.
(c) $D_1 \notin SP(U_3^M)$.
(d) Let $\mathcal{S}$ be a collection of semigroups. Then $cascade(\mathcal{S}^M) \supseteq SP(\mathcal{S}^M)$.
(e) Let $\mathcal{S}$ be a collection of semigroups. Then

$$cascade(\mathcal{S}^M \cup \{U_k^M\}) = SP(\mathcal{S}^M \cup \{U_k^M, D_1\})$$

for $k = 1$ or 3.

Proof. (a) follows immediately from construction (5.67) since clearly $A^{rM} \mid \prod U_1^M$ (ℓp) and $D_A \mid \prod D_1$ (ℓp) for any finite set A where $\prod f$ denotes some finite direct product of f with itself, i.e., $f \times \cdots \times f$ (n times) for some finite n.

(b) Let $\mathbf{C} = (\{a, b\}, \{a, b, *\}, U_1^M, U_1^M, U_1^M, U_1^M, N_4, N_3, N_2, N_1, j)$ with $N_1 : \{a, b\} \to \{a, b\}$ defined by $N_1(x) = x$; $N_2 : \{a, b\} \times \{a, b\} \to \{a, b\}$ defined by $N_2(x, y) = y$; $N_3 : \{a, b\} \times \{a, b\} \times \{a, b\} \to \{a, b\}$ defined by $N_3(x, y, z) = y$; $N_4 : \{a, b\} \times \{a, b\} \times \{a, b\} \times \{a, b\} \to \{a, b\}$ defined by $j(w, x, y, z) = *$ iff $w \neq x$ and w when $w = x$. Finally let $\bar{q} = (a, a, a, b)$. Then it is easily verified that $C_{\bar{q}} = D_1$ proving (b).

(c) For the proof we refer the reader to [MATMI] or [M-L-S], chapter 5.

(d) is Proposition (5.66).

(e) follows from (a), (b) and (d) since $U_1^M \mid U_3^M$ (ℓp). This proves Proposition (5.71). $\qquad\square$

Remark 5.72. The Prime Decomposition Theorem for Machines, Theorem (5.17), can be stated for cascade using the previous proposition as follows (all machines f considered have finite basic input and output sets and are realizable by finite state circuits so f^S is a finite semigroup):

(a) Let f be a machine, then

$$f \in \text{cascade}(\mathcal{S}^M \cup \{U_3^M\}) \text{ iff } \textbf{PRIMES}(f^S) \subseteq \textbf{PRIMES}(\mathcal{S}) .$$

In particular, $f \in \text{cascade}(\textbf{PRIMES}(f^S) \cup \{U_3^M\})$.

(b) The following two statements are equivalent:

(i) $P \in \textbf{PRIMES} \cup \textbf{UNITS}$
(ii) If $P \mid f^S$ and $f \in \text{cascade}(\mathcal{M})$, then $P \mid g^S$ for some $g \in \mathcal{M}$.

(See below concerning (b).)

Reduction of the Prime Decomposition Theorem for Machines, Theorem (5.17), to the Prime Decomposition Theorem for Finite Semigroups, Theorem (3.4).

To prove the 'if' part of (5.17a) simply use (3.5), (5.53b) and then construction (5.67).

To prove the 'only if' part of (5.17a) and to prove (5.17b) we use (3.6) and the following

Lemma 5.73. $f \in SP(\mathcal{M})$ *implies*

$$(f^Q, f^S) \mid (g_n^Q, g_n^S) \wr \cdots \wr (g_1^Q, g_1^S)$$

for some $n \geq 1$ and $g_1, \ldots, g_n \in \mathcal{M}$.

Proof. By applying (5.48c) to $f = M_{\bar{q}}$ where M is the state-output-dependent circuit associated with the series-parallel combination defining f and then using (5.63c) and (5.64), Lemma (5.73) follows. □

The cascade version, (5.72), is also reduced to (5.17) in a very similar manner. □

Infinite State Machines. We next prove a proposition showing that any obvious attempt to generalize the Prime Decomposition Theorem to infinite state machines will fail. In fact we show every machine with countable basic inputs can be realized by a cascade of '*two* non-negative infinite counters'. Precisely,

Proposition 5.74 (Allen). *Let* $f : A^+ \to B$ *be any machine with* A *a countable set. Then* f *is realizable by a cascade* $\mathbf{C} = (A, B, C'(\mathbb{N}^M), C'(\mathbb{N}^M), N_2, N_1, j)$ *where* $\mathbb{N}$ *is the semigroup of non-negative integers under addition.*

Proof. The proof proceeds via the following sequence of assertions:

(a) $S \mid T$ implies $(S^1, S) \mid (T^1, T)$.

Proof of (a). Easy and omitted.

(b) Every countably generated semigroup is countable and divides $\{0, 1\}^+$.

Proof of (b). Clearly a semigroup is countable iff it is countably generated. Further, the strings of length n generate a subsemigroup of $\{0, 1\}^+$ (consisting of all strings of length divisible by n) isomorphic to X^+ with $|X| = 2^n$. But $X^+ \twoheadrightarrow FG(2)$ where $|X| = 6$ and $FG(2)$ denotes the free group on two generators (by adding the relations $x_1 = x_2$, $x_1 x_i = x_i x_1 = x_i$ for $i = 1, \ldots, 6$, $x_3 x_4 = x_4 x_3 = x_1$, $x_5 x_6 = x_6 x_5 = x_1$ so $x_1 = x_2 = 1$, $x_4 = x_3^{-1}$ and $x_5 = x_6^{-1}$). But the derived group of $FG(2)$, $FG(2)'$, is a free group on a countable number of generators so $A^+ \le FG(2)'$ with A countably infinite (i.e., $|A| = |\mathbb{Z}|$). But if S is any countable semigroup $A^+ \twoheadrightarrow S$. Thus $S \mid A^+ \mid FG(2)' \mid FG(2) \mid X^+ \mid \{0, 1\}^+$ proving (5.74b).

(c) $((\{0, 1\}^+)^1, \{0, 1\}^+) \mid (\mathbb{N}, \mathbb{N}) \wr (\mathbb{N}, \mathbb{N})$, where $\mathbb{N}$ denotes the non-negative integers under addition.

Proof of (c). For $t = (x_1, \ldots, x_n) \in (\{0, 1\}^+)^1$ the length $|t|$ of the t is n (and $|\epsilon| = 0$ where we write ϵ for the identity of $(\{0, 1\}^+)^1$ to distinguish it from the word 1 of length 1). Let $r(t) = (x_n, \ldots, x_1)$ and $b(t)$ be the non-negative integer represented by $r(t)$ in binary where $b(\epsilon) = 0$. Thus $b(0001) = 8$, $b(100) = 1$, etc.

Now let $X = \{(b(t), |t|) : t \in (\{0, 1\}^+)^1\} \subseteq \mathbb{N} \times \mathbb{N}$ and for each $t \in (\{0, 1\}^+)^1$ let $\hat{t} \in (\mathbb{N}, \mathbb{N}) \wr (\mathbb{N}, \mathbb{N})$ be defined by

$$(n_2, n_1)\,\hat{t} = (n_2 + 2^{n_1} b(t), n_1 + |t|) .$$

Let $\hat{T}$ be the subsemigroup of $(\mathbb{N}, \mathbb{N}) \mathbin{\text{w}} (\mathbb{N}, \mathbb{N})$ generated by $\{\hat{t} : t \in (\{0, 1\}^+)^1\}$.

Then we claim $\theta : X \to (\{0, 1\}^+)^1$ defined by $\theta(b(t), |t|) = t$ is a well defined one-to-one onto map with $\theta(x\hat{t}) = \theta(x)t$ for all $x \in X$,

$t \in (\{0,1\}^+)^1$. This follows since $|rt| = |r| + |t|$ and $b(rt) = b(r) + 2^{|r|}b(t)$. Thus $\rho : \hat{T} \twoheadrightarrow (\{0,1\}^+)^1$ defined by $\rho(\hat{t}_1 \ldots \hat{t}_k) = t_1 \ldots t_k$ is a well defined surmorphism, since letting $\overline{1} = (0,0)$, $\theta(\overline{1}\hat{t}_1 \ldots \hat{t}_k) = \theta(\overline{1}\hat{t}_1 \ldots \hat{t}_{k-1})t_k = \theta(\overline{1}\hat{t}_1 \ldots \hat{t}_{k-2})t_{k-1}t_{k-2} = \theta(\overline{1})t_1 \ldots t_k = \epsilon t_1 \ldots t_k = t_1 \ldots t_k$ showing $\hat{r}_1 \ldots \hat{r}_m = \hat{t}_1 \ldots \hat{t}_n$ implies $r_1 \ldots r_m = t_1 \ldots t_n$ for all $t_j, r_k \in (\{0,1\}^+)^1$.

Now a similar calculation shows $\theta(x\alpha) = \theta(x)\rho(\alpha)$ for all $x \in X$, $\alpha \in \hat{T}$, proving (5.74c).

(d) Let S be a countably generated semigroup. Then

$$(S^1, \dot{S}) \mid (\mathbb{N}, \mathbb{N}) \wr (\mathbb{N}, \mathbb{N}).$$

Proof of (d). Use (5.74a–c).

(e) Let $f : A^+ \to B$ be a machine with A a countable set. Then f^S and f^Q are countable and

$$(f^Q, f^S) \mid (f^{S1}, f^S).$$

Proof of (e). f^S is generated by $\{\cdot a : a \in A\}$ so f^S is countable. Further if $C'(f)_{\overline{q}} = f$, then $\{\overline{q} \cdot s : s \in f^S\} = f^Q$ so f^Q is countable. Finally, $\theta : f^{S1} \twoheadrightarrow f^Q$ defined by $\theta(s) = \overline{q} \cdot s$ is such that $\theta(s \cdot t) = \theta(s)t$ for $s \in f^{S1}$, $t \in f^S$, proving (5.74e).

(f) Let $f : A^+ \to B$ be a machine with A a countable set. Then

$$(f^Q, f^S) \mid (\mathbb{N}, \mathbb{N}) \wr (\mathbb{N}, \mathbb{N}).$$

Proof of (f). Use (5.74d) and (5.74e).

Now Proposition (5.74) follows from (5.74f) and Proposition (5.53b). $\qquad \square$

Finite State Machines whose Semigroups are Groups.

Proposition 5.75. *Let $f : A^+ \to B$ be a machine with A and B finite sets. Then f^S is a finite group iff $C'(f)$ is a finite state-output-dependent circuit, each input of which permutes the states (i.e., if $C'(f) = (A, B, Q', \lambda', \delta')$ then for each $a \in A$, $q' \mapsto \lambda'(q', a) = q' \cdot a$ is a one-to-one map) iff (f^Q, f^S) is a finite transitive permutation group.*[*]

---

[*]Proposition 5.75 is by no means obvious and even perhaps surprising since if G is a group and $X = G \times \{0,1\}$ with $(g,x) \cdot g' = (gg', 0)$ for all $g \in G$, $x \in \{0,1\}$, $g' \in G$, then $\cdot g'$ is not a permutation of X, so (X, G) is *not* a permutation group even though G is a group. -CLN

Proof. The middle condition is sufficient to make f^S a finite group by Proposition (5.44a). Thus assume $f^S = G$ is a finite group and $f = j_f f^{SM} \overline{h_f}$ is given by (5.15), so $f(a_1, \ldots, a_m) = j_f(h_f(a_1) \cdots \cdot h_f(a_n))$. Then define the state-output-dependent circuit $C' = (A, B, C, \lambda_1, \delta_1)$ as follows: $\lambda_1(g, a) = g \cdot h_f(a)$ and $\delta_1 = j_f$. Then $C'_1 = f$ since $C'_1(a_1, \ldots, a_m) = j_f(1 \cdot h_f(a_1) \ldots h_j(a_m)) = f(a_1, \ldots, a_m)$. Now Proposition (5.48c) implies $(f^Q, f^S = G) \,|\, (G, (C')^S) = (G, G)$ which immediately implies (f^Q, f^S) is a transitive permutation group proving Proposition (5.75).   □

Circuits with Feedback. We next briefly consider *circuits with feedback*. Intuitively, given the circuit $C = (A_2, B, Q, \lambda, \delta)$ we introduce an additional basic input set A_1 which will receive the actual inputs and a feedback map $\alpha : B \times A_1 \to A_2$ where α (output $b_t \in B$ at time t, input $(a_1)_{t+1} \in A_1$ at time $t + 1) = (a_2)_{t+1} \in A_2$ the input of C at time $t + 1$. A diagram like

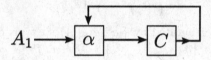

is sometimes drawn. Also some 'starting conditions' must be given. We make all this precise via the following

Definition 5.76. $F = (A_1, A_2, B, Q, \lambda, \delta, \alpha, m)$ is a *feedback circuit over the circuit* $C = (A_2, B, Q, \lambda, \delta)$ (so $\lambda : Q \times A \to Q$ and $\delta : Q \times A \to B$) with basic inputs A_1, feedback α and starting condition m iff C is a circuit and A_1 is a non-empty set and $\alpha : B \times A_1 \to A_2$ and $m : A_1 \to A_2$.

For $q \in Q$, $F_q : A_1^+ \to B$ is defined inductively as follows: Let $q_0 = q$. Let $F_{q_0}(a_1) = C_{q_0}(m(a_1)) = \delta(q_0, m(a_1)) = b_1$ and let $q_1 = q_0 \cdot m(a_1)$ (with $\cdot$ taken with respect to λ). Now assume $F_q(a_1, \ldots, a_{m-1}) = b_{m-1}$ and q_{m-1} has been defined and define

$$F_q(a_1, \ldots, a_{m-1}) = C_{q_{m-1}}(a'_m = \alpha(b_{m-1}, a_m)) = \delta(q_{m-1}, a'_m) = b_m$$

and $q_m = q_{m-1} \cdot a'_m$ (with $\cdot$ taken with respect to λ). By induction this completes the definition of F_q.

We next show that every machine can be 'realized' by some feedback over a *one-state* circuit showing 'circuits are already in feedback form'.

Proposition 5.77. Let $C = (A, B, Q, \lambda, \delta)$ be a circuit and assume $f = C_{q_0}$. Then consider the one-state circuit $D = (Q \times A, Q \times A, \{x\}, \lambda^*, \delta^*)$

with $\lambda^* : \{x\} \times (Q \times A) \to \{x\}$ *defined by* $\lambda^*(x, (q, a)) = x$ *and*
$\delta^* : \{x\} \times (Q \times A) \to Q \times A$ *defined by* $\delta^*(x, (q, a)) = (q, a)$. *Then define a feedback circuit F over D by* $F = (A, Q \times A, Q \times A, \{x\}, \lambda^*, \delta^*, \alpha, m)$
with $\alpha : (Q \times A) \times A \to Q \times A$ *defined by* $\alpha((q, a_1), a_2) = (\lambda(q, a_1), a_2)$ *and*
$m : A \to Q \times A$ *defined by* $m(a) = (q_0, a)$.
   Thus $F_x : A^+ \to Q \times A$. *We claim*

(5.78) $$\delta F_x = f .$$

Proof. Let $(a_1, \ldots, a_m) \in A^+$ be given. Define $q_0 = q$, $q_1 = q_0 \cdot a_1, \ldots, q_m = q_{m-1} \cdot a_m$ where $\cdot$ is taken with respect to λ. Then we claim

(5.79) $$F_x(a_1, \ldots, a_m) = (q_{m-1}, a_m) .$$

We verify (5.79) by induction on m. For $m = 1$, $F_x(a_1) = \delta^*(x, m(a_1)) = \delta^*(x, (q_0, a_1)) = (q_0, a_1)$. Now assuming $F_x(a_1, \ldots, a_{m-1}) = (q_{m-2}, a_{m-1}) = b_{m-1}$ we find $F_x(a_1, \ldots, a_m) = D_x(a'_m = \alpha(b_{m-1}, a_m)) = D_x(\alpha((q_{m-2}, a_{m-1}), a_m)) = (D_x(q_{m-2} \cdot a_{m-1}, a_m)) = D_x(q_{m-1}, a_m) = (q_{m-1}, a_m)$ (since $D_x(x_1, \ldots, x_m) = x_m$), proving (5.79).
   Now $f(a_1, \ldots, a_m) = C_{q_0}(a_1, \ldots, a_m) = \delta(q_0 \cdot a_1 \ldots a_{m-1}, a_m) = \delta(q_{m-1}, a_m) = \delta F_x(a_1, \ldots, a_m)$ proving (5.78) and hence also Proposition (5.77). □

5.80. State Assignment Problem. We next make some brief comments regarding the state-assignment problem discussed in (5.24).

Formally, the *state-assignment problem* for the machine $f : A^+ \to B$ (where for ease we assume $A = B = \{0, 1\}$, but this is not essential) is to find a state-output-dependent (say) circuit C' with $C' = (A, B, Q', \lambda', \delta')$ such that $f = C'_{q'}$ for some $q' \in Q'$ and to choose a state-assignment function (see (5.24b)) $j : Q' \to \{0, 1\}^n$ so that functions B_a and B defined in (5.24c) are as 'simple', 'least costly', etc. as possible. Notice since $B_a : \{0, 1\}^{n+1} \to \{0, 1\}^n$, B_a consists of n Boolean functions of $n + 1$ variables and $B : \{0, 1\}^n \to \{0, 1\}$ with $B(j(q)) = \delta'(q)$ so B is one Boolean function of n variables.

Now suppose f is realized by the cascade $\mathbf{C} = (A, B, C'_m, \ldots, C'_1, N_m, \ldots, N_1, j)$ with $C'_k = (A_k, B_k, Q'_k, \lambda'_k, \delta'_k)$ so $C'(\mathbf{C})_{\overline{q}} = f$ (see Definition (5.50)). Then $C'(\mathbf{C}) = (A, B, Q'_m \times \cdots \times Q'_1, \lambda', \delta')$. Now for each k with $1 \leq k \leq m$ choose (in an arbitrary but fixed way) $i_k : Q'_k \to \{0, 1\}^{\pi(k)}$ (where $\pi(k)$ is the smallest integer such that $2^{\pi(k)} \geq |Q'_k|$). Then define $i : Q' = Q'_m \times \cdots \times Q'_1 \to \{0,1\}^n$ with $n = \pi(1) + \cdots + \pi(m)$ by $i(q'_m, \ldots, q'_1) = (i_m(q'_m), \ldots, i_1(q'_1))$ so $i = i_m \times \cdots \times i_1$. Now if $|Q_k| = 2$ for all k, and since by (5.51a) the λ' action is in *triangular* form, B_a becomes (one Boolean function of one variable, one Boolean function of two variables, $\ldots$, one Boolean function of $n + 1$ variables), quite a savings from n Boolean functions of $n + 1$ variables. When $|Q'_k| > 2$ but m is not too small, then similar simplifications in the B_a's occur. The main cloud on the horizon is that $|Q'_m \times \cdots \times Q'_1|$ may be quite a bit larger than the states of $C(f)$ (the minimal possible by Proposition (5.9)) and thus the simplifications given by triangular action are lost in the increased size of $Q'_m \times \cdots \times Q'_1$.

If f^S is a finite *group*, then the above method of *triangular state-assignments* is 'good'. See [MinGroup]. In general, much more work needs to be done regarding the 'minimal solutions' (see last part of Chapter 4 for precise definitions) of $(f^Q, f^S) \mid (Q'_m, S'_m) \wr \cdots \wr (Q'_1, S'_1)$ before the triangular state-assignments can be fairly observed. This is a promising area of research. See also [Haring] regarding the state assignment problem.

5.81. Complexity Axioms. Proof of Theorems (5.30) and (5.32e).

(In the following proof we omit some details.)

We first note that it is not difficult to verify that $\alpha = \beta$ by using (5.61), (5.65) and (5.67). Also it is not difficult to verify that $\beta(f) = \#_G(f^S)$ by using (5.53). Thus $\alpha(f) = \beta(f) = \#_G(f^S)$ for all $f \in \mathcal{M}_{Fin}$.

Now it is very easy to verify that α satisfies Axioms B and C. Also α satisfies Continuity of Axiom B, since if $\alpha(t) = n$ and f is decomposed as in (5.32b) and $n = a + b$ with a and b non-negative integers, then let m_1 be the first $2a$ terms and m_2 the remaining terms. Then f divides m_2 series m_1. Now clearly $\alpha(m_1) = a$ and $\alpha(m_2) = b$. Now using the fundamental expansion (5.15) we find f divides m_2^{SM} series m_1^{SM}. But since $\alpha(f) = \#_G(f^S)$, $\alpha(m_k) = \alpha(m_k^{SM})$ so $\alpha(f) = n = \alpha(m_2^{SM}) + \alpha(m_1^{SM}) = a + b$. But S^M is nice for all S since if $C(S) = (S, S, S^1, \cdot, \cdot)$, then $C(S)_1 = S^M$. This proves Continuity of Axiom B for α.

Now since $f^S \twoheadrightarrow (fL_t)^S$, $f \mid g$ implies $f^S \mid g^S$ and $(f \times g)^S \leq\leq f^S \times g^S$, Axiom A follows from $\alpha(f) = \#_G(f^S)$ and Axiom I, (3.17), for $\#_G$.

Thus $\alpha = \beta = \#_G(\cdot^S)$ satisfies Axioms A, B, Continuity of B, and C.*

---

*The reader may like to observe that if Axiom D is omitted then the constant zero function $\Theta_0(f) = 0$ for all $f \in \mathcal{M}_{Fin}$ also satisfies these other complexity axioms. Also, observe that if Continuity of Axiom B is omitted then the analogue of (5.82b) need not hold. Indeed, if Continuity of Axiom B is omitted and Θ satisfies Axioms A, B, and C,

To continue the proof we require the following

Lemma 5.82. *(a) Since* $\Theta : \mathcal{M}_{Fin} \to \mathbb{N}$ *be a function which satisfies Axioms A, B, Continuity of B, and C. Then* $\Theta(f) = \Theta(f^{SM})$ *for all* $f \in \mathcal{M}_{Fin}$.

(b) Let $\Theta : \mathcal{M}_{Fin} \to \mathbb{N}$ *be a function which satisfies Axioms A, B, Continuity of B, and C. Then* $\Theta(f) \leq 1$ *whenever* $f^S = G$ *is a group.*

Proof. The proof proceeds via the following statements:

(1) If Θ satisfies Axiom A and f is nice, then $\Theta(f) = \Theta(f^{SM})$.

Proof of (1): By definition, f nice means $C(f) = (A, B, Q, \lambda, \lambda)$; (see (5.6)). Thus $C'(f) = (A, B, Q', \lambda, \text{id})$; (see (5.42)). Thus for $t \in A^+$, $fL_t(a_1, \ldots, a_n) = q_t \cdot a_1 \ldots a_n$ where $q_t = \hat{f}\hat{L}(t)$. Further, by (5.44), $f^S = \{\cdot t : t \in A^+\}$.

Let $g : A^+ \to f^S$ be defined by $g(a_1, \ldots, a_n) = \cdot a_1 \circ \cdots \circ \cdot a_n = \cdot a_1 \ldots a_n \in f^S$. Then let $t_1, \ldots, t_m$ be such that $\{\hat{f}\hat{L}(t_i) : i = 1, \ldots, m\} = Q'$. Then clearly

$$g = j(fL_{t_1} \times \cdots \times fL_{t_m})$$

where by definition $j(q_1, q_2, \ldots, q_m) = s$ iff $q_{t_i} \cdot s = q_i$ for each $i = 1, \ldots, m$.

But define $H : f^S \to A^+$ so that $H(s) = a_1 \ldots a_n$ implies $\cdot a_1 \ldots a_n = s$. Then extending H to a homomorphism* gives

$$f^{SM} = gH .$$

Now by Axiom A, $\Theta(f^{SM}) \leq \Theta(g) \leq \max\{\Theta(fL_t)\} \leq \Theta(f)$. But $f \mid f^{SM}$ by (5.15) so $\Theta(f) \leq \Theta(f^{SM})$. Thus $\Theta(f) = \Theta(f^{SM})$. This proves (1).

(2) Axioms A, B, C and Continuity of B imply that $\Theta(f) = 0$ if f^S is combinatorial and $\Theta(f) = n > 0$ implies

$$f \text{ divides } S_n^M \text{ series } \cdots \text{ series } S_1^M$$

with $\Theta(S_j^M) = 1$ for $j = 1, \ldots, n$.

Proof of (2): That $\Theta(f) = 0$ if f^S combinatorial follows from (5.19) and Axioms A, B and C. The assertion when $\Theta(f) = n > 0$ follows from (1),

---

then so does $\Theta'(f) = 2\Theta(f)$; it would follow in this case that no non-zero function could satisfy the modified Axiom D (in the second version, D$_2$, but with Continuity of Axiom B not mentioned). -CLN

*That is, extending H in the unique natural way, as in (5.10), to a trivial code $H : (f^S)^+ \to A^+$. -CLN

(5.15) and Continuity of Axiom B with $a = n - 1$, $b = 1$ and induction. This proves (2).

(3) $(S^1, S) \mid \prod(Q, S)$ where $\prod$ denotes a finite product of (Q, S) with itself.

Proof of (3): Fact (2.14c) of [M-L-S, Chapter 5].

Proof of (5.82a). Let $S = f^S$. From (2), (5.48d) and (5.44) follows that

(4) $(Q', S) \mid (S_n^1, S_n) \wr \cdots \wr (S_1^1, S_1)$.

From (3) and (4) follows

(5) $(S^1, S) \mid \prod((S_n^1, S_n) \wr \cdots \wr (S_1^1, S_1))$.

But $\prod((X_2, S_2) \wr (X_1, S_1)) \mid \prod(X_2, S_2) \wr \prod(X_1, S_1)$ by [M-L-S], proof of Fact 5.2.2. Thus (5) yields

(6) $(S^1, S) \mid (\prod(S_n^1, S_n)) \wr \cdots \wr (\prod(S_1^1, S_1))$.

Now (6) and (5.53b) (with $f = f^{SM}$ so $Q' = S^1$) and (5.67) implies (by (5.68)) that

(7) $S^M = f^{SM}$ divides g_n series $g_{n-1,2}$ series $g_{n-1,1}$ series $\cdots$ series g_{12} series g_{11} with $\Theta(g_{11}) = \Theta(S_1^M)$, $\Theta(g_{12}) = 0, \ldots, \Theta(g_{n-1,1}) = \Theta(S_{n-1}^M)$, $\Theta(g_{n-1,2}) = 0$, $\Theta(g_n) = \Theta(S_n^M)$.

Thus $\Theta(f^{SM}) \leq \sum_{i=1}^{n} \Theta(S_i^M) = n = \Theta(f)$. But $f \mid f^{SM}$ by (5.15), so $\Theta(f) \leq \Theta(f^{SM})$ so $\Theta(f) = \Theta(f^{SM})$. This proves Lemma (5.82a).

Proof of (5.82b). We begin with

(8) Let G be a finite group of order n, then $G \leq A_{2n}$. Also trivially $n \leq k$ implies $A_n \leq A_k$.

Proof of (8): Let $(G_1 + G_2, G)$ be the permutation group with $G_1 + G_2$ denoting two disjoint copies of G and $g_1 \cdot g = g_1 g$ for $g_1 \in G_1$ and $g_2 \cdot g = g_2 g$ for $g_2 \in G_2$. Then clearly G is a subgroup of the even permutations on $G_1 + G_2$.

We now complete the proof of (b). Suppose Θ satisfies Continuity of Axiom B. Then $\Theta(f) = \Theta(f^{SM})$ for all $f \in \mathcal{M}_{Fin}$ by (a). Now suppose $n \geq 5$ and $\Theta(A_n^M) = k \geq 2$. Then

$$A_n^M \text{ divides } f_2 \text{ series } f_1$$

with $\Theta(f_2) = k-1$ and $\Theta(f_1) = 1$. But $A_n^{MS} = A_n$ and by elementary group theory (see [Hall]), $A_n \in \mathbf{PRIMES}$ so by (5.17b), $A_n^M \mid f_2^{SM}$ or $A_n^M \mid f_1^{MS}$. Now since $\Theta(f_j) = \Theta(f_j^{SM})$ it follows that $k \leq \Theta(f_j^{SM}) = \Theta(f_j)$ for $j = 1$ or $j = 2$. But $\Theta(f_2) = k-1$ and $\Theta(f_1) = 1$ so this is a contradiction. Thus $\Theta(A_n^M) \leq 1$ for $n \geq 5$. But by (8), if $f^S = G$ is a group, $G \leq A_n$ for some $n \geq 5$ so $f \mid f^{SM} \mid A_n^M$ so $\Theta(f) \leq \Theta(A_n^M) \leq 1$. This proves (b) and hence Lemma (5.82). $\qquad\Box$

We next check α satisfies D_2, (5.29b). Suppose $\alpha(f) = n$ and f is expanded by (5.32b). Then by Axioms A and B,

$$\Theta(f) \leq \sum_{j=0}^{n} \Theta(f_j) + \sum_{j=1}^{n} \Theta(g_j) .$$

But by Axioms A, B and C, $\Theta(f_j) = 0$ for $j = 0, \ldots, n$. Further, by Axiom A, $\Theta(g_k) = \Theta(g_{k1})$ for $k = 1, \ldots, n$. However $g_{k1}^S = G_k$ is a group so by Lemma (5.82b), $\Theta(g_{k1}) \leq 1$. Thus

$$\Theta(f) \leq \sum_{n=0}^{n} \Theta(f_j) + \sum_{j=1}^{n} \Theta(g_j) = n = \alpha(f) .$$

Thus D_2 holds for α.

Now obviously by the nature of D_2, at most one function satisfies Axioms A, B, Continuity of B, C, and D_2. Thus Theorem (5.32e) has been proved if $D=D_2$.

In the following let $\mathfrak{A}$ denote Axioms A, B, Continuity of B, and C. Then we must show $\Theta \models \mathfrak{A}+D_2$ iff $\Theta \models \mathfrak{A}+D_1$.*

(9) However, before we can do this we must make D_1, (5.29a), completely precise. We assume the reader is familiar with some universal algebra concepts of equational theory. See for example [Cohn]. For our purposes here we will say an equation holds for the machine f iff the equation holds for the semigroup f^S. (We could directly define equations for f, but since the foundations of equational theories for semigroups are well known, our method seems the easiest.) Now by an equation E_n holding among the

---

*Notation: '$\models$' means 'is a model of' or, equivalently, 'satisfies'. -CLN

values of Θ we mean an equation (for semigroups) which holds for all f (i.e. for f^S) such that $\Theta(f) \leq n$. Thus E_n holds for Θ iff $\Theta(f) \leq n$ implies E_n holds for f.

Now since by (5.82a) and Axiom A, if $\Theta \models \mathfrak{A}$, then

$$\{f^S : \Theta(f) \leq n\} = \{S : \Theta(S^M) \leq n\}$$

is a variety of semigroups (i.e. closed under division and finite direct products). We denote this variety by Θ_n. Thus E_n holds among the values of Θ iff E_n holds for Θ_n.

Let W_1, W_2 be collections of finite semigroups (but W_j can be infinite). Then by the Theorem of G. Birkhoff (see [Cohn]),

(10) $W_1 \subseteq W_2$ iff every equation for W_2 holds for W_1; and

(11) $W_1 \subsetneqq W_2$ iff every equation for W_2 holds for W_1 but *not* conversely.

Now clearly $\Theta \leq \Theta'$ iff $\Theta_n \supseteq \Theta'_n$ for all n and $\Theta \lneqq \Theta'$ iff $\Theta_n \supseteq \Theta'_n$ and $\Theta_{n_0} \neq \Theta'_{n_0}$ for some n_0.

Now that $\mathfrak{A} + D_2$ implies $\mathfrak{A} + D_1$ and that $\mathfrak{A} + D_2$ implies $\mathfrak{A} + D_1$ follow immediately from (10) and (11).

This proves Theorems (5.30) and (5.32e).   □

5.83. Remarks. Θ denotes the unique function satisfies Axioms A, B, Continuity of B, C, and D.

(a) "Is everything true about Θ provable?" "Globally, how continuous is $\Theta : \mathcal{M}_{Fin} \to \mathrm{N}$?" Making these questions precise and attempting to answer them is a good area for future research.

(b) If one uses cascade in Axiom B, and Continuity of B, instead of series, then $f \mid g$ implies $\Theta(f) \leq \Theta(g)$, Axiom B, Continuity of B, C, and D_2 are axioms for Θ, i.e., parts of Axiom A can be omitted.

(c) If $C(f) = (A, B, Q, \lambda, \delta)$ and $C'(f) = (A, B, Q', \lambda', \delta')$, then

$$\Theta(f) = \#_G(f^S) = \#_G((\lambda')^S) = \#_G(\lambda^S).$$

Proof of (c). Use Proposition (5.44) and Axiom II for complexity, (3.20). □

(d) The proofs of Remark (5.31c) and (5.31d) follow from (c), (3.16g) and that in (5.31d), $C = C(f)$ for $f = C_q$ and $C(f)^S = F_R(X_n)$.

5.84. Computation Time. A Proof of Proposition (5.35). Reconsider Definition (5.34). For a (d, r)-circuit let $h_j(y)$ be the value on the jth output line when the overall output configuration is $h(y)$.

Let $\phi : X_1 \times \cdots \times X_n \to Y$ and let C compute ϕ. Then $R \subseteq X_m$ is called an h_j-*separable set* for C in the mth argument of ϕ if whenever r_1 and r_2 are distinct elements of R we can find $x_1, \ldots, x_{m-1}, x_{m+1}, \ldots, x_n$ with $x_i \in X_i$ such that

$$h_j(\phi(x_1, \ldots, x_{m-1}, r_1, x_{m+1}, \ldots, x_n)) \neq h_j(\phi(x_1, \ldots, x_{m-1}, r_2, x_{m+1}, \ldots, x_n)) .$$

The following important observation was first made by Winograd [Win].

5.85. In a (d, r)-circuit the output of an element at time τ can depend upon at most r^τ input lines.

Proof. Consider the fan-in with modules having r input lines to the height of τ. □

The following basic lemma is due to Spira [Spira].

Lemma 5.86. *Let C be a (d, r)-circuit which computes ϕ in time τ. Then*

$$(5.87) \qquad \tau \geq \max_j \{ \log_r (\log_d |S_1(j)| + \cdots + \log_d |S_n(j)|) \}$$

where $S_i(j)$ is an h_j-separable set for C in the ith argument of ϕ.

Proof. The jth output line at time τ must depend upon at least $\log_d |S_i(j)|$ input lines from $I_{c,i}$ or else there would be two elements of $S_i(j)$ which are not h_j-separable. Thus the jth output depends upon at least $\log_d |S_1(j)| + \cdots + \log_d |S_n(j)|$ input lines and this number is at most r^τ. □

Now we specialize (5.86) to the case where ϕ is multiplication in a finite semigroup so $n = 2$, $X_1 = X_2 = S$ and $\phi = \cdot : S \times S \to S$. In this case what can $S_1(j)$ be? The answer follows

5.88. Let $h_j : S \to Y$. Let R_j be the equivalence relation on S given by $s_1 R_j s_2$ iff $h_j(s_1 s) = h_j(s_2 s)$ for all $s \in S^1$. Further, let $\equiv_j$ be the equivalence relation on S given by $s_1 \equiv_j s_2$ iff $h_j(ts_1 s) = h_j(ts_2 s)$ for all $t, s \in S^1$. Let S/R_j denote the equivalence classes of R_j, etc. Then $(S/R_j, S/\equiv_j)$ is a right mapping semigroup under the following operations:

multiplication in $S/\equiv_j$ is defined by $[s]_{\equiv_j} \cdot [t]_{\equiv_j} = [st]_{\equiv_j}$ and the action is defined by $[s]_{R_j} \cdot [t]_{\equiv_j} = [st]_{R_j}$. It is trivial to verify that all the operations are well defined and $(S/R_j, S/\equiv_j)$ is a right mapping semigroup (so $S/\equiv_j$ acts faithfully). Also $s \mapsto [s]_{\equiv_j}$ is a surmorphism of S onto $S/\equiv_j$, denoted by N_j.

5.89. X is an h_j-separable set for $\cdot : S \times S \to S$ in the first argument of $\cdot$ iff no two members of X are R_j equivalent. Thus $|X| \le |S/R_j|$ and the bound is obtained if X is a set of representations of R_j.

Proof. Trivial! $\qquad\qquad\square$

Combining (5.87) and (5.89) for the case $\phi = \cdot : S \times S \to S$ yields

5.90. Let C be a circuit which computes $\phi = \cdot : S \times S \to S$ in time τ. Then

$$(5.91) \qquad \tau \ge \log_r \log_d \max_j\{|S/R_j|\}$$

where $h_j(s)$ is the value on the jth output line when the overall output configuration of C is $h(s)$.

We next relate the bound of (5.91) to $\#_G(S)$. Given $h_j : S \to Y$ for $j = 1, \ldots, m$ we say $\{h_j\}_{i=1}^m$ *separates points* of S iff $h_j(s_1) = h_j(s_2)$ for $j = 1, \ldots, m$ implies $s_1 = s_2$ (i.e., $h_1 \times \cdots \times h_m$ is one-to-one). Let $\mathcal{H} = \{\{h_j : j = 1, \ldots, m\} : h_j : S \to Y$ and $h_1, \ldots, h_m$ separates points of $S\}$. Then define

$$(5.92) \qquad \beta(S) = \min_{\mathcal{H}}\{\max_j\{|S/R_j| : \{h_j\} \in \mathcal{H}\}\}.$$

We now complete the proof (5.36).

5.93. Let τ be defined as in (5.90). Then

(a) $\tau \ge \log_r \log_d \beta(S)$.
(b) $\beta(S) \ge \#_G(S) + 1$.

Proof. Clearly (5.93a) follows from (5.91) and (5.92). Let $\{h_1, \ldots, h_m\} \in \mathcal{H}$. Then $s \mapsto (N_1(s), \ldots, N_m(s))$ (see the end of (5.88)) yields $S \le\le S/\equiv_1 \times \cdots \times S/\equiv_m$ so $\#_G(S) = \max_j\{\#_G(S/\equiv_j)\}$ by Axiom I, (3.17a), for $\#_G$. But the fact that $(S/R_j, S/\equiv_j)$ is a right mapping semigroup implies $\#_G(S/\equiv_j) \le |S/R_j| - 1$ by (3.16g). Thus $\#_G(S) \le \beta(S) - 1$, proving (5.93b). $\qquad\qquad\square$

Clearly (5.93) implies (5.36), proving Proposition (5.35).

Computing with SNAGs. In [Simple] the following 'Stone-Weierstrass' theorem characterizing SNAGs is proved.

Theorem 5.94 ([Simple]). *Let G be a finite group of order ≥ 3, and let A be a finite set of order ≥ 3. Let $F(A,G)$ be the group of all functions $f : A \to G$ under pointwise multiplication $(f_1 \cdot f_2)(a) = f_1(a) \cdot f_2(a)$, and let H be a subgroup of $F(A,G)$ such that*

(a) $C_g \in H$, *where* $C_g(a) = g$ *for all a, for each $g \in G$. That is, H contains the constant maps.*

(b) *If $x \neq y$, $x, y \in A$, then there exists $f \in H$ with $f(x) \neq f(y)$. That is, H separates points of H.*

Then H is necessarily equal to $F(A,G)$ iff G is a SNAG (simple non-abelian group).

The proof is not hard. See [Simple].

If we set $A = G^n$, the direct product of G taken with itself n times, we obtain a result useful in the algebraic theory of machines. Let $V_n = \{v_1, \ldots, v_n\}$ where $v_1, \ldots, v_n$ are n distinct points or 'variables'. Let $t \in (G \cup V_n)^+$. Then $f_t : G^n \to G$ is defined by 'replacing the variable v_i by the group element g_i and multiplying'. Precisely, if $t = (a_1, \ldots, a_m) \in (G \cup V_n)^+$, then $f_t(g_1, \ldots, g_n) = a'_1 a'_2 \ldots a'_m$ where $a'_i = a_i$ if $a_i \in G$ and $a'_i = g_j$ if $a_i = v_j$. Clearly $t \mapsto f_t$ is a homomorphism of $(G \cup V_n)^+$ into $F(G^n, G)$. Its image, being a subsemigroup of a finite group, is a group denoted by $F_n(G)$. Then as a corollary of (5.94) we have

Corollary 5.95. *Every function $f : G^n \to G$ belongs to $F_n(G)$ iff G is a SNAG (or $G = 1$).*

Notice (5.95) says given *any* function $k : G^n \to G$ one can construct a finite sequence $(a_1, a_2, \ldots, a_m) \in (G \cup V_n)^+$ such that when the variables are evaluated (i.e., given $(g_1, \ldots, g_n)$ replace v_i by g_i) and the evaluated sequence $(a'_1, \ldots, a'_m)$ is presented to G^M, then $G^M(a'_1, \ldots, a'_m) = k(g_1, \ldots, g_n)$. Thus by having the circuit $C(G^M)$ and a 'storage-tube' containing $(a_1, \ldots, a_m)$, one can realize k. For an exposition of this as a novel method of realizing the Boolean functions, see [Real].

Philosophically (5.94) and (5.95) may shed some light on the question of why there are SNAGs — why not just $\mathbb{Z}_p$'s? Well, suppose one tries to compute modulo p in the presence of noise (as all biological systems must). Things will go random if no 'error correction' is used. Now for the machine G^M, the groups $F_n(G)$ may be some measure of how good an 'error correction' G^M can be. (Let the v_i denote places for errors and the constants in $(a_1, \ldots, a_m)$ denote places without errors. Now if each set of errors give a distinct output, the errors can be corrected, and the likelihood of this is proportional to $|F_n(G)|$ for n errors.) *Thus SNAGs are Nature's answer to the problem of performing arithmetic in the presence of noise.*

Finally,

notice $G^{M+} : G^+ \to G^+$ so $G^{M+}(g_1, \ldots, g_n) = (g_1, g_1 g_2, \ldots, g_1 \cdots g_n)$ has an inverse given by

$$(x_1, \ldots, x_n) \mapsto (x_1, x_1^{-1} x_2, \ldots, x_{n-1}^{-1} x_n).$$

Thus we conjecture

$$(g_1, \ldots, g_n) \mapsto$$
$$(5.96) \quad (G^{M+}(g_1, \ldots, g_n), (G^{M+})^2(g_1, \ldots, g_n), \ldots, (G^{M+})^k(g_1, \ldots, g_n))$$
$$= (g_1, g_1 g_2, \ldots, g_1 \cdots g_n, g_1, g_1 g_1 g_2, g_1 g_1 g_2 g_1 g_2 g_3, \ldots)$$

might be a good error correcting code (for suitably chosen k) when G is a SNAG, like M_{11} say, which is easy to decode.

Bibliography

[Auto Th.] J. Rhodes. *Automata Theory.* unpublished. Class Notes, Math Dept., University of California, Berkeley.

[AxII Weak] J. Rhodes. A proof of the fundamental lemma of complexity [weak version] for arbitrary finite semigroups. *Journal of Combinatorial Theory, Series A*, 10:22–73, 1971.

[Cohn] P. M. Cohn. *Universal Algebra.* (first published 1965 by Harper & Row) revised edition, D. Reidel Publishing Company, Dordrecht, 1980.

[Hall] M. Hall. *Theory of Groups.* Macmillan, New York, 1959.

[Haring] D. R. Haring. *Sequential-Circuit Synthesis.* Research Monograph No. 31. Massachusetts Institute of Technology Press, Cambridge, Massachusetts, 1966.

[M-L-S] M. A. Arbib, editor. *Algebraic Theory of Machines, Languages and Semigroups.* Academic Press, 1968.

[MATMI] K. Krohn, R. Mateosian, and J. Rhodes. Methods of the algebraic theory of machines. I: Decomposition theorem for generalized machines; properties preserved under series and parallel compositions of machines. *Journal of Computer and System Sciences*, 1(1):55–85, 1967.

[MinGroup] W. D. Maurer and J. Rhodes. Minimal decompositions of group machines. unpublished.

[q-theory] John Rhodes and Benjamin Steinberg. *The q-theory of Finite Semi-groups*. Springer Monographs in Mathematics, Springer Verlag, 2009. [The most complete treatment of group complexity to date].

[Real] K. Krohn, W. D. Maurer, and J. Rhodes. Realizing complex boolean functions with simple groups. *Information and Control*, 9(2):190–195, 1966.

[Simple] W. D. Maurer and J. Rhodes. A property of finite simple nonabelian groups. *Proceedings of the American Mathematical Society*, 16(3):552–554, 1965.

[Spira] P. M. Spira. The time required for group multiplication. *Journal of the Association for Computing Machinery*, 16(2):235–243, 1969.

[Trans] K. Krohn and J. Rhodes. Algebraic theory of machines. *Transactions of the American Mathematical Society*, 116:450–464, 1965.

[Win] S. Winograd. On the time required to perform addition. *Journal of the Association for Computing Machinery*, 12(2):277–285, 1965.

Chapter 6

Applications

"A new scientific truth does not triumph by convincing its opponents and making them see the light, but rather because its opponents eventually die, and a new generation grows up that is familiar with it."
— Max Planck

Introduction

In this section we give five applications of the preceding theory of complexity.

Our first application in Part I is to the analysis of metabolic reactions occurring in biology leading to a Mendeleev type table for the classification of biological reactions (e.g., bacterial intermediary metabolism) isomorphic to the classification scheme (à la Dickson, Chevalley, Thompson, etc.) for simple non-abelian groups (SNAGs).

The second application in Part II is devoted to precisely stating and developing the important relations imposed upon the complexity of evolved organisms (in our case the cell) by the Darwin-Wallace theory of evolution. We then proceed to derive some properties of the evolved organism from the 'Evolution-Complexity Relations' which include the following:

- proving the possibility of a 'genetics' in any evolved organism (even one on Mars (say), which we have never observed) and throwing some new light and perspectives on Mendelian genetics;
- finding strong restraints on the number of reactions an enzyme can catalyze in a cell in the presence of adequate coenzymes, substrates and inorganic ion concentrations;
- proving the impossibility of simplifying the complete set of data for the

cell's reactions beyond certain limits, which then suggests reasonable procedures for organizing this immense amount of data.

In Part III(A), a definition of a global Lagrangian is given from which Thom's program for studying morphogenesis (i.e., the origin, development, and evolution of biological structures) via the topological methods of the theory of structural stability might possibly begin. While Part II is devoted to studying the conditions that complexity must satisfy in *evolved* organisms (a static law), Part III is devoted to the laws the complexity of *evolving* organisms must satisfy (a dynamical law). In Part III(A) we have an example in mind like a human fetus developing in the womb.* However the theory will apply in other situations as, for example, to the emotional development of an individual during his or her own lifetime. In Part III(B) we define a Lagrangian for emotional development and relate this to studies of mental illness, namely, Freud's early work on neurosis and R. D. Laing's more recent work on schizophrenia. We also give a precise definition of emotion.

In Part IV we develop a theory of complexity for certain games (in the sense of von Neumann) which includes Chess, Checkers, Go, Hex, etc. This allows one to analyze games in greater depth and to compare the complexity of Chess, Go, etc.

Before plunging into the details we make a few comments and observations.

First, as was explained after Definition (5.1), the best examples of (sequential) machines are organisms which must respond in real time with their environment just to stay alive (e.g. a bull fighting a bear). Second, just as physics for several centuries got along quite well by approximating much of nature by *linear* differential equations, we plan to approximate organisms with finite state (sequential) machines. Parts II, III and IV all

---

*John Rhodes' notion of *evolution* encompasses but is not limited to the Darwinian sense. The term also includes more general developmental senses and will entail a progressive viewpoint as will become clear in the discussions below relating 'contact' to complexity. See the discussions of the Evolution-Complexity Relations herein where a Darwin-Wallace adaptationist argument is applied to organismal complexity, development, etc. Richard Dawkins gives non-mathematical arguments for progression in the evolution of complexity (R. Dawkins, *The Blind Watchmaker: Why the Evidence of Evolution Reveals a Universe without Design*, W.W. Norton, 1996). See also John Tyler Bonner's *Evolution of Complexity by Means of Natural Selection* (Princeton, 1988), and, for an information-theoretic approach, C. Adami, C. Ofria, and T. C. Collier, "Evolution of Biology Complexity", *Proceedings of the National Academy of Sciences U.S.A.*, 97(9):4463-4468, 2000. -CLN

proceed by applying these two ideas.

We note philosophically in passing that applying a theory of complexity is not so dependent on the exact details of the theory as might first be suspected. Otherwise said, creating a model and detailed theory of complexity for (say) finite state machines and applying a model of complexity to evolution, games, etc., are almost separate problems. In Parts II, III and IV any reasonable model of complexity could be exchanged for our model with similar results (e.g. Winograd's, Kolmogorov's, etc.). (See the discussion in Chapter 5, Part II.)

Of course, a "reasonable" theory means a theory of complexity for "finite-type" objects satisfying Axiom A (5.25), and Axiom B (5.26).

Thus the main problems of complexity theory fall into two parts. One is to determine all reasonable complexity theories for finite-type objects (e.g. finite semigroups, Boolean functions, finite state machines, finite strings of zeros and ones) and develop detailed mathematical theories for each and then to determine the relation among the various theories. (Our conjecture at present is that all theories will satisfy Axiom A (5.25) and Axiom B (5.26), and more or less differ only in the choice of complexity zero and in the "units" of complexity.) The second main problem is the application of complexity theory. For example, as we shall see, complexity and evolution are intimately wedded. In fact, biology and complexity are deeply related in many areas. However, the knowledge of almost any phenomenon can be deepened by developing a theory of complexity for the phenomenon. Complexity theory is the first level of sophistication after the basic fundamental facts and laws are known.

The reader may be disappointed with the presentation which follows. We sluff many details. We plan to present each application in greater detail in other papers. Here we are primarily interested in presenting our viewpoint and hope to enlist the aid of others in the hard work which must precede final results.

As was kindly pointed out to me by M. W. Hirsch, I subconsciously owe very much of the following to that excellent early book *Design for A Brain* by W. Ross Ashby, which I recommend to the reader for a valuable viewpoint. It is that viewpoint, coupled with powerful algebraic techniques, which we present in the following.

Part I. Analysis and Classification of Biochemical Reactions

"In other words, life merely becomes, to use Haldane's apt phrase, 'a self-perpetuating pattern of chemical reaction'."
— J. Singh, quoting J. B. S. Haldane

The model we will start with is essentially the one we already introduced for cellular metabolism in Chapter 4. Thus the reader should peruse again Chapter 4 up to Definition 4.1. However, it is very important that the mathematics allows the models to be varied relatively freely so that the final model both fits the experimental facts and is in a form suitable for significant applications. Thus we will start with a relatively simple example (Krebs' citric acid cycle) to demonstrate the procedures. For references consult [Watson], [Rose] and [B-F] from which the following is taken.

Consider a cell in the human body. The cell receives most of its energy by oxidizing foodstuffs. We want to consider a latter part of this process in detail. Oxygen from the air enters the bloodstream via the lungs to form a combination with the red, iron-containing blood pigment, hemoglobin. When the circulating blood reaches the body tissues the hemoglobin-oxygen complex dissociates, releasing oxygen into solution whence it can diffuse into the cell. Thus the cell is kept constantly supplied with enough oxygen to help produce the energy it requires. Food is broken down in the mouth, stomach and intestines by a series of enzymes (amylases in the saliva and pancreatic juice for carbohydrates, pepsin in the stomach and trypsin and chymotrypsin in the intestine for proteins, etc.). The resulting mixture includes simple sugars and these are directly absorbed through the intestinal wall and enter the bloodstream and are thus carried to the cell. Under normal circumstances the major portion of the cell's energy requirements are met by the breakdown of sugars.

The principal source of energy for the cell is glucose and the complete burning of glucose is given by

(6.1) $C_6H_{12}O_6 + 6O_2 \rightarrow 6CO_2 + 6H_2O + 673,000$ cal/mol.

Notice both constituents on the left-hand side are brought to the cell by the bloodstream. Also the carbon dioxide (CO_2) on the right-hand side is carried away by the bloodstream.

However, in the cell, glucose is converted to CO_2 and water by a process involving nearly 30 different steps where at certain steps a small amount of energy is released. How does the cell store this energy released by glucose oxidation to carbon dioxide? It stores it in the energy-rich bonds of ATP! ATP denotes the compound adenosine triphosphate. Adenosine (denoted A) is the nucleotide formed from adenine and ribose and it can be phosphorylated to yield first adenosine monophosphate (AMP), then adenosine diphosphate (ADP), and finally ATP. Thus ATP is

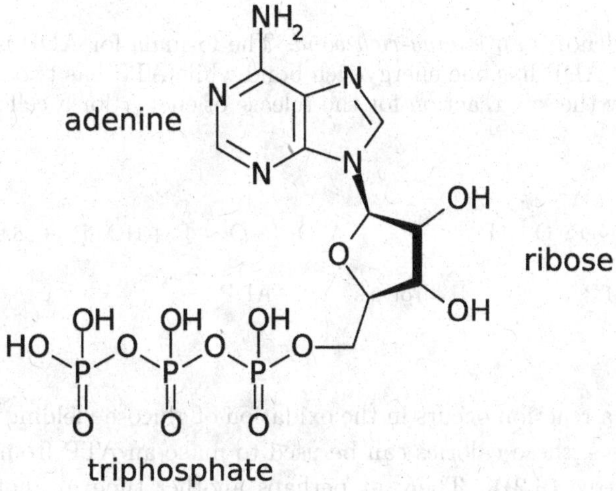

We write ATP as A–O–℗–O ~℗–O~℗ where –℗ terminally represents

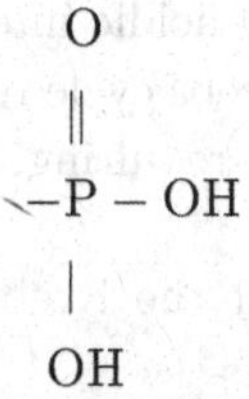

$$
\begin{array}{c}
\text{O} \\
\parallel \\
\text{--P--OH} \\
\mid \\
\text{OH}
\end{array}
$$

and internally

$$
\begin{array}{c}
O \\
\| \\
-P- \\
| \\
OH\,.
\end{array}
$$

The $\sim$ denotes an *energy-rich bond*. The formula for ADP is A–O–℗–O$\sim$℗; thus ADP has one energy-rich bond while ATP has two energy-rich bonds. Now the key reaction for the release of energy'for a cell's use is

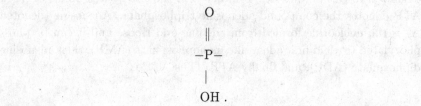

$$\underset{\text{ATP}}{\text{A–O–℗–O} \sim \text{℗–O} \sim \text{℗}} \quad \underset{\text{Hydrolysis}}{\overset{\text{H}_2\text{O}}{\longrightarrow}} \quad \underset{\text{ADP}}{\text{A–O–℗–O} \sim \text{℗}} + \text{HO–℗} \; + \; 8{,}000 \text{ cal/mol.}$$

Whenever a reaction occurs in the oxidation of glucose yielding more than 8,000 calories, these calories can be used to make an ATP from ADP and phosphate (see (6.2)). Then, at perhaps another time in another place, ATP can be hydrolysed to ADP releasing 8,000 calories.

The breakdown of glucose is accomplished in two parts. The first part is through a sequence of reactions (called the Embden-Meyerhof pathway) splits the 6-carbon glucose molecule into two 3-carbon fragments of pyruvic acid. This first part releases only a little energy and can be carried out in the absence of oxygen (as was discovered by Pasteur) and is called *glycolysis*.

The second part starts with pyruvic acid and completely oxidizes it to CO_2 and water via a series of acidic intermediates called the *citric acid cycle* or the *Krebs cycle*. The Krebs cycle releases a great amount of energy (in the form of ATP), is oxygen-requiring, and is sometimes called *glucose oxidation*.

Now that we have said what the Krebs cycle does, we must next say how it does it. See Figure 1.

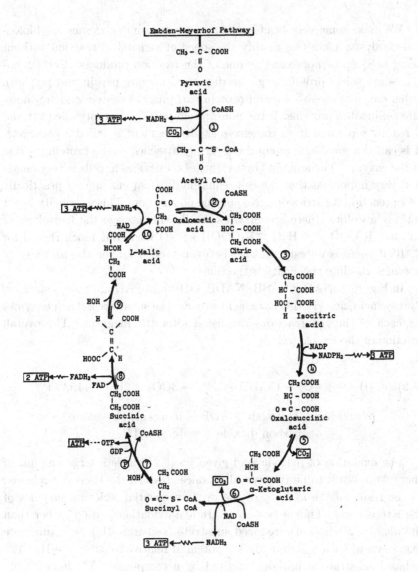

Figure 1. Krebs' citric acid cycle.

The enzymes involved in the reactions as numbered in the diagram are: (1) pyruvate oxidase, (2) citrate synthase, (3) aconitase, (4) and (5) isocitrate dehydrogenase, (6) α-ketoglutarate dehydrogenase, (7) succinyl thiokinase, (8) succinate dehydrogenase, (9) fumarase, (10) malate dehydrogenase.

We make some very brief explanatory comments. *Enzymes* are biological catalysts. Catalysts modify the speed of a chemical reaction without being used up or appearing as one of the reaction products. Certain enzymes are solely protein (e.g. the digestive enzymes pepsin and trypsin). Other enzymes consist of a non-protein part plus the protein and are therefore conjugated proteins. If the non-protein portion is an organic part and is readily separated from the enzyme, this fragment is called a *coenzyme*. It is called a *prosthetic group* if it is firmly attached to the protein portion of the enzyme. The amount (by weight) of coenzymes in cells is very small. It is very important that the cell require a different enzyme for practically every reaction it carries out. No enzyme no reaction. This specificity is not entirely absolute (there are enzymes which will catalyze the hydrolysis of an ester $R\text{-}COO\text{-}R' + H_2O \rightleftarrows R\text{-}COOH + R'\text{-}OH$ without much regard for R, R' if there is an ester linkage between them) but for the majority of enzymes absolute specificity is the rule.

In Figure 1, NAD, CoASH, NADP, GDP and FAD are the coenzymes. Pyruvic acid, acetyl CoA, citric acid, etc. are the substrates. Then enzymes for each of the ten reactions are listed after the diagram. The overall reaction of the Krebs cycle is

$$(6.2) \quad C_3H_4O_3 + \tfrac{5}{2}O_2 + 15ADP + 15P_i \rightarrow 3CO_2 + 2H_2O + 15ATP$$

$$\text{pyruvic acid + oxygen + ADP + inorganic phosphate} \rightarrow$$
$$\text{carbon dioxide + water + ATP}$$

The oxidation of pyruvic acid given by (6.2) yields a large amount of energy, too much to be released all at once. Instead the energy is released as the result of the activity of the cyclic sequential series of enzymes of the Krebs cycle. Thus a system of gradual oxidation is used rather than the direct oxidation of a reduced substrate (substrate-H_2) by a substrate (e.g. oxygen) with a high oxidation potential relative to substrate-H_2. The reduced substrate is initially oxidized by a compound (X') having only a slightly higher oxidation potential. Thus a small amount of energy is released. X' becomes reduced ($X'H_2$) and is then re-oxidized by a second oxidizing agent (X''), etc. At each stage oxidizable energy becomes available which can be transferred to the energy-rich ATP bonds.

We can say this another way. The overall reaction is of the form $B + O_2 \rightarrow BO_2$, i.e. B is oxygenated to BO_2. However this overall reaction is performed by a series of removal of hydrogen (not oxygenation) according to the reactions

$$BH_2 + X \rightarrow B + XH_2$$

where we remove hydrogen (i.e. oxidize) B at the expense of X which has been correspondingly reduced. Substances like X are hydrogen carriers and the enzymes which catalyze the reactions are dehydrogenases. The enzymes are specific but the hydrogen carriers are not specific. The hydrogen carriers are arranged in order of decreasing potential energy. The hydrogen carriers are NAD, NADP, etc.

For example,

(6.3)

$$\text{L-malic acid} \xrightarrow[\text{malate dehydrogenase}]{\text{NAD} \searrow \text{NADH}_2} \text{Oxaloacetic acid}$$

Subsequent oxidation of $NADH_2$ (through the so-called respiratory chain or cytochrome system) gives rise to three molecules of ATP per molecule of $NADH_2$ by

(6.4) $NADH_2 + \frac{1}{2} O_2 + 3\textcircled{P}OH + 3ADP \rightarrow NAD + 3ATP + H_2O$

We write (6.4) as $NADH_2 \rightsquigarrow 3ATP$. In the Krebs cycle there is no direct involvement of molecular oxygen, but oxygen is involved in (6.4).

Things are very much more complicated than as we have presented them here. For example, in (6.3) a series of oxidation-reduction reactions occur that involve a series of closely linked enzymes, all of which contain iron atoms (called the respiratory enzymes). In the reaction labeled ① in Figure 1, two other cofactors and several enzymes are involved, one of the cofactors being vitamin B_1, etc. However, the basic facts are valid and they will adequately serve as an example to which we can apply our algebraic analysis.

We want to apply Principle I (4.1), to the Krebs cycle. Thus we must define the phase space, the inputs and the action.

With reference to Figure 1 we define the set of inputs to be

$$\{NAD + CoASH, CoASH, NADP, GDP, FAD, NAD\}.$$

For ease in defining the states we list

1_1	Pyruvic acid	7_1	Succinyl CoA
2_1	Acetyl CoA	8_1	Succinic acid
3_1	Citric acid	9_1	Fumaric acid
4_1	Isocitric acid	10_1	L-malic acid
5_1	Oxalosuccinic acid	11_1	Oxaloacetic acid
6_1	α-ketoglutaric acid		

We let all the relevant enzymes (listed in Figure 1), the ADP, P_i, O_2, etc., i.e. everything required for the reactions but not listed in the inputs or states, to be considered in the 'soup' and present in the amount required and at the time and place required.

Now by the reaction ③ of Figure 1, citric acid is converted to its isomer (chemical composition the same but structure or shape differing) isocitric acid by way of the enzyme aconitase which functions by removing a molecule of water from the citric acid and then replacing the water molecule isomerically. However, since no cofactor (i.e. input) is required for this (just members of the 'soup') we identify 3_1 and 4_1.

Similarly in the reaction ⑤ of Figure 1, oxalosuccinic acid is decarboxylated to give α-ketoglutaric acid, so we identify 5_1 and 6_1. Similarly we also identify 9_1 and 10_1.

Thus the states are formally

$$\{1_1, 2_1, \{3_1, 4_1\}, \{5_1, 6_1\}, 7_1, 8_1, \{9_1, 10_1\}, 11_1\}.$$

Now we define the action in a natural way from Figure 1 by, for example, (6.3) giving $10_1 \cdot \text{NAD} = 11_1$.

We introduce the following elementary

Definition 6.5. Let X be a set and let $a, b \in X$. Then T_{ab} is by definition the mapping $T_{ab} : X \to X$ with $T(a) = b$ and $T(x) = x$ for all $x \in X$, $x \neq a$. We often symbolically denote T_{ab} by ⓐ → ⓑ.

6.6. Now to be completely formal, the first model K_1 of the Krebs cycle is the following:

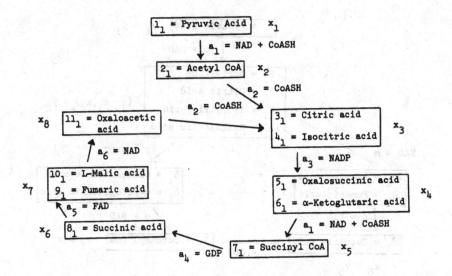

Figure 2. Model K_1 of the Krebs cycle.

Here states $X_1 = \{1_1 = x_1,\ 2_1 = x_2,\ \{3_1, 4_1\} = x_3,\ \{5_1, 6_1\} = x_4,\ 7_1 = x_5,\ 8_1 = x_6,\ \{9_1, 10_1\} = x_7,\ 11_1 = x_8\}$; inputs $A_1 = \{$NAD+CoASH$= a_1$, CoASH$= a_2$, NADP$= a_3$, GDP$= a_4$, FAD$= a_5$, NAD$= a_6\}$ and the action $\lambda_1 : X_1 \times A_1 \to X_1$ is defined by $\lambda_1(x, a_1) = (x)T_{x_1 x_2} T_{x_4 x_5}$, $\lambda_1(x, a_2) = (x)T_{x_2 x_3} T_{x_8 x_3}$, $\lambda_1(x, a_3) = (x)T_{x_3 x_4}$, $\lambda_1(x, a_4) = (x)T_{x_5 x_6}$, $\lambda_1(x, a_5) = (x)T_{x_6 x_7}$, and $\lambda_1(x, a_6) = (x)T_{x_7 x_8}$.[1] Thus our first model of the Krebs cycle yields (X_1, A_1, λ_1) with *semigroup* S_1 and right mapping semigroup (X_1, S_1) in accordance with (4.1), (4.4) and (4.6).

Now suppose we want to vary the model. We might want to further simplify it or we might wish to expand it to put back in many of the glossed over details.

One important way models are changed is to put some inputs into the 'soup' or, conversely, take away some members of the 'soup' and make them inputs. In the following we will demonstrate both of these procedures.

First we modify K_1 by putting the coenzyme CoASH into the 'soup' and identifying NAD and NAD+CoASH, yielding

---

[1]From (6.5), $(x)T_{x_1 x_2} T_{x_4 x_5}$ equals x if $x \neq x_1, x_4$ and $(x_1)T_{x_1 x_2} T_{x_4 x_5} = x_2$ and $(x_4)T_{x_1 x_2} T_{x_4 x_5} = x_5$. $T_{x_1 x_2}$ and $T_{x_4 x_5}$ occur because a_1 occurs from x_1 to x_2 and from x_4 to x_5.

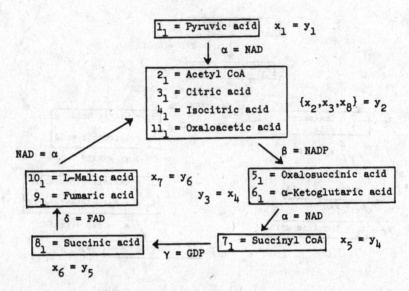

Figure 3. Model K_2 of the Krebs cycle derived from K_1 with CoASH put 'in the soup' and the effects of NAD and NAD+CoASH identified (See (6.8)).

The intuition is clear, if $S_1 \to S_2$ under CoASH plus the old 'soup', then we identify S_1 and S_2. If $S_3 \to S_4$ under CoASH+C in the old model then we define $S_3 \xrightarrow{C} S_4$ in the new model. Precisely

Definition 6.7. Suppose (X, A, λ) is given with $\lambda : X \times A \to X$. (See (4.1)). For $a \in A$ define $\lambda_a : X \to X$ by $\lambda_a(x) = \lambda(x, a) \equiv x \cdot a$.

(a) Assume $\lambda_{a_0}\lambda_{a_0} = \lambda_{a_0}$ for some $a_0 \in A$ (so $x \cdot a_0 \cdot a_0 = x \cdot a_0$ for all $x \in X$). Then (X', A', λ'), read *"putting a_0 into the 'soup' of (X, A, λ)"*, is defined by $X' = \lambda_{a_0}(X) = \{x \cdot a_0 : x \in X\} = \{x \in X : x \cdot a_0 = x\}$ and $A' = A - \{a_0\}$ and $\lambda' : X' \times A' \to X'$ with $\lambda'(x', a) = x' \cdot a_0 \cdot a \cdot a_0$ (with $\cdot$ taken with respect to λ) so $\lambda'_a = \lambda_{a_0}\lambda_a\lambda_{a_0}$ restricted to X'.

(b) Assume $a_1, a_2 \in A$ and $\lambda_{a_1}\lambda_{a_2} = \lambda_{a_2}\lambda_{a_1}$ (so $x \cdot a_1 \cdot a_2 = x \cdot a_2 \cdot a_1$ for all $x \in X$). Then (X, A^*, λ^*), read *"a_1 is identified with a_2 in (X, A, λ)"*, is defined by $A^* = (A - \{a_1, a_2\}) + \{b\}$ and $\lambda^* : X \times A^* \to X$ is defined with $\lambda^*(x, a) = \lambda(x, a)$ for $a \in A - \{a_1, a_2\}$ and $\lambda^*(x, b) = x \cdot a_1 \cdot a_2 = x \cdot a_2 \cdot a_1$ (so $\lambda_a^* = \lambda_a$) for $a \in A - \{a_1, a_2\}$ and $\lambda_b^* = \lambda_{a_1}\lambda_{a_2} = \lambda_{a_2}\lambda_{a_1}$.

Then the model described by Figure 3 is precisely defined as given K_1 (defined in (6.6) and sketched in Figure 2) put CoASH into the 'soup' and identify the inputs NAD+CoASH and NAD.*

6.8. (See Figure 3.) To be formal, the second model K_2 of the Krebs cycle is the following: the states $X_2 = \{y_1, y_2, \ldots, y_6\}$, the inputs $A_2 = \{$NAD$= \alpha$, NADP$= \beta$, GDP$= \gamma$, FAD$= \delta\}$ and $\lambda_2 : X_2 \times A_2 \to X_2$ is given by $\lambda_2(x_2, \alpha) = (x_2)T_{y_1y_2}T_{y_3y_4}T_{y_6y_2}$, $\lambda_2(x_2, \beta) = (x_2)T_{y_2y_3}$, $\lambda_2(x_2, \gamma) = (x_2)T_{y_4y_5}$ and $\lambda_2(x_2, \delta) = (x_2)T_{y_5y_6}$. Thus our second model of the Krebs cycle yields (X_2, A_2, λ_2) with semigroup S_2 in accordance with (4.1), (4.4) and (4.6).

But how is S_2 related to S_1?

Fact 6.9. *Let (X_1, A_1, λ_1) be given with semigroup S_1 (via (4.6)). Suppose (X_2, A_2, λ_2) is derived from (X_1, A_1, λ_1) by a sequence of finite steps, each of which is given by (6.7a) (putting something in the 'soup') or by (6.7b) (identifying some commuting inputs). Let S_2 be the semigroup of (X_2, A_2, λ_2) (via (4.6)).*

Then S_2 divides S_1 and (X_2, S_2) divides (X_1, S_1). (See Definition (3.1).)

Proof. Assume (X_2, A_2, λ_2) equals putting a_0 in the 'soups' of (X_1, A_1, λ_1). For notational ease we write λ_i as $\lambda^{(i)}$. Then S_1 is the semigroup of maps of X_1 into X_1 generated under composition (writing the variable on the left) by $\{\lambda_a^{(1)} : a \in A_1\}$. Let S_1' be the subsemigroup of S_1 generated by $\{\lambda_{a_0}^{(1)} \lambda_a^{(1)} \lambda_{a_0}^{(1)} : a \in A_1 - \{a_0\}\}$. Let $S_1' \twoheadrightarrow S_2$ be defined by $\rho(f_1')$ equals the mapping f_1' restricted to X_2. Then ρ is a well-defined surmorphism since $\lambda_a^{(2)}$ equals $\lambda_{a_0}^{(1)} \lambda_a^{(1)} \lambda_{a_0}^{(1)}$ restricted to X_2. Thus $X_2 S_1' \subseteq X_2$ and defining $\theta : X_2 \to X_2$ to be the identity map so $\theta(x_2) = x_2$ we have $\theta(x_2 \cdot s_1') = \theta(x_2) \cdot \rho(s_1')$ proving (X_2, S_2) divides (X_1, S_1).

Next assume (X, A_2, λ_2) equals a_1 and a_2 being identified in (X, A_1, λ_1). Then again writing λ_i as $\lambda^{(i)}$ we find that S_2 is the subsemigroup of S_1

---

*In detail, putting CoASH into the 'soup' causes substrate states x_2 and x_8 to be identified with x_3 since in the presence of CoASH these substrates are further transformed into x_8. This results in a new state comprising $\{x_2, x_3, x_8\}$. Then identifying the input NAD+CoASH with NAD results in identifying the labels for inputs a_1 and a_6 of Figure 2 to a single label $\alpha =$NAD yielding the model in Figure 3.

There is a somewhat subtle point regarding (6.7a): one must interpret any arrow labels containing a plus-sign, like "NAD+CoASH", in a biochemically natural way: In the new model, the input $\alpha =$ NAD (in the presence of CoASH) yields a nontrivial transition whenever there is a nontrivial transition due to *at least one of* $\lambda_{\text{CoASH}} \lambda_{\text{NAD+CoASH}} \lambda_{\text{CoASH}}$ or $\lambda_{\text{CoASH}} \lambda_{\text{NAD}} \lambda_{\text{CoASH}}$ in the old model, e.g. as occurs in Figure 2 at x_4 and at x_7. -CLN generated by $\{\lambda_a^{(1)} : a \in A - \{a_1, a_2\}\} \cup \{\lambda_{a_1}\lambda_{a_2} = \lambda_{a_2}\lambda_{a_1}\}$ so $S_2 \leq S_1$ so $(X, S_2) \leq (X, S_1)$ so clearly (X, S_2) divides (X, S_1).

Now Fact (6.9) clearly follows.   □

We can also take things out of the 'soup' or, pedantically,

Notation 6.10. *We say (X', A', λ') arises from (X, A, λ) by "taking some-thing out of the 'soup'" iff by a sequence of finite steps each of which is given by (6.7a) (putting something in the 'soup') or by (6.7b) (identifying some commuting inputs) we can pass from (X', A', λ') to (X, A, λ).*

6.11. Thus by Fact (6.9) we have: if (X_1, A_1, λ_1) arises from (X_2, A_2, λ_2) by taking something out of the 'soup' of (X_2, A_2, λ_2) then (X_2, S_2) divides (X_1, S_1) (where S_j is the semigroup of (X_j, A_j, λ_j) defined by (4.6)).

Since enzymes are (very nearly always) specific, they are a natural thing to take out of the 'soup'. For example, considering Figure 1, if we take all the relevant enzymes out of the 'soup' we obtain

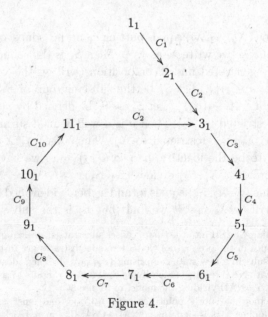

Figure 4.

where C_1 = pyruvate oxidase + NAD + CoASH, C_2 = citrate synthase + CoASH, C_3 = aconitase, C_4 = isocitrate dehydrogenase + NADP, etc.

Similarly, if we consider the model K_2 given by (6.8) and Figure 3 and bring the enzymes up out of the 'soup' we can obtain

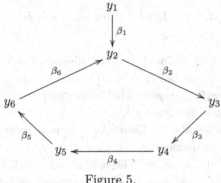

Figure 5.

where the y_j's are as in Figure 3 and β_1=NAD + pyruvate oxidase, β_2 = isocitrate dehydrogenase + NADP, β_3 = α-ketoglutarate dehydrogenase + NAD, β_4 = succinyl thiokinase + GDP, β_5 = succinate dehydrogenase + FAD, and β_6 = malate dehydrogenase + NAD. Here we assume y_2 is at isocitric acid (the 'soup' driving it there), y_3 is at α-ketoglutaric acid, and y_6 is at L-malic acid. Thus we pull only the enzymes associated with the old inputs A_2 out of the 'soup'.

Given certain labeled directed graphs H we can associate an (X, A, λ). We have been doing this implicitly but we now formalize this.

Definition 6.12. H is a *labeled directed graph* over the *points* X, *arrows* E and *labels* L iff $H = (X,\ E \subseteq X \times X,\ \rho : E \to L)$.

Example: Let $X = \{1, 2, 3, 4\}$, $E = \{(1, 2), (2, 3), (3, 4), (4, 1)\}$, $L = \{\alpha, \beta\}$, $\rho(1, 2) = \rho(3, 4) = \alpha$, $\rho(2, 3) = \rho(4, 1) = \beta$. Then the *picture* of $H(X, E, \rho)$ in this case is

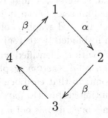

In general the *picture* or *figure* of $H = (X, E, \rho)$ is constructed by taking X to be a subset of points of the plane and drawing $a \xrightarrow{\alpha} b$ iff $(a, b) \in E$

and $\rho(a,b) = \alpha$. Clearly H and its picture uniquely determine one another. Notice only one label per arrow is allowed.

Definition 6.13. Let the labeled directed graph $H = (X, E, \rho : E \to L)$ be given.

(a) H is *single-valued* iff $(a, a) \notin E$ for all a, and $(a, b_1), (a, b_2) \in E$ and $\rho(a, b_1) = \rho(a, b_2)$ implies $b_1 = b_2$. (Otherwise said, in the picture of H no two arrows with the same label leave any given point, and no arrow starts and ends at the same point.)

(b) Let H be single-valued. Then (X, L, λ) is *associated with* H, where $\lambda : X \times L \to X$ is defined by $\lambda(x, y) = x'$ if $(x, x') \in E$ and $\rho(x, x') = y$ and otherwise $\lambda(x, y) = x$. The semigroup of (X, L, λ) (see (4.6)) is termed the *right mapping semigroup associated with* H.

(c) A single-valued labeled directed graph H is called a *reaction graph* if $\rho(a, b) = \rho(b, c)$ never holds in H.

Note that the condition on graphs in (6.13c) corresponds to the convention that reactions run to completion (see the metabolic model discussed in Chapter 4 and the example of the Krebs cycle above). [*,†]

In the previous example $X = \{1, 2, 3, 4\}$, $L = \{\alpha, \beta\}$, and $\lambda(1, \alpha) = 2$, $\lambda(2, \alpha) = 2$, $\lambda(3, \alpha) = 4$, $\lambda(4, \alpha) = 4$, $\lambda(1, \beta) = 1$, $\lambda(2, \beta) = 3$, $\lambda(3, \beta) = 3$, $\lambda(4, \beta) = 4$.

Clearly K_1 defined in (6.6) is associated with the graph whose picture is Figure 2, similarly K_2 defined in (6.8) is associated with the graph whose picture is Figure 3, etc.

Another way to expand the model is to *add substrates*. Basically there are two ways to do this. One is to *expand the substrates already occurring*.

---

[*]If $a \xrightarrow{x} b$ and $b \xrightarrow{y} c$ were reactions in the graph with $x = y$, this would violate the convention since in the presence of x the reactions take a to c. -CLN

[†]While the text considers general single-valued labeled directed graphs throughout this Part, in the analysis of biochemical systems one may restrict one's attention to reaction graphs. This is not an essential limitation since it is clear how one can simplify (if necessary) any given finite labeled directed graph to obtain a reaction graph. To obtain a reaction graph in this situation forbidden by (6.13c) the two arrows would need to be replaced by a single arrow $a \xrightarrow{x} c'$, with c' identified with b and c. However, note that if $a = c$ this would not yet result in a valid reaction graph, since by (6.13a), no arrow may start and leave from the same point; in that case one could eliminate this difficulty by removing the arrow from the resulting graph. (In particular a digraph containing $a \xrightarrow{x} b$ and $b \xrightarrow{x} a$, or any cycle having only one label, is not a reaction graph.) The process may need several iterations to finally yield a reaction graph (but always terminates since it reduces the size of the graph at each iteration). The restriction to reaction graphs (which include e.g. all elementary reactions) is only needed in Proposition (6.49a). -CLN

For example, in Figure 1 the reaction labeled ① is not as simple as our diagram would indicate; lipoic acid is alternately oxidized and reduced during the reaction. Thus we might on occasion wish to add lipoic acid to the substrates. Also then we would add additional inputs (e.g. new enzymes). However, assume the original model (X, A, λ) was simplified but not false and let $(X_1, A_1, \lambda^{(1)})$ denote (X, A, λ) with all the enzymes pulled out of the 'soup'. Let $(X_2, A_2, \lambda^{(2)})$ denote the expanded model so $A_1 \subseteq A_2$ and let (X_i, S_i) denote the (right mapping) semigroup associated with $(X_i, A_i, \lambda^{(i)})$ (via (4.6)). Let $\theta : X_2 \twoheadrightarrow X_1$ be the map defined by $\theta(x_2) = x_1$ where x_2 is in the expansion of the substrate x_1. Let $\rho : S_2 \twoheadrightarrow S_1^1$ be defined by $\rho(\lambda_{a_2}^{(2)}) = \lambda_{a_2}^{(1)}$ for $a_2 \in A_1$ and let $\rho(\lambda_{a_2}^{(2)}) = 1$ for $a_2 \in A_2 - A_1$ (where $x_1 \cdot 1 = x_1$ for all $x_1 \in X_1$). Then we would expect that ρ is a well-defined surmorphism and

$$\theta(x_2 \cdot \lambda_{a_2}^{(2)}) = \theta(x_2)\rho(\lambda_{a_2}^{(2)})$$

for all $x_2 \in X_2$, $a_2 \in A_2$, i.e. (using the notation of Definition (3.1)),

(6.14) $$(\theta, \rho) : (X_2, S_2) \twoheadrightarrow (X_1, S_1^1) .$$

This simply asserts that $A_1 \subseteq A_2$ and if x_2, x_2' are expansions of x_1 (i.e. $x_2 =$ pyruvic acid, $x_2' =$ lipoic acid, $x_1' =$ pyruvic acid, with reference to reaction ① of Figure 1), then for $a_2 \in A_1$, $\lambda^{(2)}(x_2, a_2)$ and $\lambda^{(2)}(x_2', a_2)$ are expansions of $\lambda^{(1)}(x_1, a_2)$. Also if $a_2 \in A_2 - A_1$ and x_2 is an expansion of x_1, then $\lambda^{(2)}(x_2, a_2)$ is also an expansion of x_1.

In fact, if we start with (X, A, λ), pass to $(X_1, A_1, \lambda^{(1)})$ and "expand it" to $(X_2, A_2, \lambda^{(2)})$ but $(X_2, S_2) \twoheadrightarrow (X_1, S_1^1)$ does *not* hold, then (assuming the second model fits the experiment facts) the first model (X, A, λ) must have contained falsifications and was not just a simplification.

For a simple example let (X, A, λ) be associated with

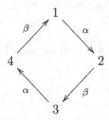

and let $(X_1, A_1, \lambda^{(1)})$ be associated with

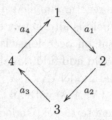

and let

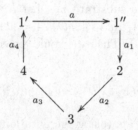

be associated with $(X_2, A_2, \lambda^{(2)})$ so $\theta(j) = j$, $j = 2, 3, 4$, $\theta(1') = \theta(1'') = 1$, $\rho(\lambda_{a_q}^{(2)}) = \lambda_{a_q}^{(1)}$ for $q = 1, 2, 3, 4$ and $\rho(\lambda_a^{(2)}) = 1$.

See Definition (6.15b).

Another way to add substrates is to *adjoin new substrates*. Intuitively this means adding substrates which feed into the old substrates or are fed from the old substrates, but do not form part of the basic reactions of the old substrates. It is easy to make this precise as follows.

Definition 6.15. (a) Let $(X_1, A_1, \lambda^{(1)})$ be given. Then $(X_2, A_2, \lambda^{(2)})$ is obtained from $(X_1, A_1, \lambda^{(1)})$ by *adjoining new substrates and. inputs* iff $X_1 \subseteq X_2$, $A_1 \subseteq A_2$ and if $x_1 \in X_1$, $a_1 \in A_1$, then $\lambda^{(2)}(x_1, a_1) \in X_1$ and $\lambda^{(1)}(x_1, a_1) = \lambda^{(2)}(x_1, a_1)$. Clearly then (X_1, S_1) divides (X_2, S_2).

(b) Let $(X_1, A_1, \lambda_1^{(1)})$ be given. Then $(X_2, A_2, \lambda_2^{(2)})$ is obtained from $(X_1, A_1, \lambda_1^{(1)})$ by *expanding substrates* iff . $A_1 \subseteq A_2$ and there exists $\theta : X_2 \to X_1$ and $\rho : S_2 \twoheadrightarrow S_1^1$ satisfying (6.14) so $(X_2, S_2) \twoheadrightarrow (X_1, S_1^1)$.

As an example, let $(X_1, A_1, \lambda^{(1)})$ be associated with

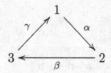

Let $(X_2, A_2, \lambda^{(2)})$ be associated with

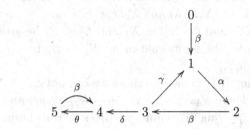

Then $(X_2, A_2, \lambda^{(2)})$ is obtained from $(X_1, A_1, \lambda^{(1)})$ by adjoining new substrates and inputs.

We now introduce the last operations to change the model. We might like to replace the two inputs a_1, a_2 by a single input whose effect is (a_1, a_2), or conversely to drop an input.

Definition 6.16. (a) Let $(X, A_1, \lambda^{(1)})$ be given. Let $a_1, a_2 \in A_1$. Then $(X, A_2, \lambda^{(2)})$ is obtained from $(X, A_1, \lambda^{(1)})$ by *introducing* (a_1, a_2) *as an input*, iff $A_2 = A_1 + \{b\}$ and for $a \in A_1$, $x \in X$, $\lambda^{(2)}(x, a) = \lambda^{(1)}(x, a)$ and $\lambda^{(2)}(x, b) = (x \cdot a_1 \cdot a_2 = \lambda^{(1)}(\lambda^{(1)}(x, a_1), a_2)$.

(b) Let $(X, A_2, \lambda^{(2)})$ be given. Let $a \in A_2$, then $(X, A_1, \lambda^{(1)})$ is obtained from $(X, A_2, \lambda^{(2)})$ by *dropping the equivalent input* a iff $A_1 = A_2 - \{a\}$ and there exists $t = (a'_1, \ldots, a'_m) \in A_1^+$ such that $\lambda_{a'_1} \ldots \lambda_{a'_m} = \lambda_a$ (i.e. $x \cdot a'_1 \ldots a'_m = x \cdot a$ for all $x \in X$), and for $x \in X$, $a_1 \in A_1$, $\lambda^{(1)}(x, a_1) = \lambda^{(2)}(x, a_1)$.

(c) Let $(X, A_1, \lambda^{(1)})$ be given. Then $(X, A_2, \lambda^{(2)})$ is obtained from $(X, A_1, \lambda^{(1)})$ by *adding an input* a iff $A_2 = A_1 + \{a\}$ and $\lambda^{(2)}(x, a_1) = \lambda^{(1)}(x, a_1)$ for all $x \in X$, $a_1 \in A_1$.

Proposition 6.17. *Let $(X_j, A_j, \lambda^{(j)})$ with X_j finite be given for $j = 1, 2$. Then the following assertions are equivalent*

(a) (X_1, S_1) *divides* (X_2, S_2)

(b) *By a finite sequence of operations, each of which is one of the following, $(X_1, A_1, \lambda^{(1)})$ can be transferred into $(X_2, A_2, \lambda^{(2)})$.*

    (i) *Introducing (a, b) as an input (6.16a).*
    (ii)$_1$ *Dropping the equivalent input a (6.16b).*
    (ii)$_2$ *Adding the input a (6.16c).*
    (iii) *Adjoining new substrates and inputs (6.15a).*
    (iv) *Expanding substrates (6.15b).*

Proof. Let (X, A, λ) be given with X a finite set. Then by applying (i) and (ii)$_1$ numerous times we can obtain (X, S, λ_S^*) where (X, S) is the (right mapping) semigroup of (X, A, λ) and $\lambda_S^*(x, s) = x \cdot s$.

Now let $X_2' \subseteq X_2$ and $\theta : X_2' \twoheadrightarrow X_1$ and $S_2' \leq S_2$ be given such that $X_2' \cdot S_2' \subseteq X_2'$ and let the surmorphism $\rho : S_2' \twoheadrightarrow S_1$ be given such that $\theta(x_2' \cdot s_2') = \theta(x_2') \cdot \rho(s_2')$.

Then by applying (ii)$_2$ numerous times we can obtain $(X_1, S_1 \cup S_2', \lambda')$ from $(X_1, S_1, \lambda_{s_1}^*)$ where $\lambda'(x_1, s_1) = x_1 \cdot s_1 = \lambda_{S_1}^*(x_1, s_1)$ and $\lambda'(x_1, s_2') = x_1 \cdot \rho(s_2')$. Then by applying (ii)$_1$ numerous times we can obtain $(X_1, S_2', \rho^\#)$ from $(X_1, S_1 \cup S_2', \lambda')$ with $\rho^\#(x_1, s_2') = x_1 \cdot \rho(s_2') = \lambda_{S_1}^*(x_1, \rho(s_2'))$. Then we can obtain $(X_2', S_2', \lambda_{S_2'}^*)$ from $(X_1, S_2', \rho^\#)$ by applying (iv) once since $\theta(x_2' \cdot s_2') = \theta(x_2') \cdot \rho(s_2')$. Now $(X_2, S_2, \lambda_{S_2}^*)$ can be obtained from $(X_2', S_2', \lambda_{S_2'}^*)$ by applying (iii) once. Finally $(X_2, A_2, \lambda^{(2)})$ can be obtained from $(X_2, S_2, \lambda_{S_2}^*)$ by applying (ii)$_1$ numerous times.

Thus (6.17a) implies (6.17b).

The converse follows easily since if (X_2, A_2, λ_2) is derived from (X_1, A_1, λ_1) by (i), (ii)$_j$, (iii) or (iv), then clearly (X_1, S_1) divides (X_2, S_2).

This proves Proposition (6.17). $\qquad\qquad\square$

6.18. **Discussion.** We briefly summarize. Given some reaction (e.g. intermediary metabolism like the Krebs cycle) we construct a (single-valued labeled directed) graph by choosing the substrates, the 'soup' and the inputs. We have seen how we can vary these choices to fit the application desired. Thus we have the important functor

$$(6.19) \qquad \text{reactions} \mapsto \text{semigroup(reactions)},$$

given by

$$\text{reactions} \mapsto \text{graph} \mapsto \text{state-space and action} \mapsto \text{semigroup of the action}$$

where the last two arrows are defined by (6.13b) and (4.6) respectively.

In the previous section we introduced some important functors on (finite) semigroups, namely

$$(6.20a) \qquad S \mapsto \mathbf{PRIMES}(S), \text{ the primes of } S$$

$$(6.20b) \qquad S \mapsto \#_G(S), \text{ the complexity of } S$$

and

$$(6.21) \qquad (X, S) \,|\, (X_n, S_n) \wr \cdots \wr (X_1, S_1)$$

where each $S_j \in$ **PRIMES**(S) or S_j is a combinatorial semigroup. (6.21) was termed a *"theory* for the reactions" in Chapter 4.

We have also seen from Fact (6.9) and Proposition (6.17) that the function (6.19) varies 'nicely' as the model of the reactions is varied. This is an extremely important property, for without it the mathematics would be of no value. Thus the functors of (6.20) and (6.21) also vary 'nicely'. One might even say that the theory of semigroups presented in Chapter 3 was created just to have the critical property that it varies 'nicely' as models of biological reactions are varied in a natural and practical way. We see again the critical role of 'divide'.

Let **PRIMES**(reactions) denote **PRIMES**(S) where S is the semigroup determined by (6.19). Thus **PRIMES**(reactions) is the composition of (6.19) and (6.20a). Now we can state our viewpoint toward classifying biological reaction data.

6.22. Viewpoint. *The useful way to classify reactions is by using the functor*

$$(6.23) \qquad \text{reactions} \mapsto \textbf{PRIMES}(\text{reactions}) \ .$$

The functors (using the obvious notation) reactions $\mapsto$ complexity of the reactions $= \#_G(\text{semigroup}(\text{reactions}))$, and reactions $\mapsto$ a "physical theory" for semigroup action of the reaction via (6.21), will be less important for reasons to be given later. We will discuss Viewpoint (6.22) at length later, but first we will work out a detailed example. We note here however that if semigroup(reactions) is defined by (6.19), then

$$(6.24) \qquad \text{reactions} \mapsto \text{semigroup}(\text{reactions})$$

vaults intermediary metabolism (say) into semigroup theory and the obvious procedure is to classify the reactions in a natural *algebraic* way.

Consider again Figure 3. Let semigroup(Figure 3) $= S$. In the following we will compute **PRIMES**(S), $\#_G(S)$ and a solution of (6.21) and then interpret the results biologically; specifically we will show

6.25. Let semigroup(Figure 3) $= S$. Then

(a) **PRIMES**$(S) = \{\mathbb{Z}_3, \mathbb{Z}_2\}$ and in fact the maximal subgroups of S are all cyclic and (isomorphic) to $\mathbb{Z}_3$, $\mathbb{Z}_2$ and $\{1\}$.

(b) $\#_G(S) = 2$ and in fact the length of a longest chain of essential dependencies is 2 (see (4.11) and Theorem (4.11)(a)). Our calculation will proceed by first considering

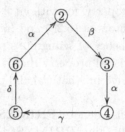

Figure 6.

We notice Figure 6 is Figure 3 with y_j written as j and 1 deleted. Let T denote the semigroup associated with Figure 6. Then we claim

Proposition 6.26. $S \underset{\gamma}{\twoheadrightarrow} T$ where the surmorphism[2] is given by restricting members of S to $\{y_2, \ldots, y_6\} = \{2, \ldots, 6\}$.

Proof. Let $G \neq 1$ be a subgroup of S. Then $(\{y_1, \ldots, y_6\}, G)$ is a right mapping semigroup. But then (see notation of (4.11)) $(X(G), G)$ is a faithful permutation group. Let $e^2 = e \in G$ be the identity of G. Then there exists $t \in \{\alpha, \beta, \gamma, \delta\}^+$ so that $x \cdot t = x$ for all $x \in X(G)$. But for no t of Figure 3 is $y_1 \cdot t = y_1$. Thus $y_1 \notin X(G)$ so $X(G) \subseteq \{2, \ldots, 6\}$ so $G \leq T$. This proves (6.26). □

Thus it suffices to verify (6.25) for T.

Now, relative to Figure 6, notice $\alpha = T_{62}T_{34} = T_{34}T_{62}$, $\beta = T_{23}$, $\gamma = T_{45}$ and $\delta = T_{56}$.

Next note

(6.27)
$$
\begin{array}{ccccc}
2 & 2 & 2 & 3 & 4 \\
4 \xrightarrow{\delta} 4 & \xrightarrow{\gamma} 5 & \xrightarrow{\beta} 5 & \xrightarrow{\alpha} 5 \\
5 & 6 & 6 & 6 & 2
\end{array}
$$

so $\delta\gamma\beta\alpha$ cyclically permutes 245. Also $\{2, 3, 4, 5, 6\} \cdot \delta\gamma\beta\alpha = \{2, 4, 5\}$.

---

[2] $S \underset{\gamma}{\twoheadrightarrow} T$ means there exists an surmorphism of S onto T which is one-to-one on each subgroup of S. See Chapter 3.

Next note

$$(6.28) \quad
\begin{array}{ccccccc}
2 & 3 & 3 & 3 & 4 & 4 & 4 \\
3 & 3 & 3 & 3 & 4 & 4 & 4 \\
4 \xrightarrow{\ \beta\ } 4 \xrightarrow{\ \delta\ } 4 \xrightarrow{\ \gamma\ } 5 \xrightarrow{\ \alpha\ } 5 \xrightarrow{\ \beta\ } 5 \xrightarrow{\ \alpha\ } 5 \\
5 & 5 & 6 & 6 & 2 & 3 & 4 \\
6 & 6 & 6 & 6 & 2 & 3 & 4
\end{array}$$

so $\beta\delta\gamma\alpha\beta\alpha \equiv r$ cyclically permutes 45 and has range 45 and partition 4,2356. Thus $\mathbb{Z}_2$ is a maximal subgroup of T.

Next every subgroup G of T is *cyclic*: If $|X(G)| \leq 2$, obviously G is cyclic. Otherwise, observe that whenever $A \subseteq \{2,3,4,5,6\}$ with $A = \{a_1 < a_2 < \cdots < a_k\}$, $k > 2$, and $x \in \{\alpha, \beta, \gamma, \delta\}$ such that $\cdot x$ is one-to-one on A, then $\cdot x$ mapping A to $A \cdot x$ preserves the cyclic ordering of $\{2 < \cdots < 6\}$. If $w = x_1 \ldots x_n \in \{\alpha, \beta, \gamma, \delta\}^+$ lies in G, it follows inductively that $\cdot x_1$, $\cdot x_1 x_2$, ..., $\cdot x_1 \ldots x_n = \cdot w$ preserve this cyclic ordering when applied to A, $A \cdot x_1$, ..., and $A \cdot x_1 \ldots x_{n-1}$, respectively. In particular $\cdot w$ as a map from $A = X(G)$ to $X(G)$ preserves the cyclic ordering on $X(G)$. Thus $w \in \mathbb{Z}_k$, and, since w was an arbitrary element of G, we conclude $G \leq \mathbb{Z}_k$.[*]

Now β, γ and δ are idempotents and β commutes with γ and δ and $\gamma\delta\gamma = \gamma\delta$ so all products of β, γ, δ can be written as $\beta^{t_1}\gamma^{t_2}\delta^{t_3}$ with $t_j = 0$ or 1. Thus the subsemigroup of T generated by β, γ, δ is combinatorial. On the other hand, α has rank (= cardinality of its range) 3. Thus all maximal subgroups are cyclic of order ≤ 3. On the other hand, $\mathbb{Z}_3$ and $\mathbb{Z}_2$ occur by (6.27) and (6.28). This shows (6.25a).

We next verify (6.25b). Using the notation of (4.11), $G_1 = \{t, t^2, t^3\}$ with $t = \delta\gamma\beta\alpha$, so

$$
\begin{array}{cc}
2 & 4 \\
3 & 4 \\
4 \xrightarrow{\ t\ } 5 \\
5 & 2 \\
6 & 2
\end{array}
$$

and $(t^3)^2 = t^3$ and $X(G_1) = \{4,5,2\}$. Let $G_2 = \{r, r^2\}$ with r defined by

---

[*]This argument generalizes to any reaction graph whose underlying directed graph is a cycle. That is, every subgroup of the semigroup of a cyclic reaction graph is cyclic. -CLN

(6.28) so

$$
\begin{array}{rcl}
2 & & 4 \\
3 & & 4 \\
4 & \xrightarrow{\ r\ } & 5 \\
5 & & 4 \\
6 & & 4
\end{array}
$$

Then $t^3 = (t^3)^2$, $r^2 = (r^2)^2$ and $rt^3 = r$ and $t^3 r = r$ so $t^3 r^2 = r^2 = r^2 t^3$ and $t^3 > r^2$ and $X(G_2) = \{4,5\} \subsetneqq \{2,4,5\}$. Also $G_1 \cong \mathbb{Z}_3$, $G_2 \cong \mathbb{Z}_2$. Further, $\{X(G_2)g_1 : g_1 \in G_1\} = \{\{4,5\}, \{2,5\}, \{2,4\}\}$. Notice

(6.29)
$$
\begin{array}{ccccccccccc}
2 & & 2 & & 2 & & 2 & & 2 & & 4 \\
3 & & 2 & & 3 & & 2 & & 3 & & 4 \\
4 & \xrightarrow{\ rt^2\ } & 4 & , & 4 & \xrightarrow{\ trt\ } & 5 & , & 4 & \xrightarrow{\ t^2 r\ } & 4 \\
5 & & 2 & & 5 & & 5 & & 5 & & 5 \\
6 & & 2 & & 6 & & 5 & & 6 & & 5
\end{array}
$$

Then $rt^2 = (rt^2)^2$ with range $(rt^2) = \{2,4\}$, $trt = (trt)^2$ with range $(trt) = \{2,5\}$ and $t^2 r = (t^2 r)^2$ with range $(t^2 r) = \{4,5\}$ and $(rt^2)(trt)(t^2 r) = (rt^3)(rt^3)r = rrr = r$. Thus $\mathcal{I}_2 \cap$ Per $(X(G_2)) \supseteq G_2 \neq \{1\}$ so $X(G_1) \supsetneqq X(G_2)$ is essential.

We next give a solution of (6.21) for $(\{2,\ldots,6\}, T)$. Essentially we use the method of Zeiger [Zeiger].

Let $X_1 = \{\{2,4,5\}, \{2,4,6\}, \{2,5,6\}, \{3,4,5\}, \{3,4,6\}, \{3,5,6\}\}$. Let $X = \{((\{s_3\}, S_2, S_1) : S_1 \in X_1,\ S_2 \subseteq S_1$ and $|S_2| = 2,\ s_3 \in S_2\}$ so $(\{2\}, \{2,4\}, \{2,4,5\}) \in X$, etc.

Members of X like $(\{2\}, \{2,4\}, \{2,4,5\})$ can be represented as
$$
\begin{array}{c}
245 \\
24 \\
2
\end{array}
\quad .
$$

See Figure 7. Thus
$$
\begin{array}{ccc}
 & 245 & \\
24 & 25 & 45 \\
2\ 4 & 2\ 5 & 4\ 5
\end{array}
$$
represents six members of X, namely

$(\{2\}, \{2,4\}, \{2,4,5\}), (\{4\}, \{2,4\}, \{2,4,5\}), \ldots, (\{5\}, \{4,5\}, \{2,4,5\})$.

X	$\hat{\alpha}$	$\hat{\beta}$	$\hat{\gamma}$	$\hat{\delta}$
245	245	345	256	246
24 25 45	24 25 45	34 35 45	25 25 25	24 26 46
2 4 2 5 4 5	2 4 2 5 4 5	3 4 3 5 4 5	2 5 2 5 5 5	2 4 2 6 4 6
246	245	346	256	246
24 26 46	24 24 24	34 36 46	25 26 56	24 26 46
2 4 2 6 4 6	2 4 2 2 4 2	3 4 3 6 4 6	2 5 2 6 5 6	2 4 2 6 4 6
256	245	356	256	246
25 26 56	25 25 25	35 36 56	25 26 56	26 26 26
2 5 2 6 5 6	2 5 2 2 5 2	3 5 3 6 5 6	2 5 2 6 5 6	2 6 2 6 6 6
345	245	345	356	346
34 35 45	45 45 45	34 35 45	35 35 35	34 36 46
3 4 3 5 4 5	4 4 4 5 4 5	3 4 3 5 4 5	3 5 3 5 5 5	3 4 3 6 4 6
346	245	346	356	346
34 36 46	24 24 24	34 36 46	35 36 56	34 36 46
3 4 3 6 4 6	4 4 4 2 4 2	3 4 3 6 4 6	3 5 3 6 5 6	3 4 3 6 4 6
356	245	356	356	346
35 36 56	45 24 25	35 36 56	35 36 56	36 36 36
3 5 3 6 5 6	4 5 4 2 5 2	3 5 3 6 5 6	3 5 3 6 5 6	3 6 3 6 6 6

Figure 7.

Let $\theta : X \twoheadrightarrow \{2,3,4,5,6\}$ be defined by $\theta(\{s\}, S_2, S_1) = s$.

Now for each $x \in \{\alpha, \beta, \gamma, \delta\}$ we define $\cdot\hat{x} : X \to X$ by Figure 7. Thus

$$(\{2\}, \{2,4\}, \{2,4,5\}) \cdot \hat{\alpha} = (\{2\}, \{2,4\}, \{2,4,5\}),$$

$$(\{5\}, \{5,6\}, \{3,5,6\}) \cdot \hat{\alpha} = (\{5\}, \{2,5\}, \{2,4,5\}),$$

etc.

X	$\alpha^\#$	$\beta^\#$	$\gamma^\#$	$\delta^\#$
245	245	345	256	246
1 2 3	1 2 3	1 2 3	1 1 1	1 2 3
1 2 1 2 1 2	1 2 1 2 1 2	1 2 1 2 1 2	1 2 1 2 2 2	1 2 1 2 1 2
246	245	346	256	246
1 2 3	1 1 1	1 2 3	1 2 3	1 2 3
1 2 1 2 1 2	1 2 1 1 2 1	1 2 1 2 1 2	1 2 1 2 1 2	1 2 1 2 1 2
256	245	356	256	246
1 2 3	2 2 2	1 2 3	1 2 3	2 2 2
1 2 1 2 1 2	1 2 1 1 2 1	1 2 1 2 1 2	1 2 1 2 1 2	1 2 1 2 2 2
345	245	345	356	346
1 2 3	3 3 3	1 2 3	1 1 1	1 2 3
1 2 1 2 1 2	1 1 1 2 1 2	1 2 1 2 1 2	1 2 1 2 2 2	1 2 1 2 1 2
346	245	346	356	346
1 2 3	1 1 1	1 2 3	1 2 3	1 2 3
1 2 1 2 1 2	2 2 2 1 2 1	1 2 1 2 1 2	1 2 1 2 1 2	1 2 1 2 1 2
356	245	356	356	346
1 2 3	3 1 2	1 2 3	1 2 3	2 2 2
1 2 1 2 1 2	1 2 2 1 2 1	1 2 1 2 1 2	1 2 1 2 1 2	1 2 1 2 2 2

Figure 8.

We note that $\cdot\hat{x}$ can also be defined as follows:

(6.30a) $\qquad (\{s_3\}, S_2, S_1) \cdot \hat{x} = (\{s_3 \cdot x\}, (S_2, S_1) \cdot x, S_1 \cdot x_1)$

where $s_3 \cdot x$ is the action in T, $S_1 \cdot x_1$ is defined via Figure 7 with $\{2,4,5\} \cdot \alpha_1 = \{2,4,5\}$, $\{2,4,5\} \cdot \beta_1 = \{3,4,5\}$, etc., and

(6.30b) $\qquad (S_2, S_1) \cdot x = \begin{cases} S_2 \cdot x & \text{when } |S_1 \cdot x| = |S_1| = 3 \\ S_1 \cdot x & \text{otherwise .} \end{cases}$

Now if $Y = \{y_3 < y_2 < y_1\}$, then $(Y)_j = Y - \{y_j\}$. If $Z = \{y_2 < y_1\}$, then $(Z)_j = Z - \{y_j\}$. Let $X_2 = \{1,2,3\}$ and let $X_3 = \{1,2\}$. Then $j : X_3 \times X_2 \times X_1 \to X$ is defined by

(6.31) $\qquad j(x_3, x_2, S_1) = (\{s_3\} = (S_2)_{x_3}, \ S_2 = (S_1)_{x_2}, S_1) .$

Clearly j^{-1} exists. Now define $x^\# : X_3 \times X_2 \times X_1 \to X_3 \times X_2 \times X_1$ by

$$(6.32) \qquad\qquad x^\# = j \cdot \hat{x} j^{-1} .$$

The values of $x^\#$ are computed in Figure 8. For example,

$$
\begin{array}{c}
245 \\ 1 \\ 1
\end{array}
\xrightarrow{j}
\begin{array}{c}
245 \\ 24 \\ 2
\end{array}
\xrightarrow{\hat{\alpha}}
\begin{array}{c}
245 \\ 24 \\ 2
\end{array}
\xrightarrow{j^{-1}}
\begin{array}{c}
245 \\ 1 \\ 1
\end{array}
\text{ or }
\begin{array}{c}
356 \\ 3 \\ 2
\end{array}
\xrightarrow{j}
\begin{array}{c}
356 \\ 56 \\ 6
\end{array}
\xrightarrow{\hat{\alpha}}
\begin{array}{c}
245 \\ 25 \\ 2
\end{array}
\xrightarrow{j^{-1}}
\begin{array}{c}
245 \\ 2 \\ 1
\end{array}
$$

so $(1,1,\{2,4,5\})\alpha^\# = (1,1,\{2,4,\bar{5}\})$, $(2,3,\{3,5,6\})\alpha^\# = (1,2,\{2,4,5\})$, etc. in agreement with Figure 8.

Now it is very easy to verify from Figure 7 that

$$(6.33) \qquad\qquad \theta(k \cdot \hat{x}) = \theta(k) \cdot x$$

for all $k \in X$ and $x \in \{\alpha, \beta, \gamma, \delta\}$. Thus defining

$$(6.34) \qquad\qquad \theta' = \theta j$$

we obtain

$$(6.35) \qquad \theta'((x_3, x_2, x_1) \cdot x^\#) = \theta'(x_3, x_2, x_1) \cdot x$$

for all $x \in \{\alpha, \beta, \gamma, \delta\}$ and $(x_3, x_2, x_1) \in X_3 \times X_2 \times X_1$. Notice θ' is easy to compute. For example,

$$
\begin{array}{c}
245 \\ 2 \\ 2
\end{array}
\xrightarrow{j}
\begin{array}{c}
245 \\ 25 \\ 5
\end{array}
\xrightarrow{\theta}
5 = \theta'(2, 2, \{2, 4, 5\}), \text{ etc.}
$$

Let $\mathbb{Z}_n = \{1, \ldots, n\}$ be the cyclic group of integers modulo n. Let $(\mathbb{Z}_n, \mathbb{Z}_n^*)$ denote $(\mathbb{Z}_n, \mathbb{Z}_n)$ with the constant maps added, i.e. $\mathbb{Z}_n^* = \mathbb{Z}_n + \{C_z : z \in \mathbb{Z}_n\}$ with $z_1 \cdot z = z_1 + z$ and $z_1 \cdot C_z = z$ so in $\mathbb{Z}_n^*$ $z_1 \cdot z_2 = z_1 + z_2$, $z_1 \cdot C_{z_2} = C_{z_2}$, $C_{z_2} \cdot z_1 = C_{z_2 + z_1}$, $C_{z_1} \cdot C_{z_2} = C_{z_2}$ for all $z_1, z_2 \in \mathbb{Z}_n$.

Let C be the semigroup generated by $\{\cdot x_1 : X_1 \to X_1 : x \in \{\alpha, \beta, \gamma, \delta\}\}$ where $\cdot x_1$ is defined by $(6.30)(a)$. Let $T^\#$ be the semigroup generated by $\{x^\# : X_3 \times X_2 \times X_1 \to X_3 \times X_2 \times X_1 : x \in \{\alpha, \beta, \gamma, \delta\}\}$. Notice $X_3 = \mathbb{Z}_2$ and $X_2 = \mathbb{Z}_3$ as sets. Then we have from (6.35) and Figure 8 that

$$(6.36)$$

$$(\{2, \ldots, 6\}, T) \xleftarrow{(\theta', \rho')} (X_3 \times X_2 \times X_1, T^\#) \leq (\mathbb{Z}_2, \mathbb{Z}_2^*) \wr (\mathbb{Z}_3, \mathbb{Z}_3^\#) \wr (X_1, C)$$

where $\rho'(x_1^\# \ldots x_k^\#) = x_1 \ldots x_k$. Further, the action of the generators of $T^\#$ is given by Figure 8 and C is a combinatorial semigroup. This is easy to see since if $c \in C$ and $c = \cdot x_1 \cdots x_k$ and $Y \subseteq X_1$ and $Y \cdot c = Y$, then

c is the identity on Y. (Clearly if some $x_j = \alpha$ this follows since $\cdot\alpha_1$ is the constant map always $\{2, 4, 5\}$. But $\cdot\beta_1$ sends $2\sim$ to $3\sim$ and is the identity on $3\sim$, γ_1 and $\cdot\delta_1$ leave $2\sim$ and $3\sim$ invariant and are a constant map restricted to either.)

Finally

$$(6.37) \qquad (G, G^*) \text{ divides } (G, G^r) \wr (G, G).$$

To prove (6.37) define $\theta : G \times G \to G$ by $\theta(g_2, g_1) = g_2 g_1$. For $g \in G$ define $\hat{g} : G \times G \to G \times G$ by $(g_2, g_1)\hat{g} = (g_2, g_1 g)$. For $g \in G$ define $\hat{C}_g : G \times G \to G \times G$ by $(g_2, g_1) \cdot \hat{C}_g = (gg_1^{-1}, g_1)$. Let $\hat{T}$ be the subsemigroup generated by $\{\hat{g} : g \in G\} \cup \{C_g : g \in G\}$. Then

$$\theta((g_2, g_1) \cdot \hat{x}) = \theta(g_2, g_1) \cdot x$$

for all $(g_2, g_1) \in G \times G$ and $x \in G \cup \{C_g : g \in G\}$. But then $\rho : \hat{T} \twoheadrightarrow G^*$ with $\rho(\hat{x}_1 \ldots \hat{x}_j) = x_1 \ldots x_j$ for $x_i \in G \cup \{C_g : g \in G\}$ is a well-defined surmorphism and $(G \times G, \hat{T}) \leq (G, G^r) \wr (G, G)$ and

$$(\theta, \rho) : (G \times G, \hat{T}) \twoheadrightarrow (G, G^*)$$

proving (6.37).

Now (6.37) implies $\#_G(G^*) = 1$ (if $G \neq 1$) and so (6.36) and (3.16)(j) implies $\#_G(T) \leq 2$. But T has a length two essential dependency so $\#_G(T) = 2$ and $\#_G(T) = \#_G(S)$ by Axiom II for complexity (3.19), and (6.26). This completes the computation (6.25).

6.38. Discussion. Starting with a standard model of the Krebs cycle given by Figure 1 we chose the inputs to be the coenzymes and placed the relevant enzymes in the 'soup' leading to Figure 2. Then we also placed Coenzyme A, CoASH, in the 'soup' because the reactions leading to the formation of ATP (via the respiratory chain) are the most important. This leads us to the model K_2 given in Figure 3 and precisely defined by (6.8).

Then in Discussion (6.18) we set forth our viewpoint (see (6.22)) and in (6.25) calculated **PRIMES**(K_2), complexity of K_2 and found a chain of essential dependencies at complexity, and determined a 'physical theory' for K_2 (actually for Figure 6) by Figure 7 and 8 and (6.36). We would next like to *interpret* these calculations in a biologically significant way.

We start with **PRIMES**$(K_2) = \{\mathbb{Z}_2, \mathbb{Z}_3\}$.

First of all, that **PRIMES**(Krebs cycle of Figure 3) are all cyclic groups is intuitively appealing. In fact our calculations show that all the subgroups of semigroup (K_2) are cyclic.

Next consider (6.27). Here $\delta = T_{56}$, $\gamma = T_{45}$, $\beta = T_{23}$ and $\alpha = T_{62}T_{34} = T_{34}T_{62}$. We recall the interpretation of $5 \xrightarrow{\delta} 6$. We place substrate(s) 5 in a beaker together with all the elements of the 'soup' (say a mole of each). Then if the reaction $5 \to 6$ does occur when a mole of δ is added we write $5 \xrightarrow{\delta} 6$. However if we replace substrate(s) 5 by substrate(s) 3 (say) and no reaction using up any of 3 occurs in the presence of the 'soup' and δ we write $3 \xrightarrow{\delta} 3$. For another example, in Figure 3, $y_3 = 3$ consists of oxalosuccinic acid and α-ketoglutaric acid. If 3 and the 'soup' is placed in a beaker, the reaction oxalosuccinic acid $\to \alpha$-ketoglutaric acid occurs, and then when $\alpha = $ NAD is added α-ketoglutaric ccid $\to$ succinyl CoA occurs. In this case we write $3 \xrightarrow{\alpha} 4$.

We now make explicit an important property of our models. Consider

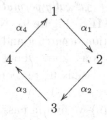

Figure 9.

so $\alpha_k = T_{k,k+1}$ ($1 \le k \le 3$) and $\alpha_4 = T_{41}$. Now chemists, biologists, etc. usually say we have a series of reactions $1 \xrightarrow{\alpha_1} 2 \xrightarrow{\alpha_2} 3 \xrightarrow{\alpha_3} 4 \xrightarrow{\alpha_4} 1$. However on the molecular level if we are (a pacified Maxwell Demon) riding a molecule of 1, it is true that we might be hit by a molecule of α_1 and be transferred into 2, and then hit by a molecule of α_2 and be transformed into 3, etc. until we are transformed once again back to 1. But this assumes starting from 1 we are hit by $\alpha_1, \alpha_2, \alpha_3, \alpha_4$ in that order. But since the molecules are in solution, a molecule of 1 might come in contact with α_1 or α_2 or α_3 or α_4 at any time. For example suppose 1 comes in contact with α_3 at time 1, α_2 at time 2, α_1 at time 3 and α_4 at time 4 (the cycle *backwards* starting at α_3) then

(6.39a) $$1 \xrightarrow{\alpha_3} 1 \xrightarrow{\alpha_2} 1 \xrightarrow{\alpha_1} 2 \xrightarrow{\alpha_4} 2.$$

In words a molecule of 1 meets α_3 and no reaction takes place, it next meets a molecule of α_2 and no reaction takes place, it thirdly meets a molecule of α_1 and is transformed into 2 and lastly meets α_4 and remains unchanged at 2. Similarly

(6.39b) $2 \xrightarrow{\alpha_3} 2 \xrightarrow{\alpha_2} 3 \xrightarrow{\alpha_1} 3 \xrightarrow{\alpha_4} 3$

with a reaction taking place only at α_2.

Finally

(6.39c) $3 \xrightarrow{\alpha_3} 4 \xrightarrow{\alpha_2} 4 \xrightarrow{\alpha_1} 4 \xrightarrow{\alpha_4} 1$

with a molecule hiting α_3 at time 1 and being transferred into 4, then meeting α_2 and α_1 and remaining unchanged and fourthly 4 is hit by α_4 and is transformed into 1.

Now writing (6.39a)–(6.39c) together yield

$$
(6.40) \qquad
\begin{array}{ccccc}
1 & 1 & 1 & 2 & 2 \\
2 \xrightarrow{\alpha_3} & 2 \xrightarrow{\alpha_2} & 3 \xrightarrow{\alpha_1} & 3 \xrightarrow{\alpha_4} & 3 \\
3 & 4 & 4 & 4 & 1
\end{array}
$$

which has the *global effect of cyclically permuting* 1 to 2 to 3 to 1, denoted (1 2 3). Thus $\mathbb{Z}_3$ is in **PRIMES** (Figure 9). Now in our model we want to take account of all the reaction occurring and not just the ones 'following the arrows'. The molecules don't know about the diagrams drawn by the biologist or chemists. Any possible finite sequence of molecules can hit a given molecule.

Recalling Discussion (6.19) we find in passing from reactions $\rightarrow$ graph $\rightarrow$ state-space, inputs and action (X, A, λ), we can define $\cdot : X \times A^+ \rightarrow X$ via (4.4), i.e. $x \cdot a_1 = \lambda(x_1 a_1)$, $x \cdot a_1 \ldots a_n = (x \cdot a_1 \ldots a_{n-1}) \cdot a_n$. The physical interpretation of the map $\cdot : X \times A^+ \rightarrow X$ is as follows. The string $t = a_1 \ldots a_m \in A^+$ denotes *the order we assume each molecule of X will be hit by inputs*. Dividing time by the contacts of the molecule $x \in X$ with the inputs in A, it must be hit by some string $a_1' \ldots a_m'$. Thus by knowledge of what happens to x under *every* string we have complete knowledge of the reactions. Now $\cdot : X \times A^+ \rightarrow X$ simply tabulates the resulting molecule $x' = \cdot(x, a_1 \ldots a_m) \equiv x \cdot a_1 \ldots a_m$ if the molecule x is hit by $a_1, \ldots, a_m$ in that order. Thus for Figure 9 and (6.39a)–(6.39c) we have $1 \cdot \alpha_3 \alpha_2 \alpha_1 \alpha_4 = 2$, $2 \cdot \alpha_3 \alpha_2 \alpha_1 \alpha_4 = 3$, $3 \cdot \alpha_3 \alpha_2 \alpha_1 \alpha_4 = 1$.

Now all the chemicals are in a 'big soup' in the cell. In our modeling we arbitrarily separate the 'big soup' into three disjoint parts, namely: substrates, inputs and 'soup'. Now molecules combine in the 'big soup'. The substrates are so labeled so that if $s_1 \rightarrow s_2$ and no input is required, only members of the 'soup', we identify s_1 and s_2 under a single label. On the other hand if $s_1 + \text{'soup'} + \alpha \rightarrow s_2 + \ldots$, then we write $s_1 \xrightarrow{\alpha} s_2$ or $s_1 \cdot \alpha = s_2$. In short given a molecule $s \in \text{'big soup'}$

suppose $s_1 \xrightarrow{} s_1 \rightarrow s_2 \xrightarrow{\alpha} s_3 \xrightarrow{\beta} s_4 \rightarrow s_5 \rightarrow s_6 \xrightarrow{\gamma} s_7$ occurs in the 'big soup', then in our model $x_1 = \{s_1, s_2\}$, $x_2 = \{s_3\}$, $x_3 = \{s_4, s_5, s_6\}$, $x_4 = \{s_7\}$ and $x_1 \xrightarrow{\alpha} x_2 \xrightarrow{\beta} x_3 \xrightarrow{\gamma} x_4$ and so $x_1 \cdot \alpha\beta\gamma = x_4$, $x_2 \cdot \alpha\beta\gamma = x_4$, etc. Thus the map $\cdot : X \times A^+ \rightarrow X$ tells us everything that can possibly happen in the 'big soup' relative to the conventions of our model (i.e. relative to our division of 'big soup' = inputs + substrates + 'soup').

The semigroup of the reaction can be constructed from (X, A, λ) as either the set of all maps of X into X of the form $x \rightarrow x \cdot t$ for $t \in A^+$ under composition (i.e. $f, g : X \rightarrow X$, then $f \cdot g(x) = gf(x)$) or as the equivalence relation $\equiv$ on A^+ where for $s, t \in A^+$, $s \equiv t$ iff $x \cdot s = x \cdot t$ for all $x \in X$. Then letting $[t]_\equiv = \{s \in A^+ : s \equiv t\}$, $S = \{[t]_\equiv : t \in A^+\}$ with $[t]_\equiv \cdot [r]_\equiv = [tr]_\equiv$. Thus the semigroup of the reactions identifies $t = a_1 \ldots a_m$ with $r = a_1' \ldots a_n'$ iff a molecule of X hits $a_1, \ldots, a_m$ in that order or $a_1' \ldots a_n'$ in that order, the result is always the same (i.e. $x \cdot a_1' \ldots a_n' = x \cdot a_1 \ldots a_m$ for all $x \in X$).

6.41. Cross-over Groups. Now consider (6.27) for Figure 3 or 6 and (6.39a)–(6.39c) and (6.40) for Figure 9. In (6.40) $\alpha_3 \alpha_2 \alpha_1 \alpha_4$ cyclically permutes 123 to 231, denoted (123).

Now assume substrates 1, 2 and 3 plus the 'soup' are in a beaker. Further suppose an amount of α_1, α_2, α_3 and α_4 is added. Then nothing will seem to happen, 1, 2, 3 just sit there. *But if we ride a molecule of 1 we will trace out the cyclic group* $\mathbb{Z}_3$ (for if $t = \alpha_3 \alpha_2 \alpha_1 \alpha_4$ $1 \cdot t = 2$, $2 \cdot t = 3$, $3 \cdot t = 1$ but $\{1, 2, 3\} \cdot t = \{1, 2, 3\}$).

Of course the successful experimental method for 'riding molecules' around the metabolic pathways is that of *radioisotopes*. Using the isotope carbon 14 (say) we can observe the molecules tracing out various permutations as for example the cyclic permutation (1 2 3) or (6.40).

In the following when we use *macroscopic* we mean the substrates present and easily detected by color, standard chemical tests, etc. When we use *microscopic* we mean activity at the molecular level as would be discovered (say) by tracing with carbon 14.

Thus (6.40) is *macroscopically stable* under $\alpha_3 \alpha_2 \alpha_1 \alpha_4$ since 1, 2, 3 is always present and just 'sits there' but microscopically (6.40) is cyclically permuted under $\alpha_3 \alpha_2 \alpha_1 \alpha_4$.

Using the notation of (4.11) we can easily generalize the previous to arbitrary groups. Let (X, S) be the semigroup of the reaction. Let $G \leq S$

be a non-trivial subgroup of S. Let $X(G) \subseteq X$ as defined in (4.11). Then

$$X(G) \cdot G = X(G)$$
$$X(G) \cdot g = X(G) \text{ for all } g \in G$$

and $(X(G), G)$ is a permutation group.

Let $g \in G$ and $x \cdot g = x \cdot a_1 \ldots a_m$ for $x \in X(G)$. Then if we place substrates $X(G)$ plus the 'soup' in a beaker and then add $a_1, a_2, \ldots, a_m$, we will find no change in substrates at the beginning and end of the process, namely $X(G)$ starts and ends, but if we travel on the molecule $x \in X(G)$ we will trace out $x, x \cdot g, x \cdot g^2, \ldots$.

In other words by the reactions driven by $a_1, \ldots, a_m$, the substrates $X(G)$ are not changed, macroscopically the substrates $X(G)$ are fixed, but microscopically, riding an individual molecule in $X(G)$, the molecules of $X(G)$ are permuted among themselves according to the permutation group $(X(G), G)$.

Because the phenomenon is macroscopically fixed ($X(G)$ is static) but microscopically the molecules of $X(G)$ are permuted or crossed over among themselves, we term $(X(G), G)$ a *cross-over group* of the reactions.

From (6.25), (6.27) and (6.28) we find cross-over groups $\mathbb{Z}_3$ and $\mathbb{Z}_2$ for our model K_2 (see (6.8) and Figure 3) of the Krebs Cycle.

With reference to Discussion (6.38), we next interpret the lower bound for complexity of Theorem (4.11a), namely the longest chain of essential dependencies, which by (6.25b) is perfect for the model K_2.

From the definitions presented in (4.11) we see that roughly the length k of the longest chain of essential dependencies of the reaction R (with semigroup and action (X, S)) is the longest chain of cross-over groups $G_1, G_2, \ldots, G_k$ whose associated substrates $X(G_1) = X_1, \ldots, X(G_k) = X_k$ are properly contained in one another $X_1 \supsetneq X_2 \supsetneq \cdots \supsetneq X_k$ and such that microscopic knowledge of the jth cross-over group (X_j, G_j) gives information regarding the macroscopic movements of X_{j+1} in X_j, but gives *no information* regarding the microscopic movement of the molecules within X_{j+1} via G_{j+1}.

First of all it is not too difficult to show (but we omit the proof) that given $G_1, \ldots, G_k$ such that $X(G_1) \supset X(G_2) \supset \cdots \supset X(G_k)$, then there exists groups $G'_1, \ldots, G'_k$ with G'_j isomorphic with G_j and $X(G'_j) = X(G_j)$ and, if $e'_j = (e'_j)^2 \in G'_j$ is the identity of G'_j, then $e'_1 > e'_2 > \cdots > e'_k$.[*] Thus

---

[*]Recall that this is the natural partial order on idempotents (elements satisfying $e^2 = e$) where $e_1 \geq e_2$ is defined by the condition $e_1 e_2 = e_2 = e_2 e_1$. -CLN

essentially for our purposes $X(G_1) \supset X(G_2)$ and $e_1 > e_2$ are equivalent biological concepts.

As before we write $X(G_j)$ as X_j. Now $\mathcal{X}_{j+1} = \{X_{j+1} \cdot g_j : g_j \in G_j\}$ is easy to interpret. Under the inputs G_j (meaning strings $a_1 \ldots a_n$ so that the map $\cdot a_1 \ldots a_n$ lies in G_j) the substrates X_j are macroscopically static (i.e. $X_j \cdot g_j = X_j$ for all $g_j \in G_j$) but the subset of substrates $X_{j+1} \subset X_j$ is moved around from X_{j+1} to $X_{j+1} \cdot g_j$ under g_j etc. Clearly (X_j, G_j) and X_{j+1} determines $\mathcal{X}_{j+1}$ with no knowledge of G_{j+1} being required. Thus (X_j, G_j) gives information regarding the macroscopic movements of X_{j+1} in X_j.

Finally we want to show $X_j \supset X_{j+1}$ being essential (in the sense of (4.11)) implies (X_j, G_j) and X_{j+1} gives *no information* about the microscopic movements of molecules within X_{j+1} (i.e. *no information* about G_{j+1}). First of all let us give a situation where (X_j, G_j) and X_{j+1} *does* determine some microscopic movements of X_{j+1}. Let $(G_j)_{X_{j+1}} = \{g_j \in G_j : X_{j+1} \cdot g_j = X_{j+1}\}$. Then $(X_{j+1}, (G_j)_{X_{j+1}})$ is a permutation group derived from (X_j, G_j) and X_{j+1}. To exclude these movements (since they are already known from (X_j, G_j) and X_{j+1}) we notice the set of idempotents (element e with $e^2 = e$) of $(G_j)_{X_{j+1}}$ is just $\{e_j\}$, since $(G_j)_{X_{j+1}}$ is a group. Further $x_{j+1} \cdot e_j = x_{j+1}$ for all $x_{j+1} \in X_{j+1}$ so e_j just gives the trivial identity movement.

The *critical idea* is the following. To exclude microscopic movements of X_{j+1} which can be derived from the previous larger cross-over groups we demand that the movements of G_{j+1} be products of idempotents. Since the only idempotents in the previous groups are trivial identity movements, the movements derived in this manner cannot be deduced from the knowledge of the larger microscopic cross-over groups.

We make this critical idea more precise and clearer. Let $t = a_1, \ldots, a_n \in A^+$ be a string of inputs. Then t is an *idempotent* iff $(x \cdot t) \cdot t = x \cdot t$ for all $x \in X$. Now chemically this means $x \xmapsto{t} x \cdot t$ but then $x \cdot t$ does not change under another application of t. Pictorially

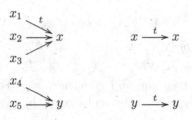

Thus t transforms x_1, x_2, x_3 into x and preserves x (i.e. $x \cdot t = x$) and t transforms x_4, x_5 into y and then preserves y (i.e. $y \cdot t = y$).

Also by definition the *range* of the idempotent $t = t^2$ is $\{x \in X : x \cdot t = x\}$, thus the range of t is all substrates stable under application of t. Clearly Range$(t) = \{x \cdot t : x \in X\}$, when $t = t^2$ is an idempotent.

Now suppose $A, B \subseteq X$ and $t = t^2$ is an idempotent. Then we write $A \overset{t}{\Rightarrow} B$ iff $t^2 = t$ is an idempotent and $B = $ Range(t) and $a \to t \cdot a$ is a one-to-one map of A onto B. In our example above, $\{x_1, x_4\} \overset{t}{\Rightarrow} \{x, y\}$, $\{x_2, x_5\} \overset{t}{\Rightarrow} \{x, y\}$, or $\{x_i, x_j\} \overset{t}{\Rightarrow} \{x, y\}$ iff either $1 \le i \le 3$, $4 \le j \le 5$ or $1 \le j \le 3$, $4 \le i \le 5$. Thus $A \overset{t}{\Rightarrow} B$ implies

$$(6.42) \qquad \begin{array}{c} a_1 \longrightarrow a_1 \cdot t = b_1 \\ \vdots \\ a_n \longrightarrow a_n \cdot t = b_n \end{array} \qquad b_j \overset{t}{\longrightarrow} b_j$$

$$\begin{array}{l} b_i = b_j \quad \Longrightarrow \quad i = j \\ x \cdot t = x \quad \Longrightarrow \quad x = b_j \text{ for some } j. \end{array}$$

Now intuitively since $B = $ Range(t) are *all* the stable substrates under t and $a_i \to a_i \cdot t = b_i = b_i \cdot t$ is a one-to-one map of A onto B, *the microscopic action $a_i \to a_i \cdot t = b_i = b_i \cdot t$ is not given by the previous cross-over groups, not even from those which give the macroscopic action $A \cdot g = B$.* Intuitively $A \overset{t}{\Rightarrow} B$ means 'locally no cross-over' from A to B via t.

Now $\mathcal{X}_{j+1} = \{X_{j+1} \cdot g_j : g_j \in G_j\}$. We write $A \overset{g}{\longrightarrow} B$ iff $\overset{}{a} \to a \cdot g$ is a one-to-one map of A onto B. Then $\mathcal{X}_{j+1} = \{B : X_{j+1} \overset{g}{\longrightarrow} B$ for some $g \in G_j\}$.

Now suppose (X_j, G_j) and X_{j+1} are given. Consider first all chains

$$(6.43) \qquad X_{j+1} = A_1 \overset{g_1}{\longrightarrow} A_2 \overset{g_2}{\longrightarrow} \ldots \overset{g_n}{\longrightarrow} A_{n+1} = X_{j+1}$$

with $g_1, \ldots, g_n \in G_j$ and each $A_i \subseteq X_j$. Thus (6.43) represents a *macroscopic movement of $X_{j+1} = A_1$ to $A_2 \ldots$ to A_n to $A_{n+1} = X_{j+1}$ under $g_1, \ldots, g_n \in G_j$.* Now choose idempotents $t_1 = t_1^2, \ldots, t_n = t_n^2$, when *possible*, so that

$$(6.44) \qquad X_{j+1} = A_1 \overset{t_1}{\Rightarrow} A_2 \overset{t_2}{\Rightarrow} \cdots \overset{t_n}{\Rightarrow} A_{n+1} = X_{j+1}.$$

Thus *macroscopically $X_{j+1} = A_1 \overset{g_1}{\longrightarrow} A_2 \overset{g_2}{\longrightarrow} \ldots \overset{g_n}{\longrightarrow} A_{n+1} = X_{j+1}$ does the same thing as $X_{j+1} = A_1 \overset{t_1}{\Rightarrow} A_2 \overset{t_2}{\Rightarrow} \cdots \overset{t_n}{\Rightarrow} A_{n+1} = X_{j+1}$ but microscopically they may differ considerably.* For example in (6.43) $X_{j+1} \cdot g_1 \ldots g_n = X_{j+1}$. Thus by replacing g_n by $g_n(g_1 \ldots g_n)^{-1}$ in (6.43) we can assume $g_1 \ldots g_n = 1$ or $x_{j+1} \mapsto x_{j+1} \cdot g_1 \ldots g_n = x_{j+1}$ is the identity map. On the other hand in (6.44), $x \mapsto x \cdot t_1 t_2 \ldots t_n$ is a permutation of X_{j+1} which may not be the identity.

In fact we demand that G_{j+1} is contained that set of mappings on X_{j+1} of the form $x \mapsto x \cdot t_1 \ldots t_n$, where $t_1, \ldots, t_n$ satisfies (6.43) and (6.44). Thus G_{j+1} is all the microscopic movements of molecules of X_j which arise from a macroscopic pattern (6.43) which is possible from G_j, but where the movements microscopically are made by idempotents of which all their stable states are present and whose macroscopic behavior is the same as (6.43).

Thus the lower bound k to complexity of Theorem 4.11a is the longest chain of cross-over groups whose substrates form a decreasing nested set and such that the microscopic behavior at step $j + 1$ can be mimicked *macroscopically* by the jth step, but *cannot be microscopically* mimicked by any of the other cross-over groups including the jth step.

We close with an example from K_2 (see (6.8) and Figure 3) already computed. Let (X, S) be given by (6.8). Delete y_1 from X and write y_j as j leading to Figure 6. Let $t = \delta\gamma\beta\alpha$ and $r = \beta\delta\gamma\alpha\beta\alpha$ as in (6.27) and (6.28). Thus

$$
\begin{array}{ccccc}
2 & 4 & 2 & & 4 \\
3 & 4 & 3 & & 4 \\
4 & \xrightarrow{t} 5 & 4 & \xrightarrow{r} & 5 \\
5 & 2 & 5 & & 4 \\
6 & 2 & 6 & & 4
\end{array}
$$

Let $G_1 = \{t, t^2, t^3\}$, $G_2 = \{r, r^2\}$, $X(G_1) \equiv X_1 = \{2, 4, 5\}$ and $X(G_2) = X_2 = \{4, 5\}$. $\mathcal{X}_2 = \{\{4, 5\} \cdot t^j : j = 1, 2, 3\} = \{\{4, 5\}, \{2, 4\}, \{2, 5\}\}$. Then $X_2 = \{4, 5\} \xrightarrow{t^2} \{2, 4\} \xrightarrow{t^2} \{2, 5\} \xrightarrow{t^2} \{4, 5\} = X_2$. Consider the idempotents $rt^2, trt, t^2 r$ tabulated in (6.29). Then

$$
X_2 = \{4, 5\} \overset{rt^2}{\Rightarrow} \{2, 4\} \overset{trt}{\Rightarrow} \{2, 5\} \overset{t^2 r}{\Rightarrow} \{4, 5\} = X_2.
$$

But

$$
rt^2 trtt^2 r = rt^3 rt^3 r = r^3 = r
$$

so

$$
\begin{array}{ccc}
4 & \xrightarrow{ rt^2 \cdot trt \cdot t^2 r } & 5 \\
5 & & 4
\end{array}
$$

so $\{2, 4, 5\} \supset \{4, 5\}$ is essential.

Finally we wish to interpret (6.35) given by Figure 8 (and Figure 7). In Chapter 4 we had explained at length that (6.36) is a physical theory for K_2 (or more precisely for Figure 6). We now want to examine this claim in this special case in more detail.

To do this we adopt the viewpoint of Zeiger. See [Zeiger]. Given (X, S), the action on the state space of the reactions, we would like to compute $x \cdot a_1 \ldots a_n$ given substrate x and inputs $a_1, \ldots, a_n$. We do this by a sequence of *approximations*, defined as follows. If X were a metric space (space equipped with a distance function) then given $\epsilon > 0$ we might compute $\bar{x}$ instead of $x \cdot a_1 \ldots a_n$ but with the proviso that the distance from $x \cdot a_1 \ldots a_n$ to $\bar{x}$ is $< \epsilon$. Since X is finite there is no natural metric so we substitute the idea of cover analogous to the set of all open ϵ-spheres of X if X were a metric space.

Formally $\mathcal{C}$ is a *cover* of X iff $\mathcal{C} = \{C_a : a \in J\}$, where J is some index set, and $C_a \subseteq X$ for each $a \in J$ and $\bigcup\{C_a : a \in J\} = X$. Thus $\mathcal{C}$ is a cover for X iff $\mathcal{C}$ is a collection of subsets of X whose union is X. To tie $\mathcal{C}$ up with the input action we define $\mathcal{C}$ to be a *weakly preserved* (*w.p.*) *cover* for (X, S) iff $\mathcal{C} = \{C_a : a \in J\}$ is a cover of X and for each $s \in S$, and each $a \in J$ there exists $b \in J$ (which depends on a and s) such that

$$C_a \cdot s = \{c \cdot s : c \in C_a\} \subseteq C_b.$$

For example the set of all subsets of X having exactly k elements is a w.p. cover for any semigroup on X.

Now given (X, S) and a w.p. cover $\mathcal{C}$ for (X, S) we say $x_1 \in X$ *approximates* x_2 *in* X *with respect to* $\mathcal{C}$ iff there exists a $C_a \in \mathcal{C}$ so that $x_1, x_2 \in C_a$. Now the following definition is very natural. Let (X, A, λ) be given (say arising from reactions R). Then the machine $f : A^+ \to X$ *approximates* (X, A, λ) (with respect to the w.p. cover $\mathcal{C}$ and starting state $x_0 \in X$) iff for each $(a_1, \ldots, a_n) \in A^+$, $f(a_1, \ldots, a_n)$ approximates $x_0 \cdot a_1 \ldots a_n$ with respect to $\mathcal{C}$.

Now the approximations can 'get better'. Precisely, let $\mathcal{C}_1, \mathcal{C}_2$ be two w.p. covers of (X, S). Then $\mathcal{C}_1 \leq \mathcal{C}_2$ iff for each $C_a \in \mathcal{C}_1$, there exists $C_b \in \mathcal{C}_2$ (with b depending on a) such that $C_a \subseteq C_b$. Clearly if the machine f approximates (X, A, λ) with respect to $\mathcal{C}_1$ and $\mathcal{C}_1 \leq \mathcal{C}_2$, then f approximates with respect to $\mathcal{C}_2$. Also let I be the w.p. cover $\{\{x\} : x \in X\}$. Then the machine f approximates with respect to I iff $f(a_1, \ldots, a_n) = x_0 \cdot a_1 \ldots a_n$ for all $(a_1, \ldots, a_n) \in A^+$.

Now let us consider our special case. Let $X = \{2, 3, 4, 5, 6\}$ and let T be as defined just below (6.25) and after Figure 6. Then define the w.p.

covers with respect to (X, T) or $(X, \{\alpha, \beta, \gamma, \delta\}, \lambda)$ by

$$\mathcal{C}_1 = \{\{2,4,5\}, \{2,4,6\}, \{2,5,6\}, \{3,4,5\}, \{3,4,6\}, \{3,5,6\}\}$$

$\mathcal{C}_2 = $ all subsets of X having exactly 2 elements (except $\{2,3\}$)

$\mathcal{C}_3 = I = $ all subsets of X having exactly 1 element

so $\mathcal{C}_3 < \mathcal{C}_2 < \mathcal{C}_1$.

Now first we claim $(X, \{\alpha, \beta, \gamma, \delta\}, \lambda)$ can be approximated with respect to $\mathcal{C}_1$ by a *combinatorial* machine. For consider the circuit (see earlier chapters and the Appendix of Chapter 5 for terminology) $C' = (\{\alpha, \beta, \gamma, \delta\}, X = \{2, \ldots, 6\}, \mathcal{C}_1, \lambda_1, \delta_1)$ with inputs $\{\alpha, \beta, \gamma, \delta\}$, states X, output alphabet $\mathcal{C}_1$ with transition function $\lambda_1(D, a) = D \cdot a_1$ for $D \in \mathcal{C}_1$, $a \in \{\alpha, \beta, \gamma, \delta\}$ and $\cdot a_1$ is defined as above (6.30b) via Figure 7. Finally the output function $\delta_{1\prime} : \mathcal{C}_1 \twoheadrightarrow X$ is defined in any arbitrary but fixed manner subject to $\delta_1(D) \in D$ for all $D \in \mathcal{C}_1$. Then it is easily verified that if the machine $f_D : \{\alpha, \beta, \gamma, \delta\}^+ \to X$ by definition equals C'_D for $D \in \mathcal{C}_1$, then f_D approximates $(X, \{\alpha, \beta, \gamma, \delta\}, \lambda)$ (or (X, T)) with respect to the w.p. cover $\mathcal{C}_1$ and starting state $x_0 \in D$. The verification is very easy since $\mathcal{C}_1$ is a w.p. cover and $D \cdot a_1 \ldots a_n \subset D * a_1 \ldots a_n$ for all $(a_1, \ldots, a_n) \in \{\alpha, \beta, \gamma, \delta\}^+$ and $\cdot$ is multiplication with respect to λ of the state space and $*$ is multiplication with respect to λ_1 of the circuit C'_1.

Also it is trivial to verify that if $C = f_D^S$, then this is the same C as the one appearing in (6.36), which was shown to be combinatorial.

Thus with reference to Figure 6 we can say (and have proved) that of the five substrates, $2, \ldots, 6$, *without using any arithmetic* we can assert up to three substrates (or more strictly up to a member of $\mathcal{C}_1$) what substrate will result if x_0 is hit by $a_1, \ldots, a_n$ (with $a_k \in \{\alpha, \beta, \gamma, \delta\}$) in that order.

Now suppose we want to know up to two substrates what $x_0 \cdot a_1 \ldots a_n$ is. Thus we want to solve

(6.45) \qquad\qquad ? series $f_1 = f_2$

where f_j approximates with respect to $\mathcal{C}_j$ and starting state x_0.

We look at (6.36) and define the circuit

$$C'_2 = (\{\alpha, \beta, \gamma, \delta\}, X = \{2,3,4,5,6\}, X_2 \times X_1, \lambda_2, \delta_2),$$

where $X_2 = \{1,2,3\}$, $X_1 = \mathcal{C}_1$ and $\lambda_2((x_2, x_1) \cdot a)$ is defined by restricting the action of $T^\#$ of (6.36) to the first two coordinates $X_2 \times X_1$. Finally δ_2 is chosen to satisfy $\delta_2(x_2, x_1) \in j(X_3 \times \{x_2\} \times \{x_1\})$ with j defined by (6.31). Then C'_2 is a circuit which computes the cover $\mathcal{C}_2$ or precisely if $f_2 = (C'_2)_{(x_2, x_1)} : \{\alpha, \beta, \gamma, \delta\}^+ \to X$ then f_2 approximates $(X, \{\alpha, \beta, \gamma, \delta\}, \lambda)$ with respect to $\mathcal{C}_2$ and starting state $x_0 \in j(X_3 \times \{x_2\} \times \{x_1\})$.

Now the semigroup of the circuit C'_2 lies in $(X_2, \mathbb{Z}_3^*) \wr (\mathcal{C}_1, C)$ so we can solve (6.45) by

$$\mathbb{Z}_3^{*M} \text{ series } f_1 = f_2.$$

Now coupling this with the lower bounds already proved we find—*to known within two substrates what $x_0 \cdot a_1 \ldots a_n$ is for Figure 6 we must use arithmetic and in particular one $\mathbb{Z}_3$ counter.*

Similarly to know $x_0 \cdot a_1 \ldots a_n$ perfectly (i.e. with respect to $I = \mathcal{C}_3$) we must use arithmetic at complexity 2. In fact $\mathbb{Z}_2^{*M}$ series $\mathbb{Z}_3^{*M}$ series C^M will suffice and nothing less. C^M computes the approximation with respect to $\mathcal{C}_1, \mathbb{Z}_3^{*M}$ computes $\mathcal{C}_2$ given $\mathcal{C}_1, \mathbb{Z}_2^{*M}$ computes $I (= \mathcal{C}_3)$ *given* $\mathcal{C}_2$.

Essentially (6.27) forces a $\mathbb{Z}_3$ in computing $\mathcal{C}_2$, and (6.28) and (6.29) force a $\mathbb{Z}_2$ in computing $I = \mathcal{C}_3$ given $\mathcal{C}_2$. The lower bound theorem (4.11a) of Rhodes and Tilson make this precise. See [LB I, LB II]. However, heuristically (6.27) says that going from knowledge up to subsets of order 3 to knowledge up to subsets of order 2 costs a $\mathbb{Z}_3$. Similarly (6.28) and (6.29) say going from subsets of order 2 to perfect information cost a $\mathbb{Z}_2$.

Thus we see that the 'physical theory' (6.36) is a sequence of three approximations to the reactions, first with respect to 3 substrates, then up to 2 substrates and then perfect.

6.46.   Discussion. We now return to Discussion (6.18) and Viewpoint (6.22).

In the following we will show Viewpoint (6.22) is a plausible and reasonable way of classifying reactions. Then we will show that the complexity and the triangularizations of the state (e.g. like (6.21) or (6.36)) are less important for intermediary metabolism essentially because the global feedback loops through the genetic control of enzymes are ignored. However in Part II we will concentrate on just this global feedback.

In other words, modeling intermediary metabolism (say in the manner previously indicated) accounts for local feedback (e.g. the cycle behavior of Krebs cycle) but ignores global feedback (e.g. genetic enzyme control). Now the SNAGs are manifestations of local feedback while the complexity and triangularizations are directly effected by global feedback.

Now we will explicate and expand these remarks in detail.

We first demonstrate that Viewpoint (6.22) is plausible, reasonable and important.

Let R_1 and R_2 be two sets of reactions perhaps stemming from widely varying sources (e.g. the Krebs cycle and the hormonal control of the menstrual cycle). Suppose (X_j, A_j, λ_j) is the state space and action derived from R_j with semigroup (X_j, S_j) via Principle I (4.1) and (4.6) for $j = 1, 2$.

First suppose G_j is a subgroup of S_j so $(Y_j = X(G_j), G_j)$ is a cross-over group of R_j.

Now if (Y_1, G_1) is isomorphic to (Y_2, G_2) as permutation groups (meaning there exists $j : Y_1 \to Y_2$ and surmorphism $\rho : G_1 \to G_2$ so that both j^{-1} and ρ^{-1} exists and $j(y_1 \cdot g_1) = j(y_1) \cdot \rho(g_1)$ for all $y_1 \in Y_1$, $g_1 \in G_1$) then the movements of the substrates Y_1 under $a_1 \ldots a_n$ when $\cdot a_1 \ldots a_n \in G_1$ are exactly the same as the movements of $j(Y_1) = Y_2$ under $\rho(\cdot a_1 \ldots a_n) = \rho(\cdot a_1) \ldots \rho(\cdot a_n) \equiv \cdot b_1 \ldots b_m \in G_2$. Thus

$$y_1 \xrightarrow{a_1 \ldots a_n} y_2 \quad \text{iff} \quad j(y_1) \xrightarrow{b_1 \ldots b_m} j(y_2).$$

Now since the movements of Y_1 under G_1 are exactly similar to the movements of Y_2 under G_2 any theory, insight, intuition for either can be immediately applied to the other. Thus *if the divisor (Y_1, G_1) of semigroup (R_1) is isomorphic as permutation groups with the divisor (Y_2, G_2) of semigroup (R_2), then this allows one to identify these two parts biologically and physically.*

Next suppose G_1 is isomorphic with G_2 as groups (i.e. there exists a surmorphism $\rho : G_1 \to G_2$ with ρ^{-1} existing) but (Y_1, G_1) is not isomorphic to (Y_2, G_2) as permutation groups (e.g. Y_1 and Y_2 have different number of elements). In this case we appeal to the following Fact.

Fact 6.47. *Let (X, S) be a right mapping semigroup with X a finite set. Let $(X \times \ldots \times X, S, *)$ denote $X \times \ldots \times X$ taken $|X|$-times (i.e. the Cartesian product of X with itself as many times as X has distinct elements) with $(x_1, \ldots, x_n) * s = (x_1 \cdot s, \ldots, x_n \cdot s)$. Then there exists $Y \subseteq X \times \ldots \times X$ such that $Y \cdot S \subseteq Y$ and (Y, S) is isomorphic as mapping semigroups with (S^1, S).*

Proof. Let $X = \{x_1, \ldots, x_n\}$. Let $y(s) = (x_1 \cdot s, \ldots, x_n \cdot s) \in X \times \ldots \times X$. Let $y(1) = (x_1, \ldots, x_n)$. Let $Y = \{y(s) : s \in S^1\}$. Then $y(s_1) * s_2 = y(s_1 s_2)$ so taking $j : S^1 \to Y$ to be $j(s) = y(s)$ yields $j(s_1) * s_2 = j(s_1 s_2)$ so $(Y, S) \xrightarrow{(j, \text{id})} (S^1, S)$ is the desired isomorphism proving Fact (6.47). $\square$

The interpretation of $(Y_j \times \ldots \times Y_j, G_j)$ is simply that we take $|Y_j| = \pi_j$ copies of the reactions R_j (i.e. take π_j cells) and choose $a_1, \ldots, a_m$

so $\cdot a_1 \ldots a_m \in G_j$ and take (perhaps different) starting substrates $(x_1, \ldots, x_{\pi_j})$ and observe the effect in each system $(x_1 \cdot a_1 \ldots a_m, \ldots, x_{\pi_j} \cdot a_1 \ldots a_m)$. This is very similar to Gibbs's idea in Statistical Thermodynamics. For example see the first few chapters of that excellent little book [Sch.].

Assuming G_1 is isomorphic to G_2 as groups and applying Fact (6.47) with $X = Y_j$ and $S = G_j$ we obtain $(G_j, G_j) \leq (Y_j \times \ldots \times Y_j, G_j) \leq (X \times \ldots \times X(\pi_j \text{ times}), S_j)$. But since G_1 is isomorphic to G_2 as groups, (G_1, G_1) is isomorphic to (G_2, G_2) as permutation groups so we are in the situation previously considered.

Summarizing, if G_1 is isomorphic to G_2 as groups, then by taking $|Y_1|$ copies of G_1 (i.e. $|Y_1|$ cells containing the reactions R_1) and $|Y_2|$ copies of G_2, then we obtain two systems $R'_1 = R_1 \times \ldots \times R_1$ $|Y_1|$-times and $R'_2 = R_2 \times \ldots \times R_2$ $|Y_2|$-times such that $(G, G) = (G_1, G_1) = (G_2, G_2)$ is a subsystem of both R'_1, and R'_2 so we have the situation previously analyzed.

Next suppose G_1 is (perhaps) not isomorphic to G_2 but G_1 and G_2 have a common Jordan–Hölder factor P (so **PRIMES**$(G_1) \cap$ **PRIMES**(G_2) is not empty). Then as above by taking $|Y_j| = |X(G_j)|$ cells containing reactions R_j we obtain (G_1, G_1) and (G_2, G_2). Now applying (2.4) and (2.5) and the Jordan–Hölder Theorem for finite groups we obtain a triangularization of (G_1, G_1) and (G_2, G_2) which have a common component, i.e.

$$(G_j, G_j) \leq (S_{n_j}^{(j)}, S_{n_j}^{(j)}) \wr \cdots \wr (S_1^{(j)}, S_1^{(j)})$$

with $S_1^{(j)}, \ldots, S_{n_j}^{(j)}$ the Jordan–Hölder factors of G_j and $P = S_{x_1}^{(1)} = S_{x_2}^{(2)}$ for some x_1 and x_2, $1 \leq x_j \leq n_j$. Then as before the component action (see (2.2b) or (2.3)) $(S_{x_1}^{(1)}, S_{x_1}^{(1)}) = (P, P)$ can be biologically and physically identified with the component action $(S_{x_2}^{(2)}, S_{x_2}^{(2)}) = (P, P)$.

It is not difficult to show that for any group **PRIMES**$(G) = \{P : P \in$ **PRIMES** and P divides a Jordan–Hölder factor of $G\}$. Also $G_1 \mid G_2$ iff $(G_1, G_1) \mid (G_2, G_2)$. Thus a similar argument to the above implies—*if* $P \in$ **PRIMES**$(S_1) \cap$ **PRIMES**(S_2), *then by taking a suitable number of cells containing the reactions R_1 (call the parallel complex Π_1) and taking a suitable number of cells containing the reactions R_2, we obtain P as a component action of a group (H_1, H_1) of Π_1 and of a group (H_2, H_2) of Π_2.*

Up to now we are on very strong grounds for asserting the similarity of parts of two reactions if P belongs to both **PRIMES**(R_1) and **PRIMES**(R_2), since we have seen that components of each move exactly in the same manner microscopically.

However we must make one last step. First $\mathbb{Z}_p$'s being in common are not too important since $\mathbb{Z}_p$'s can only act by cyclically permuting p states. Thus SNAGs in common are the important means of classification since their actions are complicated and usually large. However for a useful classification of biological reactions a meaningful number of reactions must be classified together so as to bring about useful simplifications and insight. Thus demanding **PRIMES**(R_1) and **PRIMES**(R_2) to have a common member SNAG is too strong. Not enough reactions will satisfy this. Fortunately modern group theory has placed the known SNAGs in classes (e.g. A_n, $n \geq 5$) mainly derived from the Chevalley method and variations thereof. This classification goes back to Dickson and Artin and major recent developments are due to Thompson, etc. Essentially most of the SNAGs arise from constructions from the simple Lie algebras and the classifications stem from the classifications of the simple Lie algebras. However, there appear to be a few SNAGs who do not fit in any such scheme to date.

Thus given reactions R_1 and R_2 if $P_j \in$ **PRIMES**(R_j) and P_1 and P_2 are both SNAGs (i.e. $P_j \neq \mathbb{Z}_p$) and both P_1 and P_2 are classified together by the group theorists then we assert the components P_1 of Π_1 and P_2 of Π_2 are similar biologically and physically. Basically this follows since both P_1 and P_2 stem from (say) an orthogonal group of varying dimensions. However to go into this would take us into the classification scheme for SNAGs which we cannot go into here.

Thus the promised Mendeleev type table for the classification of biological reactions (e.g. bacterial intermediary metabolism) proceeds as follows. Given reactions R_1 and R_2, consider **PRIMES**(R_1) and **PRIMES**(R_2). Ignore the $\mathbb{Z}_p$'s and determine the number of SNAGs of **PRIMES**(R_1) which are classified by modern finite group theory in the same category as the SNAGs of **PRIMES**(R_2). The more SNAGs of both which are classified similarly in group theory, the closer (by definition) R_1 and R_2 are in the biological Mendeleev type classification system and *thus the more similar will be their respective biological properties* (for the reason presented previously).

It is interesting to speculate the continuous Lie groups and Lie algebras which are involved in continuous biological models (e.g. Thom's program) are the 'limits' of the Lie algebras involved with the SNAGs of **PRIMES**(R).

We next want to make it clearer why the complexity and triangularizations of the reactions R are not so important as **PRIMES**(R) in classifying the reactions R. In doing this we also develop a little mathematics.

Let R be some reactions which give rise to the (single-valued labeled directed) graph $H = (X, E, \rho : E \to L)$. See Definitions (6.12)–(6.13).

We next want to show that the complexity of H depends only on the 'components' of H and not how the 'components' are connected together.

We define the (directed) components of H (which depend on just (X, E) and not on the labels) in the standard way as follows. We say there exists a directed *path* from $x_1 \in X$ to $x_2 \in X$, denoted $P(x_1, x_2) = 1$, iff there exists $y_1, y_2, \ldots, y_n \in X$ $(n \geq 1)$ with $x_1 = y_1$, $(y_1, y_2) \in E$, $(y_2, y_3) \in E, \ldots, (y_{n-1}, y_n) \in E$, $x_2 = y_n$. Notice $P(x, x) = 1$ for all $x \in X$. Then by definition $C(x)$, the *component of H containing x*, equals $\{y \in X : P(x, y) = 1 = P(y, x)\}$. It is very easy to verify that xCy iff $C(x) = C(y)$ is an equivalence relation C on X whose equivalence classes are $\{C(x) : x \in X\}$, read "*the components of H*". For example given the directed graph (where the labels are not indicated),

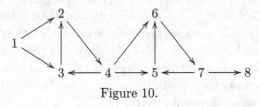

Figure 10.

the components are $\{1\}, \{2, 3, 4\}, \{5, 6, 7\}, \{8\}$. Notice if however an arrow from 8 to 1 is added the components become just $\{1, \ldots, 8\}$, i.e. all one component.

Now let (X, S) be the semigroup associated with H (see (6.13b) and (4.6)). Now using the definitions of S and the definition of the components we easily find that: (∗) if $y \in C(x)$ and $s \in S$ and $y \cdot s$ does not lie in $C(x)$, then $y \cdot s \cdot t$ also does not lie in $C(x)$ for all $t \in S$. Now using (∗) we can let S act on $C(x)^0 \equiv C(x) \cup \{0\}$, where 0 is an extra added point, by $[C(x)^0, S, *]$ (where the square brackets indicate that the action ∗ of S on $C(x)^0$ may not be faithful) with $y * s = y \cdot s$ when $y \cdot s \in C(x)$ and otherwise $y * s = 0$. Also $0 * s = 0$ all s. Then $y * s * t = y * st$, as can be verified from (∗). Let S_x denote the action $[C(x)^0, S, *]$ made faithful, i.e. $S_x = S/\equiv_x$ where $s_1 \equiv_x s_2$ iff $*s_1 = *s_2$. Thus $\theta_x : S \twoheadrightarrow S_x$ with $\theta_x[s] = [s]_{\equiv_x}$.

If G is a subgroup of $S, G \leq S$, it is easy to verify that the transitive components of $(X(G), G)$ (see (4.11)) each lie in a component of H and thus $g \mapsto \{\theta_x(g) : x \in X\}$ is a one-to-one homomorphism (monomorphism) of G into $\Pi\{S_x : x \in X\}$.

Thus $x \mapsto \{\theta_x(s) : x \in X\}$ is a γ-homomorphism of S onto a subdirect product of $\Pi\{S_x : x \in X\}$ and so using Axioms I and II for complexity of Chapter 3 we find

$$(6.48) \qquad \#_G(S) = \max\{\#_G(S_x) : x \in X\}.$$

From Figure 10 we have $C(1) = \{1\}$, $C(2) = C(3) = C(4) = \{2,3,4\}$, $C(5) = C(6) = C(7) = \{5,6,7\}$, $C(8) = \{8\}$. Thus

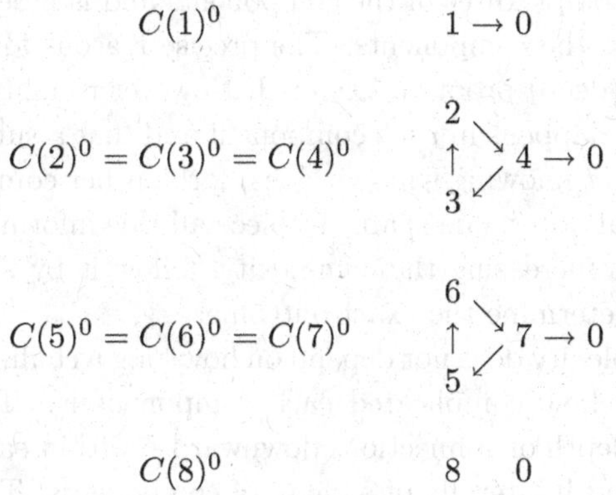

$$C(1)^0 \qquad\qquad\qquad 1 \to 0$$

$$C(2)^0 = C(3)^0 = C(4)^0 \qquad \begin{array}{c} 2 \\ \uparrow \\ 3 \end{array} \searrow 4 \to 0$$

$$C(5)^0 = C(6)^0 = C(7)^0 \qquad \begin{array}{c} 6 \\ \uparrow \\ 5 \end{array} \searrow 7 \to 0$$

$$C(8)^0 \qquad\qquad\qquad 8 \quad 0$$

Figure 11.

In fact given H it is easy to verify that $(C(x)^0, S_x)$ is associated with the graph H_x where H_x is derived from H by restricting the points to $C(x)$ then adding the point zero, leaving intact the arrows between points of $C(x)$, replacing the arrows (n, y), $n \in C(x)$, $y \notin C(x)$ by $(n, 0)$, ignoring the other arrows and defining the new ρ' in the natural way, i.e. $\rho'(n, 0) = \bigcup\{\rho(n, y) : n \in C(x), y \in X - C(x)\}$ where we are now allowing several labels on the arrows going to zero (slightly generalizing the previous definition). Also in applications the arrows going to zero can be omitted, together with zero.

Summarizing, given some reactions R whose graph (ignoring the labels) is as in Figure 10 (for example), the complexity of R just depends on the component fragments (as illustrated in Figure 11) and *not at all on the connections between the components*. In short, the complexity depends on the form of the connections within each component and is independent of the manner in which the components are connected.

Thus with respect to **PRIMES**(R), complexity(R) and even (by Axiom II for complexity) triangularizations of R, we need only consider graphs having at most two components $\{0\}$ and $\{1, \ldots, n\}$ (where 0 is a point from which no labeled arrow leaves) by passing from (X, S) to the components $(C(X)^0, S_x)$.

A well-defined partial order can be placed on the components of the graph H by defining $C(x) \geq C(y)$ iff $P(x, y) = 1$. Then $C(x_1) \geq C(x_2)$ and $C(x_2) \geq C(x_3)$ implies $C(x_1) \geq C(x_3)$. Also $C(x_1) \geq C(x_2)$ and $C(x_2) \geq C(x_1)$ implies $C(x_1) = C(x_2)$.

The complexity of the semigroup associated with H is by (6.48) the maximum of the complexities of the components and is independent of the partial ordering on the components. The precise reasons for this depends, of course, on the (deep) proof of Axiom II. However roughly, given the information of what happens in *each* component and that a substrate leaves a component (but *not* knowing where it goes) for another component further down in the partial order, one is able to piece all this information together, expand it without increasing the complexity, follow it by a combinatorial calculation, and determine the exact outcome.

Thus the complexity does not depend on how long a chain of components there are, but on how complicated each component is. The complexity depends on the "depth of connections downwards" within each component, and not on the length laterally of a chain of components. Thus

$$C_1 \to C_2 \to C_3 \to \cdots \to C_m.$$

Figure 12.

with the C_i as its components is no more complicated than $\max \#_G(C_i)$.

However, if in Figure 12 an arrow is adjoined from C_m to C_1, $C_m \to C_1$, then the complexity can rise fantastically above $\max \#_G(C_i)$ (up to total number of points minus 1).

Now in studying intermediary bacterial metabolism (say) diagrams like Figure 12 arise. Now C_m can effect C_1 but only by passing through systems not under consideration in metabolism (e.g. genetic controls).

Now the idea is that roughly placing an arrow from C_m to C_1, $C_m \to C_1$ will, more or less, leave intact the SNAGs of **PRIMES**(Figure 12) add some new ones, but radically change (upward) the complexity and vastly change the triangularizations.

Thus since intermediary bacterial metabolism is a somewhat local study, the SNAGs of **PRIMES**(reaction) are the proper means of classification, since they too are local properties of the relevant semigroups.

Next we examine in a little detail the complexities of single components. The following Proposition shows if the labels of a reaction graph H are changed so each arrow receives a distinct label, then the semigroup becomes larger (contains the old semigroup as a subsemigroup).[*,†] Second the complexity and **PRIMES** of the most complicated directed graph on n points is determined. Finally, the complexity and **PRIMES** of the cyclic reaction on n substrates is determined generalizing our previous specific computations.

Proposition 6.49. (a) *Let $H = (X, E, \rho : E \to L)$ be a reaction graph with associated semigroup S. Let $H' = (X, E, \rho' : E \to E)$ with $\rho'(e) = e$ (so each member of E is given a distinct label, thus H' is also a reaction graph). Let the semigroup of H' be S'. Then $S \leq S'$, i.e. S is a subsemigroup of S'.*

(b) *Let $n \geq 2$ and let $X_n = \{1, \ldots, n\}$. Let $H_n = (X_n, E_n, \rho_n : E_n \to E_n)$ where $E_n = \{(x_1, x_2) \in X_n \times X_n : x_1 \neq x_2\}$ and $\rho_n(e) = e$. Thus for example the picture of H_3 is given by Figure 13. Let the semigroup of H_n be denoted by T_n. Then T_n is all maps on n letters which are not one-to-one (i.e. not permutations).*

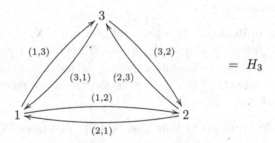

Figure 13.

Thus the semigroup of any reaction graph on n points is a subsemigroup of T_n. Further $\#_G(T_n) = n - 2$ and $\mathbf{PRIMES}(T_n) = \{P \in \mathbf{PRIMES} : P$ has a faithful permutation representation on $n - 2$ or less points$\}$.

---

[*]The conclusion of (6.49a) need not hold if H is *not* a reaction graph: For example, let H a directed n-cycle with all arrows labeled with the same label α, then α acts cyclically on the n vertices, and the semigroup of this graph is $\mathbb{Z}_n$, but this group cannot be embedded in the semigroup of corresponding directed graph $H' = C_n$ with all arrows given distinct labels. The semigroup of H' must lie in T_n by (6.49c), and hence contains no cyclic permutation of more than $n - 1$ points, and so contains no $\mathbb{Z}_n$. -CLN

[†]The formulation and proof of Proposition (6.49a) and the proof of Lemma (6.50) given here are joint work of the author and C. L. Nehaniv.

Equivalently $\textbf{PRIMES}(T_n) = \{P \mid A_{n-2} : P \in \textbf{PRIMES}\}$ *where* A_{n-2}
is the alternating group on $n - 2$ *letters.*

(c) *Let* $n \geq 2$ *and let* $X_n = \{1, \ldots, n\}$. *Let* $C_n = (X_n, E_n, \rho : E_n \to E_n)$
with $E_n = \{(1,2), (2,3), \ldots, (n-1,n), (n,1)\}$ *and let* $\rho(e) = e$. *Thus,*
for example, the picture of C_4 *is given by Figure* 14. *Let the semigroup*
of C_n *be denoted by* R_n.

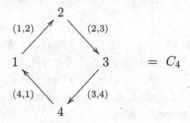

Figure 14.

Then $\#_G(R_n) = n - 2$ *and* $\textbf{PRIMES}(R_n) = \{\mathbb{Z}_p : p \leq n - 1 \text{ and } p \text{ is}$
a prime integer}.

To render the proof of Proposition (6.49) we require the following standard lemma.

Lemma 6.50. *Let* X *be a finite set. For* $a, b \in X$, $a \neq b$, *let* T_{ab} *be defined*
as in Definition (6.5) *(i.e.* $T_{ab}(a) = b$, $T_{ab}(x) = x$, $x \neq a$).

Let $f : X \to X$ *and suppose* f *is not one-to-one (or equivalently*
f *is not onto). Then* f *can be written as a product of members of*
$\{T_{ab} : a, b \in X, a \neq b\}$.

Proof. We proceed by induction on $|X|$. The cases $|X| = 1$ or 2 are very
easy. Thus assume the assertion is valid for all $|X| \leq n$ and let $|X| = n + 1$
and $f : X \to X$. Then by assumption $f(X)$ is a proper subset of X. Let
$y \in X - f(X)$. Then f restricts to a well-defined map $h : X - \{y\} \to X - \{y\}$.
Case 1: h is not a permuation of $X - \{y\}$. By induction hypothesis, h is a
product of $T_{a,b}$, with $a, b \in X - \{y\}$, and so $f = hT_{y,f(y)}$ can be written in
the required form (where the function argument is written on the left).
Case 2: h is a permutation of $X - \{y\}$. Since $f(y) \in X - \{y\}$, there
is a $z \in X - \{y\}$ with $h(z) = f(y)$. Clearly $f = T_{yz}h$. Now let h be
written as the product of disjoint cycles with entries in $X - \{y\}$. That is,
$h = \sigma_1 \ldots \sigma_n$, where each σ is of the form $(x_1, \ldots, x_{n(\sigma)})$, with $x_i \in X - \{y\}$
$(1 \leq i \leq n(\sigma))$. Let C_σ be the product $T_{x_{n(\sigma)}y}T_{x_{n-1}x_n} \ldots T_{x_2x_3}T_{x_1x_2}T_{yx_1}$.
Clearly $x \cdot C_{\sigma_i} = x \cdot \sigma_i$ for all $x \in X - \{y\}$. Therefore $f = T_{yz}h = T_{yz}\sigma_1 \ldots \sigma_n$

can be written in the desired form as $f = T_{yz}C_{\sigma_1}\ldots C_{\sigma_n}$, where each C_{σ_i} is a product of $T_{a,b}$'s as explicitly given above. $\qquad\square$

Proof of Proposition (6.49).

(a) By considering Definition (6.13a), (6.13b) and (6.13c), we find for $y \in L$ that $\cdot y = \Pi\{T_{ab} : a, b \in X, a \neq b, (a, b) \in E, y = \rho(a, b)\}$. Since $T_{a_1 b_1}T_{a_2 b_2} = T_{a_2 b_2}T_{a_1 b_1}$ when $b_1 \neq a_2$ and $b_2 \neq a_1$, we find that members of $\{T_{ab}\}$ labeled by y commute pairwise since H is a reaction graph*, so $\Pi\{T_{ab}\}$ is well-defined. Now since S' is generated by $\{T_{ab} : (a, b) \in E\}$, (a) follows.

(b) By Lemma (6.50), T_n is all maps on n letters which are not permutations. By the results on complexity of Chapter 3, $\#_G(T_n) \leq n - 2$. By (4.11a) $\#_G(T_n) \geq n - 2$. Thus $\#_G(T_n) = n - 2$. Clearly the maximal subgroups of T_n are the symmetric groups on $n - 1, \ldots, 1$ letters so (b) follows.

(c) We write $T_{i,i+1(\mathrm{mod}\ n)}$ as α_i so C_4 becomes as in Figure 15. In the following we apply functions on the right of their arguments and compute the maximal subgroups of R_n. Then if $f_k : X_n \to X_n$ is defined by for $1 \leq k < n$ by $f_k(x) = x$ for $x \leq k$ and $f_k(x) = 1$ for $x > k$, then it is easily verified that

$$f_k = \alpha_n(\alpha_{n-1}\alpha_n)(\alpha_{n-2}\alpha_{n-1}\alpha_n)\ldots(\alpha_{k+1}\alpha_{k+2}\ldots\alpha_n).$$

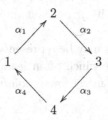

Figure 15.

Further, $t_k = \alpha_k\alpha_{k-1}\ldots\alpha_1\alpha_{k+1}\ldots\alpha_n$ cyclically permutes $(1\ldots k)$.

Thus $f_k t_k$ generates a cyclic subgroup of R_n of order k.

Next it is not difficult to verify if G is a subgroup of R_n, $G \leq R_n$, and $X(G) = \{x(1) < \cdots < x(m)\}$ and $\Pi \in G$ with $x(1) \cdot \Pi = x(j + 1)$, then $x(a) \cdot \Pi = x(a + j \ (\mathrm{mod}\ m))$ for all $1 \leq a \leq m$ (in other words the 'cyclic order' $x(1) < x(2) < \ldots x(m) < x(1)$ is preserved). Thus G is cyclic.

---

*Recall that in a reaction graph, (6.13a) implies there is a most one arrow (a, b) with label y leaving any vertex a, and (6.13c) implies a label y never occurs on consecutive arrows (a, b) and (b, c). -CLN

Thus we have proved

PRIMES$(R_n) = \{\mathbb{Z}_p : p$ is a prime integer $\leq n - 1\}$.

Using the Axioms for complexity of Chapter 3 it is not difficult to show that $\#_G(R_n) \leq n - 2$.

On the other hand we claim Theorem $(4.11a)$ implies $\#_G(R_n) \geq n - 2$ since we can construct a chain of essential dependencies of length $n - 2$ as follows.

Let $r_q = f_{q+1}t_{q+1}$. let $1 \leq q \leq n - 1$. Then consider $A = \{r_{q+1}^i r_q :$ $i = 1, \ldots, q + 1\}$ and $B = \{r_q r_{q+1}^j : j = 1, \ldots, q + 1\}$. Then form $A \cdot B = \{a \cdot b : a \in A, b \in B\}$. Then we claim r_q *is a product of idempotents of* $A \cdot B$. We omit the proof of this. Thus $\mathbb{Z}_{q+1} \cong \{r_q^m : m = 1, \ldots, q + 1\}$ is a subsemigroup of the subsemigroup generated by the idempotents of the semigroup generated by r_{q+1} and r_q.

Now using this it is easy to verify that R_n possesses a chain of essential dependencies of length $n - 2$. This proves (c) and hence proves Proposition (6.49).   □

Now any finite group G is the semigroup of some single-valued labeled directed graph. This is seen by defining $C(G) = (G, G \times G, \rho : G \times G \to G)$ with $\rho(g_1, g_2) = g_1^{-1}g_2$. Then it is trivial to verify that (in the notation of $(6.13b)$) $\lambda(g_1, x) = g_1 x$ for all $g_1, x \in G$. Thus the semigroup associated with $C(G)$ is G (acting on itself by right multiplication). $C(G)$ is called the *Cayley graph of* G.

Of course specific groups may be represented as permutation groups on less letters than their order which often leads to a representation as the semigroup of a graph with many fewer points than group elements. For example,

$$a = (x_0 \ x_1 \ x_2 \ \ldots \ x_{10})$$
$$b = (x_9 \ x_3 \ x_5 \ x_4) \ (x_6 \ x_2 \ x_7 \ x_{10})$$

generate a sharp 4-fold transitive SNAG on 11 letters (any four distinct ordered points may be sent to any other four distinct ordered points, the first to the first, etc., in exactly one way), the *Mathieu group* M_{11} of order $11 \cdot 10 \cdot 9 \cdot 8$. Then consider

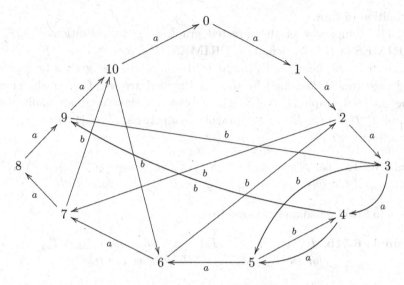

Figure 16. M_{11} a SNAG.

Then the semigroup associated the graph of Figure 16 is M_{11}. Also if a single new arrow is added to Figure 16 and labeled c, and the new graph is denoted by H and its semigroup is denoted by S, then the following holds. The maximal subgroups of S (up to isomorphism) are $M_{11}, S_{10}, S_9, \ldots, S_2$ and $\#_G(S) = 10$ (where S_j is all permutations on j letters). This follows since $\Pi^{-1} T_{ab} \Pi = T_{\Pi(a)\Pi(b)}$ for Π a permutation and M_{11} is 4-transitive (and so 2-transitive) and (6.50) and (4.11a) and (3.16g).

We next want to treat the case where distinct arrows receive distinct labels. This special case is important mathematically by (6.49a) and biologically since enzymes are (very nearly) specific. Since the definitions (6.12) and (6.13) are simpler in this case it is worthwhile to directly reformulate them.

Definition 6.51. Let $H = (X, E \subseteq X \times X)$ be a directed graph (or *digraph*). Then the *semigroup of flows on H* is by definition (X, S) where $S = \langle T_{ab} : (a, b) \in E \rangle$. We recall that $T_{ab} : X \to X$ with $T_{ab}(a) = b$ and $T_{ab}(x) = x$ for $x \neq a$, $x \in X$. Thus $f \in S$ iff $f : X \to X$ and $f = T_{a_1 b_1} \ldots T_{a_m b_m}$ with $(a_j, b_j) \in E$.

S is denoted H^S and (X, S) is denoted (X, H^S). We note H^S is the semigroup associated with $(X, E, \rho : E \to E)$ with $\rho(e) = e$ in the sense of definition (6.13b).

The complexity of the directed graph H is by definition $\#_G(H^S)$, **PRIMES** of H is by definition **PRIMES**(H^S), etc.

We can subsume graphs (undirected) under directed graphs by replacing each (undirected line) by the two directed arrows. Conversely, given the directed graph $H = (X, E)$, we can associate to it an undirected graph $U(H) = (X, E')$ in the obvious manner, i.e. $E' = \{(x,y) : (x,y) \in E \text{ or } (y,x) \in E\}$.

Fact 6.51a. *Let $H = (X, E \subseteq X \times X)$ be a digraph.*[*] *Then (X, H^S) uniquely determines the undirected graph $U(H)$ associated to H.*

Fact (6.51a) immediately follows from

Lemma 6.51b. *If $T_{ab} = T_{x_1 x_2} \ldots T_{x_{2k-1} x_{2k}}$, then either $T_{ab} = T_{x_{2j-1} x_{2j}}$ or $T_{ba} = T_{x_{2j-1} x_{2j}}$ for some $j = 1, \ldots, k$. (In fact we can take $j = 1$.)*

Proof. (In this proof we write the function variable on the left.) The partition induced by T_{ab} (namely $\{a, b\}$ and all singletons $\{x\}$ for $x \in X - \{a, b\}$) equals the partition induced by $T_{x_1 x_2}$, since Range$(T_{ab}) = X - \{a\}$ so $|\text{Range}(T_{ab})| = |\text{Range}(T_{x_1 x_2})| = |X| - 1$ and thus $T_{x_3 x_4} \ldots T_{x_{2k-1} x_{2k}}$ is one-to-one on Range$(T_{x_1 x_2})$. But Partition of T_{ab} equals Partition of T_{cd} iff $\{a, b\} = \{c, d\}$. Thus $T_{x_1 x_2} = T_{ab}$ or T_{ba}. $\qquad\square$

---

[*]This fact in its formulation here is due to the editor and John Rhodes. Note that Fact 6.51a cannot be improved to recover a directed graph from its transformation semigroup, since for example as can be immediately verified,

$$T_{ab} = (T_{ba} T_{zb} T_{az})^2,$$

which implies that the two non-isomorphic directed graphs, $H_1 = (X = \{a, b, z\}, E = \{(b, a), (z, b), (a, z)\})$ and $H_2 = (X, E \cup \{(a, b)\})$ have the same transformation semigroup $(X, H_1^S) = (X, H_2^S)$. Observe that H and H' have the same associated undirected graph,

$$U(H) = (X, E' = \{(x, y) : T_{x,y} \in H_1^S (= H_2^S)\}).$$

More generally, consider for any $n \geq 3$ a directed n-element cycle in a directed graph $H = (X, E)$, consisting of edges $(x_n, x_{n-1}), \ldots, (x_2, x_1), (x_1, x_n) \in E$. Let $W = T_{x_2 x_1} T_{x_3 x_2} \cdots T_{x_n x_{n-1}} T_{x_1 x_n}$. Clearly we have $(x_1)W = (x_2)W$ and that W cyclically permutes $C - \{x_1\}$ where $C = \{x_1, \ldots, x_n\}$ by the $(n-1)$-cycle $(x_n \, x_{n-1} \, \ldots \, x_2)$. Thus W^{n-1} maps $C - \{x_1\}$ to itself by the identity transformation. Whence $T_{x_1 x_2} = W^{n-1}$, which does not include $T_{x_1 x_2}$ as a factor. It follows by symmetry that each $T_{x_i x_{i+1}}$ $(1 \leq i < n)$ and also $T_{x_n x_1}$ are in H^S. In other words, the mapping for the inverse edge of every edge in a directed cycle occurs in the semigroup of any directed graph containing the cycle. -CLN

We next want to develop methods to compute the subgroups of the semigroup of flows of a directed graph.

Let D be a directed graph with points or vertices V so $D = (V, E \subseteq V \times V)$. Let $\{v_1, \ldots, v_s\}'$ denote $V - \{v_1, \ldots, v_s\}$ with $v_1, \ldots, v_s$ s distinct vertices. If $X \subseteq V$, we write $(X)T_{ab}$ as $X \cdot T_{ab}$.

The basis for most computations of the semigroup of flows is the following elementary

Fact 6.51c. *(1) If $x \in \{v_1, \ldots, v_s\}'$, i.e. $x \notin \{v_1, \ldots, v_s\}$, then*

$$\{v_1, \ldots, v_s\}' \cdot T_{xv_j} = \{v_1, \ldots, v_{j-1}, x, v_{j+1}, \ldots, v_s\}'$$

(2)

$$\{v_1, \ldots, v_s\}' \cdot T_{v_j a} = \{v_1, \ldots, v_s\}'$$

(3)

If $x, y \in \{v_1, \ldots, v_s\}'$, i.e. $x, y \notin \{v_1, \ldots, v_s\}$, then

$$\{v_1, \ldots, v_s\}' \cdot T_{xy} = \{v_1, \ldots, v_s, x\}'.$$

Proof. Trivial. $\qquad\qquad\qquad\qquad\qquad\qquad\qquad\qquad\qquad\qquad\qquad\square$

Definition 6.51d. (1) If $D = (V, E \subseteq V \times V)$ is a digraph, then $D^* = (V, E^* \subseteq V \times V)$ with $(v_1, v_2) \in E^*$ iff $(v_2, v_1) \in E$. D^* is called the *dual* or *reverse* digraph of D.

(2) If $D = (V, E \subseteq V \times V)$ is a digraph with p vertices, i.e. $|V| = p$, then define D_j for $1 \leq j \leq |V| - 1 = p - 1$ as follows: the vertices V_j of D_j are the subsets of V having j members, so $|V_j| = \binom{p}{j}$ and

$$\{v_1, \ldots, v_j\}' \to \{x_1, \ldots, x_j\}'$$

in D_j if $v_1 = x_1, v_2 = x_2, \ldots, v_{j-1} = x_{j-1}$, and $x_j \notin \{v_1, \ldots, v_j\}$ and $x_j \to v_j$ in D (or $v_j \to x_j$ in D^*). (Compare with (6.51c(1)).) Notice D_1 is (isomorphic with) D^*.

(3) If $\{v_1, \ldots, v_j\} \to \{v_1, \ldots, v_{j-1}, x_j\}$ is an arrow A of D_j, we say $T_{x_j v_j} \in D^S$ is *associated* with A and denote it T_A.

Now the D_j determine the maximal subgroups of D^S as follows.

Proposition 6.51e. *Let $e^2 = e \in D^S$. Let X be the range of e so $X = V \cdot e = \{v \in V : v \cdot e = v\}$. Let $V - X = \{v_1, \ldots, v_s\}$ with the s distinct elements, so $X = \{v_1, \ldots, v_s\}'$. Let G be the maximal subgroup of S containing e. Let $A_1, \ldots, A_m$ be a directed loop of D_s based at $\{v_1, \ldots, v_s\}'$. Then $eT_{A_1} \ldots T_{A_m} e \in G$ and every member of G arises in this manner.*

Proof. Obvious using Fact (6.51c).   □

We illustrate how to apply (6.51e) to a simple example. Consider Figure 16a which defines D.

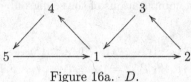

Figure 16a. D.

First let $e = T_{12}$ so $\text{Range}(T_{12}) = \{1\}'$ and let G_1 be the maximal subgroup of D^S containing e. Now $D_1 \cong D^*$.

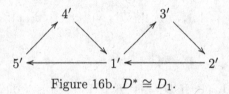

Figure 16b. $D^* \cong D_1$.

Now $(1,3), (3,2), (2,1) = \alpha$ and $(1,5), (5,4), (4,1) = \beta$ are loops of D_1 based at 1 and every loop based at 1 is a product of α's and β's. Thus

$$T_{12}T_{31}T_{23}T_{12}T_{12} = a$$

and

$$T_{12}T_{51}T_{45}T_{14}T_{12} = b$$

generate G_1.* But

---

*D_1 can be obtained by inserting primes (') on the vertex labels in Figure 16b. Unravelling this notation and the definition (6.51d) of D_1 according to Fact 6.51c, D_1 is represented as in Figure 16b with primes added and encodes:

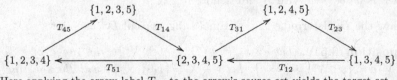

Here applying the arrow label T_{xy} to the arrow's source set yields the target set. Now the idempotent $e^2 = e = T_{12}$ has range $1' = \{2,3,4,5\}$, and the elements a and b are the two simple loops of D_1 pre- and post-fixed by e. Proposition 6.51e says they are members of the maximal subgroup of D^S having identity element $e = T_{12}$. -CLN

	a	b
1	3	2
2	3	2
3	2	3
4	4	5
5	5	4

Thus $a = T_{12}$ (2 3) and $b = T_{12}$ (4 5), so $G_1 \cong \mathbb{Z}_2 \times \mathbb{Z}_2$.

Now let $e = T_{51}T_{12}$ and let G_2 be the maximal subgroup of D^S containing e. Clearly Range$(e) = \{1, 5\}'$. Further D_2 equals*

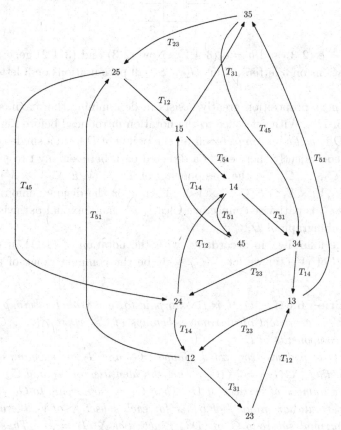

Figure 16c. D_2.

---

*For simplicity, set brackets and primes ($'$) on the vertices of D_j are dropped from now on, e.g. in Figure 16c the vertex labeled 15 denotes $\{1, 5\}' = \{2, 3, 4\}$, etc. -CLN

Thus consider the two loops of D_2 based at 15 namely

$$(15, 35), (35, 25), (25, 15) = \alpha$$

and

$$(15, 35), (35, 25), (25, 24), (24, 12), (12, 25), (25, 24), (24, 14), (14, 45), (45, 15) = \beta.$$

Then let $a = eT_{31}T_{23}T_{12}e$ and $b = eT_{31}T_{23}T_{45}T_{14}T_{51}T_{45}T_{12}T_{51}T_{14}e$ so

	a	b
1	3	3
2	3	3
3	2	4
4	4	2
5	3	3

Thus $a = e$ (2 3) and $b = e$ (3 4 2). Now (2 3) and (3 4 2) generate all permutations on 3 letters. Thus $G_2 \cong S_3$, all permutations on 3 letters.

The next proposition greatly assists in determining the maximal subgroups of D^S. With reference to the notation introduced before Figure 10 we say D is *pathwise connected* iff all the points of D form a single component or equivalently there exists a directed path between any two points.[*] Let $C_1, C_2, \ldots, C_m$ be the components of $D = (V, E \subseteq V \times V)$. Let $C_j = (V_j, V_j \times V_j \cap E \subseteq V_j \times V_j)$. Then V is the disjoint union of the V_j's and C_j is pathwise connected. Clearly C_j is a maximal pathwise connected subdigraph of D.

In the following, in accordance with the notation of (4.11), if G is a subgroup of (V, D^S) we let $X(G) \subseteq V$ be the common range of all the $g \in G$.

Proposition 6.51f. *(1) H is (isomorphic to) a maximal subgroup of D^S iff H is a product of maximal subgroups of C_j^S where $C_1, \ldots, C_m$ are the components of D.*

(2) Let D be pathwise connected. Suppose G_1 and G_2 are subgroups of D^S such that $|X(G_1)| = |X(G_2)|$, i.e. the identities of G_1 and G_2 fix the same number of vertices of D. Then G_1 is isomorphic to G_2 (in fact as permutation groups). Further for each subset X of V, there exists a maximal subgroup G of $(D^S)^1$ such that $X(G) = X$. Thus up to isomorphism there are at most $|V| - 1$ maximal subgroups of D^S.

---

[*]Sometimes this concept of pathwise connectivity for a directed graph is also called *strongly connectivity* in the literature -CLN

(3) If G_k is a maximal subgroup of D^S and $|X(G_k)| = k$ and $(X(G_k), G_k)$ is 2-transitive, then for every ℓ with $1 \le \ell < k$ and every $X \subset V$ with $|X| = \ell$, D^S has a maximal subgroup $G_\ell = G(X) \cong S_\ell$. Here S_ℓ is the symmetric group of all permutations on ℓ letters.

Proof. (Outline) (1) Since every T_{ab} with $(a, b) \in C_j$ gives the identity mapping on C_k for $k \ne j$, it follows easily that if H_j is a maximal subgroup of $(C_j^S)^1$, then $H_1 \times \cdots \times H_m = H$ is a subgroup of a maximal subgroup G of D^S. On the other hand, if $\mathcal{O}_1, \ldots, \mathcal{O}_n$ are the transitive components of the faithful permutation group $(X(G), G)$, then each $\mathcal{O}_q$ lies in some $C_{\Pi(q)}$ and $X(H_i)$ is a union of some of the $\mathcal{O}_p$'s. (Why?) Thus $\rho_i : G \twoheadrightarrow H_i$ is a surmorphism given by restricting elements of G to $X(H_i)$ and making the resulting action faithful. Further, $\rho_1 \times \cdots \times \rho_m$ is faithful. Thus $\rho_1 \times \cdots \times \rho_m$ maps G into H. Thus $H \le G \le H$ so $H = G$. This proves (1).

To prove (2) we require the following

Lemma 6.51g. *If $D = (V, E \subseteq V \times V)$ is pathwise connected then D_j is also pathwise connected for all $1 \le j \le |V| - 1$.*

Proof. A reduced (directed) path of D is a path from v_1 to v_2 all of whose vertices are distinct when $v_1 \ne v_2$ and all of whose vertices are distinct except $v_1 = v_2$ in this case. A loop is a reduced (directed) path from v_1 to v_1. $V(L)$ denotes the vertices of the path L.

Since D is pathwise connected, we will show there exists loops $L_1, \ldots, L_m$ such that

$$\begin{aligned} V(L_1) \cup \cdots \cup V(L_m) &= V \\ V(L_i) \cap V(L_{i+1}) &\ne \emptyset \text{ for } i = 1, \ldots, m - 1. \end{aligned}$$
(6.51h)

To construct such L_j we first construct reduced paths $L_1^*, \ldots, L_m^*$ as follows. Let L_1^* be any loop of D. Suppose $L_1^*, \ldots, L_{j-1}^*$ has been constructed so L_a^* is a reduced path for $1 \le a \le j - 1$, $V(L_{j-1}^*) - V(L_{j-2}^*) \ne \emptyset$ and $L_1^* \cup \cdots \cup L_{j-1}^*$ is pathwise connected. Let L_j^* be any reduced path which starts in $v_{j_1} \in V(L_{j-1}^*) - V(L_{j-2}^*)$ and ends in $v_{j_2} \in V_j^* \equiv V(L_1^*) \cup \cdots \cup V(L_{j-1}^*)$ and whose interior is disjoint from V_j^*. Clearly since D is pathwise connected there exists a reduced path between any two vertices, so such an L_j^* exists.

By continuing the process we can assume $V(L_1^*) \cup \cdots \cup V(L_m^*) = V$ for large enough m. Now $L_1^* \cup \cdots \cup L_{j-1}^*$ is pathwise connected, so we can choose a reduced path P_j^* of $L_1^* \cup \cdots \cup L_{j-1}^*$ from v_{j_1} to v_{j_2} when $v_{j_1} \ne v_{j_2}$. In this case let $L_j = L_j^* P_j^*$. If $v_{j_1} = v_{j_2}$ let $L_j = L_j^*$. Clearly $L_1, \ldots, L_m$ satisfy (6.51h).

Now with no loss of generality we can assume $D = L_1 \cup \cdots \cup L_m$. Now consecutively number the vertices of L_1, then continue consecutively numbering the remaining vertices of L_2, then the remaining ones of L_3, etc. We next show *any subset of k vertices can be moved in a path of D_k to the first k vertices.* We prove this by induction on m (since $L_1 \cup \cdots \cup L_j$ is pathwise connected for all j.) For $m = 1$, $D = L_1$ a loop so the result is obvious (why?). Now assume the assertion true for fixed m (and all $k \le m$) and suppose L_{m+1} is added satisfying (6.51h). Let A be any k subset. Let $A_1 = V \cap (V(L_1) \cup \cdots \cup V(L_m))$ and $A_2 = V - (V(L_1) \cup \cdots \cup V(L_m)) \subseteq V(L_{m+1})$. Let $|A_1| = k_1$. Then by induction A_1 can be moved to the first k_1 positions in $(L_1 \cup \cdots \cup L_m)_{k_1}$, then the result follows easily since L_{m+1} is a loop. Otherwise some member of L_m is not in A_1. Let q be the number of elements of A_1 in L_m and let v_m be a vertex where $m + 1$ enters L_m. Then by a path in $(L_m)_q$ (and thus in $(L_1 \cup \cdots \cup L_{m+1})_k$), v_m will *not* be a member of A_1.

Let p be the number of elements of A_2. Then in $(L_{m+1})_p$ by moving along a path v_m can become a member of A_2. Now $A_1 = k_1 + 1$. Now by induction A_1 can be made the first $k_1 + 1$ vertices of $L_1 \cup \cdots \cup L_m$. By continuing in this way we eventually move A to the first k positions of $L_1 \cup \cdots \cup L_{m+1}$ in $(L_1 \cup \cdots \cup L_{m+1})_k$.

Now the same argument can be applied to the dual or reverse of D. But then this gives a path of the standard set $\{1, \ldots, k\}$ to any k-set A (since this equals a path from A to $\{1, \ldots, k\}$ in $(D^*)_k$). This proves the lemma. □

We now prove (2). Assume D is pathwise connected. Let X_k be an arbitrary but fixed subset of V having k distinct members. Let X be any arbitrary subset of V having k distinct members. Now Lemma (6.51g) and Proposition (6.51e) immediately implies that there exists $\alpha, \beta \in D^S$ such that $X \cdot \alpha = X_k$ and $X_k \cdot \beta = X$. Thus $X_k \cdot \beta\alpha = X_k$ and so $X_k \cdot (\beta\alpha)^j = X_k$ for all $j \ge 1$. Now by choosing $j \ge 1$ appropriately we can assume that $(\beta\alpha)^j$ is the identity on X_k. Then by replacing α by $\alpha(\beta\alpha)^{j-1}$, we can further assume with no loss of generality that $\alpha, \beta \in D^S$ and α and β are inverse mappings of each other on X to X_k and X_k to X.

Now let G be a maximal subgroup of D^S with $X(G) = X$. Then clearly ρ given by $g \mapsto \alpha g\beta$ is a one-to-one homomorphism of G into a maximal subgroup H of D^S with $X(H) = X_k$. But if H_q for $q = 1, 2$ are maximal subgroups of D^S with $X(H_q) = X_k$ for $q = 1$ and 2 and $e_q^2 = e_q \in H_q,$, then clearly $e_a e_b = e_a$ for $a, b \in \{1, 2\}$. Then it is easy to verify that $h_1 \mapsto e_2 h_1$ with inverse $h_2 \mapsto e_1 h_2$ is an isomorphism of H_1 with H_2 (since $h_j e_j = e_j h_j e_j = e_j h_j = h_j$ and $e_a e_b = e_a$). Thus if H denotes $H_1 \cong H_2$ we have $G \leq H$. Now by symmetry, $G \cong H$. This proves the first assertion of (2).

Now to complete the proof of (2) it suffices to show that for any set $X \subseteq U$ there exists an idempotent of $D^{S1} = (D^S)^1$ with range X. But from the previous proof it follows that if an idempotent exists with k elements in its range, then for any set with k elements there is a idempotent having this range. Now $1 \in D^{S1}$ has $|V|$ elements in its range and T_{ab} with $(a, b) \in E$ has $|V| - 1$ elements in its range.

Assume any set with k elements is the range of some idempotent in D^{S1}. Choose X so $a, b \in X$ and $|X| = k$ and Range$(e = e^2) = X$. Then $T_{ab}e$ is an idempotent with range $X - \{b\}$. This now proves (2).

To prove (3) notice if Π is a permutation of a set Y and $a, b \in Y$, then $\Pi^{-1} T_{ab} \Pi = T_{\Pi(a)\Pi(b)}$. Now if $(a, b) \in E$ we can assume with no loss of generality by (2) that $a, b \in X_k$. Then $\{\Pi^{-1} T_{ab} \Pi : \Pi \in G_k\}$ restricted to $X(G_k)$ includes $\{T_{xy} : X(G_k) \to X(G_k) : x, y \in X(G_k)\}$ since G_k is 2-transitive. Now use Lemma (6.50) and (2) to prove (3).

This proves Proposition (6.51f). $\qquad\qquad\square$

As a simple application of Propositions (6.51e) and (6.51f), and our previous calculations after Figure 16a we find that the maximal subgroups of D^S where D is given by Figure 16a are $\mathbb{Z}_2 \times \mathbb{Z}_2, S_3, S_2$ and $\{1\}$.

We say digraph $D = (V, E \subseteq V \times V)$ is *anti-symmetric* iff $(a, b) \in E$ implies $(b, a) \notin E$. By Proposition (6.51f)(2) if D is pathwise connected we can associate a group with each integer $1 \leq k \leq |V| - 1$.

We say a subgroup G of D^S has '*defect*' k if $|X(G)| = |V| - k$. That is, the identity $e^2 = e$ of G maps V onto $|V| - k$ elements. Proposition 6.51f(2) implies subgroups of D^S with the same defect are isomorphic.

Conjecture 6.51i. *Let D be pathwise connected and anti-symmetric and have n vertices and not be a circle. Then*

(1) $\#_G(D^S) = n - 1$.

(2) *The maximal subgroup of D^S associated with cardinality $n - 1$ (i.e. the maximal subgroups of 'defect' one) is a product of cyclic, alternating and symmetric groups of various orders.*

(3) *The maximal subgroup of D^S associated with cardinality $n - 2$ (i.e. the maximal subgroup of 'defect' two) is S_{n-1} or A_{n-2}.*

(4) *The maximal subgroup associated with cardinality k for $1 \leq k \leq n-3$ is S_k.*

The proposed proof of the conjecture is too long to reproduce here. However, we give a very rough outline and some illustrative special cases.

Let's first consider the special case of a circle C_n (see (6.49c)) with one additional arrow added, denoted C'_n, so the resulting graph D is anti-symmetric. For example,

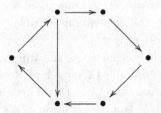

Then by applying (6.51e) we find the maximal subgroup of D^S of 'defect' one generated by $a = (1 \ldots n-1)$ and $b = (1 \ldots k)$ with $2 \leq k < n-1$. But we claim a and b generate A_m or S_m on $\{1 \ldots m\}$ with $m = n-1$. To verify this claim observe that $c = b^a b^{-1} = (aba^{-1})b^{-1} = (2\ 3\ \ldots\ k+1)(k\ \ldots\ 1) = (1\ k\ k+1)$ so $\langle a, b \rangle = G$ contains a 3-cycle. Further G is 2-transitive on $\{1, \ldots, m\}$ since if $G_m = \{\Pi \in G : m \cdot \Pi = m\}$ then $1 \cdot G_m = \{1, \ldots, m-1\}$ by considering $\{1 \cdot b^j : j = 1, \ldots, k\} = \{1, \ldots, k\}$ and $k \cdot c = k+1$, $(k+1) \cdot c^a = k+1 \cdot (1\ k\ k+1)^a = (k+1) \cdot (2\ k+1\ k+2) = k+2, \ldots, (m-2) \cdot c^{a^r} = m-1$ with $r = m-k-2$. Thus G is primitive on $\{1, \ldots, m\}$ containing a 3-cycle. Then we appeal to the following theorem of Jordan (see [Wielandt]).

Theorem 6.51j (Jordan). *A primitive group which contains a 2-cycle (transposition) is a symmetric group. A primitive group which contains a 3-cycle is either alternating or symmetric.*

Thus the group of 'defect' one for C'_n is S_{n-1} or A_{n-1}, so the groups of 'defect' j are S_{n-j} by Proposition (6.51f)(3). Then the Conjecture (6.51i) follows in this case (since $\#_G(C'_n) \geq \#_G(C_n) = n - 1$ by Proposition (6.49c)).

The rough outline of the proof of Conjecture (6.51i) now runs as follows: Write (a subgraph of) D as the union of loops $L_1, \ldots, L_m$ so that (6.51h) holds. Then one proceeds using the following

Conjectured Lemma 6.51k. *If L_1 and L_2 are intersecting loops with m_1 and m_2 vertices, respectively $(m_1, m_2 \geq 3)$ then*

(1) *If L_1 and L_2 intersect in exactly one point, then the group of defect 1 is $\mathbb{Z}_{m_1-1} \times \mathbb{Z}_{m_2-1}$ and the group of defect 2 is A_{n-2} or S_{n-2} with $n = m_1 + m_2 - 1$.*

(2) *If L_1 and L_2 intersect at two points and on the intersection have a common directed path, then the group of defect 1 is A_{n-1} or S_{n-1} with n the number of vertices of $L_1 \cup L_2$.*

Case (2) is illustrated by

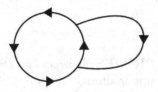

The proposed proof of (2) is a generalization of the previous proof for C_n'. The first part of (1) follows from (6.51e). The proposed proof of the second part of (1) is a generalization of the previous proof for Figure 16a which uses D_2 and (6.51e) and constructs paths in D_2 by 'running around L_1 and L_2 in all different ways' and then using (6.51j)!

Then the proof of (6.51i) is completed by induction from (6.51k) using (6.51h) and (6.51j). This completes our rough outline of a proposed proof of (6.51i).

Let us return to the viewpoint put forth in (6.22) namely that the useful way to classify reactions is by using the functor

(6.51l)   reactions $\rightarrow$ **PRIMES**(reactions).

Now the mathematics from Proposition (6.49) to this point notably Proposition (6.49), Proposition (6.51e), Proposition (6.51f) and Conjecture (6.51i) determines the maximal subgroups of the semigroups of flows of labeled graphs where each arrow receives a *distinct* label. Reactions giving rise to such graphs will be termed *elementary reactions*.

Now in (6.23) and (6.51l), **PRIMES** can be replaced by the equivalent but more useful functor **MAX PRIMES** where **MAX PRIMES(S)** = $\{P \in \textbf{PRIMES}(S) : P \mid P' \in \textbf{PRIMES}(S)$ implies $P \cong P'\}$ clearly, $\{P_1 \in \textbf{PRIMES} : P_1 \mid P \in \textbf{MAX PRIMES}(S)\} = \textbf{PRIMES}(S)$.

Thus our viewpoint is to consider

(6.51m)   reactions $\rightarrow$ **MAX PRIMES**(reactions).

Now our previous results* imply

$$(6.51\text{n}) \qquad \textbf{MAX PRIMES}(\text{elementary reactions})$$
$$\subseteq \{\mathbb{Z}_p : p \text{ is a prime}\} \cup \{A_n : n \geq 5\}.$$

In words, the **MAX PRIMES** of elementary reactions are prime order (commutative) cyclic groups or alternating groups (SNAGs). Now these are the least interesting simple groups and the information conveyed by them is essentially the maximal number of substrates in any pathwise connected component of the reaction. (See the previously mentioned propositions for a precise formulation.)

However in biology the vast majority of reactions that arise are *not* elementary reactions. This is because some important enzymes are not specific but catalyze a few (two or three) reactions, or in the models chosen the 'soup' is such that the cofactors (together with the 'soup') drive several reactions. Thus the reactions arising in this manner have some identifications (i.e. some arrows labeled the same) but not too many such identifications (e.g. two or three). Of course with arbitrary number of identifications we can obtain an arbitrary group by constructing the Cayley graph (see before Figure 16). Further with arbitrary identifications and also allowing several differing labels with each arrow we can construct the Cayley graph of an arbitrary finite semigroup and obtain an arbitrary finite subgroup. Thus reactions with arbitrary identification is just the study of finite transformation semigroups. However, in biological reactions arising as above there are in general, in important and appropriate models, *very few* identifications.

Let's consider some important reactions like for example the oxidative phosphate cycle (OPC). See [Pon].

---

*What follows appears to be predicated on the assumption that conjecture 6.51i(2) and 6.51i(3), or something similar (say with 'cyclic groups' replaced by 'solvable groups' in 6.51i(2)), can be established.   -CLN

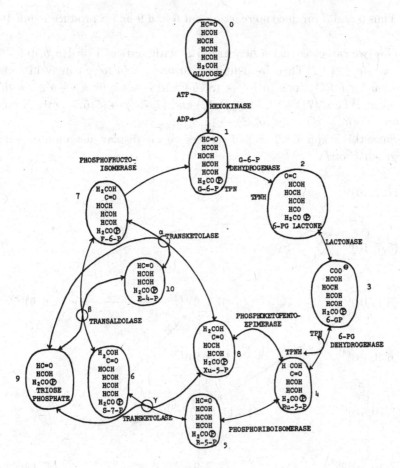

Figure 17a. The Oxidative Pentose Phosphate Cycle.

First we will identify $\{7, 0, 1, 2, 3, 4\}$ denoted also by 7. Then the reactions become

$$
\begin{aligned}
7 &\leftrightarrow 5 \\
7 &\leftrightarrow 8 \\
\{6, 9\} &\overset{\beta}{\leftrightarrow} \{7, 10\} \\
\{5, 8\} &\overset{\gamma}{\leftrightarrow} \{6, 9\} \\
\{8, 10\} &\overset{\alpha}{\leftrightarrow} \{7, 9\}
\end{aligned}
$$

Figure 17b. The reactions for OPC.

Thus 6 and β produce no reaction but 6 *and* 9 *and* β produce 7 *and* 10, etc.

Now we can construct a new graph D with vertices $V = \{(a, b, c) : 5 \leq a \leq b \leq c \leq 10\}$. Then by definition $(a, b, c) \rightarrow (a', b', c')$ in D iff some reaction for OPC sends $(a, b) \rightarrow (a', b')$ and $c = c'$, or $a \rightarrow a'$, $b = b'$, $c = c'$, etc. Thus $(7, 10, 10) \rightarrow (6, 9, 10)$ and $(7, 7, 7) \rightarrow (7, 5, 7)$, etc. Notice $(a, b, c) \rightarrow (a', b', c')$ iff $(a', b', c') \rightarrow (a, b, c)$.

Since this graph has $6^3 = 216$ vertices we will display just an interesting subgraph E of D.

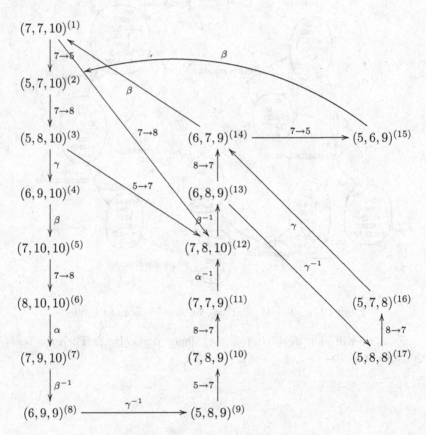

Figure 17c. E a subgraph of D for OPC.

By applying the previous theory to E (which has 17 vertices) we find that $S_{16}, S_{15}, \ldots, S_2, S = \{1\}$ are the maximal subgroups of E^S. We also make the interesting observation that under the relations $ab = ba$ (commu-

tativity) $\gamma\gamma^{-1} = \gamma^{-1}\gamma = 1$, $(7 \to 8)(8 \to 7) = (8 \to 7)(7 \to 8) = 1$, etc.
that *the symbols around every loop of E cancel to* 1!

Now from the previous theory* $\alpha = T_{14,1}T_{13,14}T_{12,13}\ldots T_{2,3}T_{1,2}\epsilon$ and
$\delta = T_{14,1}T_{13,14}T_{12,13}T_{1,12}\epsilon$ with $\epsilon = T_{17,16}T_{16,14}T_{15,2}$ generate A_{13} with
range $\{2, 3, 4, \ldots, 14\}$.

For example δ gives, $\delta = (2)(3)\ldots(11)(12\ 13\ 14)$ and $1 \to 12$, $17 \to$
14, $16 \to 14$, $15 \to 2$ and so on its range $\{2, \ldots, 14\}$, δ restricts to the
permutation $\delta' = (2)(3)\ldots(11)(12\ 13\ 14)$, i.e. δ' is the 3-cycle $(12\ 13\ 14)$.

But the truth of this assertion depends on the reactions for OPC and
the *principle of superposition*, namely, that for large sets of chemicals the
only reactions which occur are the *union* of those for smaller subsets.

Thus $(6, 7, 9) \xrightarrow{\beta} (7, 7, 10)$, $(6, 7, 9) \xrightarrow{\gamma^{-1}} (5, 7, 8)$, $(6, 7, 9) \xrightarrow{7\to5} (6, 5, 9)$,
$(6, 7, 9) \xrightarrow{7\to8} (6, 8, 9)$, $(6, 7, 9) \xrightarrow{\alpha^{-1}} (6, 8, 10)$ but $(6, 7, 9) \to (7, 8, 10)$ does not
occur.

In other words, the principle of superposition states that the basic reactions are given and all other reactions which occur are simply derived from
these basic reactions by the process of independent summation (union).

However, for the reactions known for (say) the oxidative pentose phosphate cycle the principle of superposition hold approximately, but not exactly. The present experimental methods are of such a nature that exactly
what all the many simultaneous reactions occurring (in the actual 'soup'
in the actual cell) are not observed directly, but essentially observed in
pieces by delicate and difficult experiments. What is actually happening is
deduced from these experiments by *assuming* some restricted form of the
superposition principle.

Thus for example

$$\delta' = (2)(3)\ldots(11)(12\ 13\ 14)$$

might not exactly occur (since it is composed of many reactions and the
principle of superposition might not hold exactly). Thus δ' might equal

$$(2\ 8)(3)(4)\ldots(7)(9)\ldots(11)(12\ 13\ 14)$$

because (say) the $\beta, (8 \to 7), \beta^{-1}$ reaction sequence occurring as part of
loop of reactions defining δ affects 2 and 8 in the presence of 7 and 9.

Thus the idea is that the *groups which actually do appear* are 'large'
subgroups of the alternating or symmetric *groups which theoretically appear*
under the superposition principle.

---

*Here (6.51e) can be used with idempotent $\epsilon^2 = \epsilon$ and with α and δ arising from two
loops in Figure 17c. -CLN

Thus for example M_{12} the Mathieu group (4-transitive) on 12 letters is a maximal subgroup of A_{12}.

Since the theoretically occurring groups are (usually) A_k or S_k which are $k - 2$ on k transitive, by 'large' subgroups we mean highly (2- or more) transitive groups many of which are SNAGs or closely related to SNAGs.

Of course by Proposition $(6.51f)(3)$ when the actual group with range k elements G_k is 2- or more transitive, then the actual groups G_j with $1 \le j \le k$ are all symmetric, i.e. $G_j = S_j$ for $1 \le j \le k$.

Thus the first non-trivial actual (highly transitive) group is the SNAG (or closely related to a SNAG) which classifies the reaction (the other SNAGs in **MAX PRIMES** being alternating or cyclic groups). *

For example Figure 16 with a single new arrow added is classified by the SNAG, M_{11}, the Mathieu group (4-transitive) on eleven letters.

Thus the SNAGs are the theoretically arising alternating or symmetric groups (derived by assuming the principle of superposition) 'cut down' a bit by disallowing certain permutations which in fact do not happen (e.g. because of the failure of the law of superposition in the actual cell) but leaving a highly transitive group (the SNAG). For example, M_{11} is a maximal subgroup of A_{11}, etc.

Usually in the theoretical model (with maximal subgroups H_j acting on j points – see Proposition $(6.51f)(2)$), the groups, starting with the ones of smallest defect, $H_{n-1}, \ldots, H_{k+1}$ will be commutative, but $H_k = A_k$ or S_k, and $H_j = S_j$ for $1 \le j < k$. Then in the actual model (with groups G_j), $G_n, \ldots, G_{k+1}$ are also commutative but G_k is a highly transitive subgroup of A_k and $G_j = S_j$ for $1 \le j < k$, so **MAX PRIMES** equals cyclic, alternating, and the Jordan–Hölder factors of G_k (and often G_k or G'_k is a SNAG). Thus the reactions are classified essentially by **MAX PRIMES**(G_k). For Figure 11 with another arrow added $k = n$ and $G_n = M_{11}$ and

$$\textbf{MAX PRIMES } (M_{11}) = M_{11}$$

the Mathieu group (4-transitive) on eleven letters.

Thus we see that SNAGs measure the intricacy of the interconnection of the reactions. Hence their significance in biological classification.

Remark 6.52. These ideas can be extended to (flow charts) of programs for computers leading to a theory of *classification and complexity* of pro-

---

*Hence the author's Mendeleev type classification of biochemical reactions seeks **MAX PRIMES** (for each pathwise connected component of a reaction), and regards two reactions as similar if their **MAX PRIMES** belong to classes of the same kinds (families of SNAGs) according to the classification of finite simple groups. -CLN

grams. Also we have developed much of the theory for graphs. Also the classification and complexity of finite configurations [Conway] (by passing to the obvious coresponding graph), as in Conway's Game of Life can be developed. How does the complexity of the configuration change with time?

Bibliography

Biology

[B-F] T. P. Bennett and E. Frieden. *Modern Topics in Biochemistry*. Macmillan Co., N.Y., 1966:

[Pon] N. G. Pon. Pentose phosphate cycle. *Comparative Biochemistry*, VII:3, 1964. Academic Press, Florkin and Mason Editors.

[Rose] S. Rose. *The Chemistry of Life*. Pelican Books, 1966.

[Watson] J. D. Watson. *Molecular Biology of the Gene*. Benjamin, N.Y., 1965. [Sixth edition by J. D. Watson, T. A. Baker, S. P. Bell, A. Gann, M. Levine, and R. Losick, published by Benjamin Cummings, 2008.]

(see also references for Ch. 6, Part II).

Physics

[Sch.] Erwin Schrödinger. *Statistical Thermodynamics*. Cambridge University Press, 1967.

(see also references for Ch. 4).

Mathematics

[Conway] J. H. Conway. The game of life. *Scientific American*, October 1970. February 1971.

[LB I] J. Rhodes and B. R. Tilson. Lower bounds for complexity of finite semi-groups. *Journal of Pure & Applied Algebra*, 1(1):79–95, 1971.

[LB II] J. Rhodes and B. R. Tilson. Improved lower bounds for complexity of finite semigroups. *Journal of Pure & Applied Algebra*, 2:13–71, 1972.

[Wielandt] H. Wielandt. *Finite Permutation Groups*. Academic Press, N.Y., 1964.

[Zeiger] H. Paul Zeiger. Cascade synthesis of finite state machines. *Information and Control*, 10:419–433, 1967. plus erratum.

(see also references for Ch. 2, 3, 4, and 5).

Part II. Complexity of Evolved Organisms

"... the beauty of evolution is far more striking than its purpose."
— J. B. S. Haldane

The purpose of this section is to capture some of the beauty of evolution in some beautiful equation which must necessarily hold for all *evolved* organisms.

In short, the general idea is to write down precisely some mathematical conditions which must hold for all organisms which have evolved and, for all practical purposes, stopped evolving (e.g. the cell but *not* man. We will consider evolving organisms in Part III(A).) Some elements of the Darwinian-Wallace theory of evolution seem tautological, especially the role of natural selection. Thus it should be possible to give these elements precise mathematical expression, which is our goal here.

These conditions should hold for all evolved organisms whether they are cells on earth or unobserved evolved organisms on distant planets which possibly exist in vastly different environments than on earth.

Our conditions will take the form of *relating the complexity of the evolved organism to simple physical aspects of the organism.*

The simple idea behind the derivation of these 'Evolution-Complexity Relations' is this: An evolved organism possesses a (very nearly) perfect *harmony* among its various parts. The balance among its various constituents is delicate because this balance is (very nearly) perfect.

But what do we mean by 'perfect balance'? Starting with some examples from other disciplines we would say an electrical power transformer is in 'perfect balance' when the electrical impedence equals the load. Using this practice of power engineering as an intuitive base, around 1948 at Bell Labs, C. E. Shannon developed the mathematical theory of communication. One of the important elements of his theory is the 'perfect balance' required (obtained by suitable codes) between the rate of information produced by the source (entropy of the Markov process) and the channel capacity.

In our case of evolved organisms and, in particular, the cell, we will model parts of the cell by finite state sequential machines (see Chapter 5 of this volume) and then precisely define 'perfect balance' or 'perfect harmony' in this context.

The problem then has two aspects. First, in what manner and with what success can the cell's operations be modeled by finite state machines? We take up this aspect of the problem in the Appendix at the end of Part II using the model of Jacob and Monod and the previous material of Chapter 5 and Part I.

The second problem is to define 'perfect harmony' assuming finite state machine modeling and then render in precise form the 'Evolution-Complexity Relations'. We now turn to this second problem.

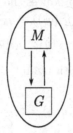

Figure 17. Cell model.

We divide the cell into two parts, metabolism and genetic control. We model the entire metabolism of the cell by a single finite state machine M (with a vast but finite number of states) and we model the genetic control with another finite state machine G. For precise details on the modeling see the Appendix at the end of Part II. However it is essentially the Jacob and Monod model. The machines M and G interact with each other. Also inputs arrive at M from outside the cell (e.g. sugars and oxygen via the bloodstream). See Figure 17. To make this precise we introduce the following definition.

Definition 6.53. Let M be a state-output-dependent circuit (see Definition 5.41) $M = (A_M, B_M, Q_M, \lambda_M, j_m)$, A_M, B_M, Q_M finite non-empty sets, $\lambda_M : Q_M \times A_M \to Q_M$, $j_M : Q_M \to B_M$. Let G be a state-output-dependent circuit $G = (A_G, B_G, Q_G, \lambda_G, j_G)$ with A_G, B_G, Q_G finite non-empty sets, $\lambda_G : Q_G \times A_G \to Q_G$, $j_M : Q_G \to B_G$. So that M and G may interact with each other we require $A_M \supseteq B_G$ and $B_M \subseteq A_G$.* We term these conditions M *interacting* with G and we denote these conditions by Figure 18.

$$\boxed{M} \; \rightleftarrows \; \boxed{G}$$

Figure 18. M and G interacting.

---

*Here outputs of genetic control are inputs to metabolism, and *vice versa*. -CLN

Given M and G interacting, let $(q_M, q_G) \in Q_M \times Q_G$. Now q_M produces the output $j_M(q_M)$ from M which enters G and makes q_G become $\lambda_G(q_G, j_M(q_M)) \equiv q_G \cdot j_M(q_M)$. We write this

(6.54a) $\qquad\qquad (q_M, q_G) \xrightarrow{-} (q_M, q_G \cdot j_M(q_M))$.

Let $q_G \cdot j_M(q_M) = q_2$. Then q_2 produces the output $j_G(q_2)$ which enters M and makes q_M become $\lambda_M(q_M, j_G(q_2)) \equiv q_M \cdot j_G(q_2)$. We denote this by

(6.54b) $\qquad\qquad (q_M, q_2) \xrightarrow{} (q_M \cdot j_G(q_2), q_2)$.

Thus given M interacting with G and $q_M \in Q_M$ and $q_G \in Q_G$ we can consider

$$(q_M, q_G) \equiv (q_{0,1}, q_{0,2}) \xrightarrow{-} (q_{0,1}, q_{1,2}) = (q_M, q_G \cdot j_M(q_M)) \xrightarrow{} (q_{1,1}, q_{1,2})$$

(6.54c) $\quad = (q_M \cdot j_G(q_{1,2}), q_{1,2}) \xrightarrow{-} (q_{1,1}, q_{2,1}) \xrightarrow{} (q_{2,1}, q_{2,2}) \ldots (q_{j,1}, q_{j,2})$

$$\xrightarrow{-} (q_{j,1}, q_{j+1,2}) \xrightarrow{} (q_{j+1,1}, q_{j+1,2}) \cdots$$

Since Q_M and Q_G are finite sets there exists a smallest nonnegative integer x so that

(6.55) $\qquad\qquad (q_{x,1}, q_{x,2}) = (q_{x+\varepsilon,1}, q_{x+\varepsilon,2})$

for some uniquely determined (smallest) positive integer ε.

Thus the repeating loop $(q_{x,1}, q_{x,2})$, $(q_{x+1,1}, q_{x+1,2})$, $\ldots, (q_{x+\varepsilon-1,1}, q_{x+\varepsilon-1,2})$, $(q_{x+\varepsilon,1}, q_{x+\varepsilon,2}) = (q_{x,1}, q_{x,2})$ is set up. This leads to the following definition.

Definition 6.56. Given M interacting with G (using the previous notation), we define $\delta : Q_M \times Q_G \to$ subsets$(Q_M \times Q_G)$ by $\delta(q_M, q_G) = \{(q_{x+j,1}, q_{x+j,2}) : 0 \leq j \leq \varepsilon$ where x and ε are determined by (6.55)$\}$.

The intuition behind Definition (6.56) is easy to understand. Given M interacting with G we might assume that due to external circumstances at time $t = 0$, M is in state $q_{0,1}$ while G is in state $q_{0,2}$. However, assuming no further external inputs or external influence the pair of states $(q_{0,1}, q_{0,2})$ may not be stable. In fact, the output of M in state $q_{0,1}$, $j_M(q_{0,1})$, enters G and carries $q_{0,2}$ to $\lambda_G(q_{0,2}, j_M(q_{0,1})) \equiv q_{0,2} \cdot j_M(q_{0,1}) \equiv q_{1,2}$. Then the output of G, $j_G(q_{1,2})$, goes into M and carries $q_{0,1}$ to $\lambda_M(q_{0,1}, j_G(q_{1,2})) \equiv q_{0,1} \cdot j_G(q_{1,2}) \equiv q_{1,1}$. And so it goes. Eventually (6.55) occurs and thus from time $t = x$ things repeat cyclically every ε units of time and this cyclic behavior is a stable phenomenon as long as there is no external interference. δ maps the initial pair of states $(q_{0,1}, q_{0,2})$ into the stable loop $\delta(q_{0,1}, q_{0,2})$.

The important thing is that δ maps the initial states (q_M, q_G) into the stable phenomenon, which will automatically result if M interacting with G is started in (q_M, q_G) and then left completely uninfluenced externally. Thus δ is indeed the *free action of M interacting with G*.

Notice δ is an idempotent map in the sense that $(q_1, q_2) \in \delta(q_M, q_G)$ implies $\delta(q_1, q_2) = \delta(q_M, q_G)$. Thus $\delta(q_1, q_2) = \delta(q_1', q_2')$ or $\delta(q_1, q_2) \cap \delta(q_1', q_2')$ is empty. By definition the *stable loops* of M interacting with G is $\{\delta(q_1, q_2) : (q_1, q_2) \in Q_M \times Q_G\}$, *Thus the collection of stable loops of M interacting with G equals the range of δ*.*

Intuitively, given M interacting with G we can also regard M as *fighting* with G, like thinking of M and G as two animals or organisms fighting one another (e.g. a bull fighting a bear) and with the states, suitably digitized, representing the positions of each. Alternatively we can consider G attempting to *control* M, where G samples the output of M and then puts a 'correction' input back into M (i.e. the usual set-up in control theory).

So far we have not considered external inputs to the interacting pair M and G. We do so now. We assume M can receive external inputs but G cannot. (We are not studying mutations here.)

Assume M and G are in a stable loop containing $(q_{0,1}, q_{0,2})$, so $(q_{0,1}, q_{0,2}) \in \delta(q_{0,1}, q_{0,2})$ and the machines are looping through a cycle which contains $(q_{0,1}, q_{0,2})$.

Suppose an input $a_1 \in A_M$ is applied to M from the outside environment when M and G are in states $(q_{0,1}, q_{0,2})$. Then $(q_{0,1}, q_{0,2})$ will pass to $(q_{0,1} \cdot a_1, q_{0,2})$ under the influence of a_1 and then $(q_{0,1} \cdot a_1, q_{0,2})$ will pass to $\delta(q_{0,1} \cdot a_1, q_{0,2})$ under the free action undisturbed by external inputs.

Under the next external input a_2, $(r_1, r_2) \in \delta(q_{0,1} \cdot a_1, q_{0,2})$ is disturbed to $(r_1 \cdot a_2, r_2)$ which passes to $\delta(r_1 \cdot a_2, r_2)$ under the free action passes, to $\delta(r_1 a_2, r_2)$ under the free action, etc.

---

*It is interesting to note that Stuart A. Kauffman, e.g. in his *The Origins of Order* (Oxford University Press, 1993) but also already in his early work by around 1970, has similarly identified stable loops in cellular dynamics in his studies of random Boolean models of genetic networks. He hypothesizes that these stable loops correspond to the set of cell types it is possible for the genome to express. John Tyler Bonner (e.g. *The Evolution of Complexity by Means of Natural Selection*, Princeton University Press, 1988) regards the number of cell types that an organism can express as a good measure for biological complexity. These approaches to biological complexity are compatible with and can find strong support from the mathematical Evolution-Complexity Relations discussed here, e.g. (6.62) below. For more details on such approaches, see C. L. Nehaniv and J. L. Rhodes, "The Evolution and Understanding of Hierarchical Complexity in Biology from an Algebraic Perspective" *Artificial Life*, 6(1):45–67, 2000, and the references cited therein. -CLN

In this model we assume that the time between external inputs $a_1, a_2, \ldots$ is of long enough duration so that a stable loop can be obtained before the next external input. By going to a somewhat more complicated model this assumption could be banished but it does not seem necessary or essential to make this addition in this presentation.

A related problem is the following. Suppose M interacting with G is in the loop $\{(q_{1,1}, q_{1,2}), \ldots, (q_{m,1}, q_{m,2})\}$ and the external input a_1 is applied to M. Then to which member of the loop is a_1 applied? Does it depend on the exact time at which a_1 is applied? Can a_1 react only upon a unique member of the loop? Is the resulting state a function of the entire loop and a_1?

All of these hinted at various models are possible. However, in our applications, the loop will usually have just a single member (see the Appendix to Part II), so in this case all models coincide. Thus for ease of exposition we will choose a representative pair of states from each stable loop and assume a_1 reacts with this pair of states when a_1 is applied to M interacting with G in this loop. Formally we have

Definition 6.57. $\bar{\delta} : Q_M \times Q_G \to Q_M \times Q_G$ is any function such that

(a) $\bar{\delta}(q_M, q_G) \in \delta(q_M, q_G)$
(b) $\bar{\delta}(q_{0,1}, q_{0,2}) = \bar{\delta}(q_{1,1}, q_{1,2})$

iff $\delta(q_{0,1}, q_{0,2}) = \delta(q_{1,1}, q_{1,2})$. $\bar{\delta}$ is termed a representation function for the free action δ, or a *representative free action*.

Clearly $\bar{\delta}(\bar{\delta}(q_{0,1}, q_{0,2})) = \bar{\delta}(q_{0,1}, q_{0,2})$, i.e. $\bar{\delta}$ is an idempotent map. Also the range of $\bar{\delta}$ is a set of representations for the stable loops.

If in the following the reader prefers one of the other models for determining how the external input a_1 affects a stable loop, he or she can vary our details to his or her liking.

All the inputs A_M may not be possible external inputs to M. Thus we assume the set A denotes the set of possible external basic stimuli to M and there exists $h : A \to A_M$ with $h(a)$ the input to M given the external input a.

Given M interacting with G we can apply a sequence of external stimuli $x_1, x_2, \ldots, x_n$ with $x_j \in A$ and obtain at the end some stable loop (or its representative). This gives us a machine, the external result of M interacting with G in the presence of external stimuli. Precisely,

Definition 6.58. Let M interact with G (using the notation of Definition

6.53) and let $\overline{\delta}$ be a representative free action. Let the set A and the map $h : A \to A_M$ be given. Then the *external result of M interacting with G in the presence of external inputs A with connection h* is the state-output dependent circuit (see Definition 5.41), $R = (A, \text{Range } \overline{\delta}, \text{Range } \overline{\delta}, \lambda, j)$ with Range $\overline{\delta} \subseteq Q_M \times Q_G$, $\lambda :$ Range $\overline{\delta} \times A \to$ Range $\overline{\delta}$ defined by $\lambda((q_1, q_2), a) = \overline{\delta}(q_1 \cdot h(a), q_2)$ where $(q_1, q_2) \in$ Range $\overline{\delta}$, $a \in A$ and $a_1 \cdot h(a)$ denotes $\lambda_M(q_1, h(a))$. Finally, $j :$ Range $\overline{\delta} \to$ Range $\overline{\delta}$ is the identity map, i.e., $j(q_1, q_2) = (q_1, q_2)$.

With reference to Definition 5.41, suppose R is initially in state $(q_{0,1}, q_{0,2})$ and $a_1, \ldots, a_n$ is presented to R. Then

$$(q_{0,1}, q_{0,2}) \xrightarrow{a_1} \overline{\delta}(q_{0,1} \cdot h(a_1), q_{0,2}) \equiv (q_{1,1}, q_{1,2})$$
$$\xrightarrow{a_2} \overline{\delta}(q_{1,1} \cdot h(a_2), q_{1,2}) \equiv (q_{2,1}, q_{2,2})$$
$$\vdots$$
$$(q_{n-1,1}, q_{n-1,2}) \xrightarrow{a_n} \overline{\delta}(q_{n-1,1} \cdot h(a_n), q_{n-1,2}) \equiv (q_{n,1}, q_{n,2})$$

and thus the final output is $(q_{n,1}, q_{n,2})$. Otherwise stated, if the cell begins in states (q_{01}, q_{02}) and external inputs $a_1, \ldots, a_n$ are given the final result is that the cell is cycling through the stable loop which contains $(q_{n,1}, q_{n,2})$.

In Definition 6.58 we took the states to be Range $\overline{\delta}$. However, Range $\overline{\delta}$ is in a natural one-to-one correspondence with the collection $\mathcal{L}$ of stable loops via $(q_1, q_2) \to \delta(q_1, q_2)$. Thus we could replace Range $\overline{\delta}$ by $\mathcal{L}$ and $\overline{\delta}$ by δ in Definition 6.58.

With reference to Figure 17, we wish to give an intuitive description of the workings of the cell following our above formal presentation.

Inside the cell we have the metabolism M and genetic control G. M and G are interacting. Let us assume initially that M and G are cycling through the stable loop L_0. Now suppose a_1 is presented externally to the cell (e.g. oxygen or sugar from the bloodstream). It penetrates the cell and $h(a_1)$ enters the metabolism M. There it disturbs the stable loop L_0 and leads to a new pair of states in M and G. In our exact model given by Definition 6.58, L_0 is denoted by its representative $\overline{L}_0$, defined by $\overline{L}_0 = \overline{\delta}(r_1, r_2)$ for $(r_1, r_2) \in L_0$. Then if $\overline{L}_0 = (q_{01}, q_{02})$ the new pair of states if $(q_{01} \cdot a_1, q_{02})$. This new pair of states moves freely (via (6.54c)) until the new stable loop L_1 is obtained with $\overline{L}_1 = \overline{\delta}(q_{01} \cdot a_1, q_{02})$. Then the external a_2 is presented which disturbs L_1 leading to a new pair of states, leading to a new stable loop L_2, etc., until a new stable loop L_n is obtained at the end of the sequence $a_1, a_2, \ldots, a_n$. The states and output of R of Definition 6.58, when started in $\overline{L}_0$ and $a_1, \ldots, a_n$ is presented as inputs, are $\overline{L}_1, \ldots, \overline{L}_n$. That is, $R_{\overline{L}_0}(a_1) = \overline{L}_1$, $R_{\overline{L}_0}(a_1, a_2) = \overline{L}_2, \ldots, R_{\overline{L}_0}(a_1, \ldots, a_n) = \overline{L}_n$. See Definition 5.41. Thus R *represents the external behavior of the cell under external inputs.* The inputs $a_1, \ldots, a_n$ are presented to the cell (say from the bloodstream) and the result is the stable condition of the cell after these inputs have terminated. (See the Appendix at the end of this Part II for details of the modeling of cell metabolism and genetic control by M and G, respectively.) Also *the states of R are (in one-to-one correspondence with) the stable configurations of the cell.* In fact, the stable configurations of the cell are the stable loops of M interacting with G of Figure 17 and Range $\overline{\delta}$ is a set of representations for these stable loops.,

In the terminology of Chapter 5 (see Definition 5.41), $R_{\overline{L}_0}$ is the *machine* $R_{\overline{L}_0} : A^+ \to$ Range $\overline{\delta}$ with $R_{\overline{L}_0}(a_1, \ldots, a_n)$ equalling the representative of the resulting stable loop obatined when the cell was initially in the stable loop $\overline{L}_0$ and external basic stimuli $a_1, \ldots, a_n$ was applied.

We are now in a position to precisely and succinctly state the 'Evolution-Complexity Relations'.

6.59.   First Version of the Evolution-Complexity Relations for the Cell. There exists a stable loop L_0 of the cell so that the complexity of the machine $R_{\overline{L}_0}$ is the maximum possible, namely, the number of states of R minus 1 which in turn equal the total number of stable loops of the cell minus one, or, symbolically,

$$(6.60) \qquad \theta(R_{\overline{L}_0}) = (\text{number of states of } R) - 1$$
$$= (\text{number of stable loops of the cell}) - 1.$$

6.61.   Extensive Remarks. (a) Since almost all of the loops have length one in practice, the stable loops of the cell are just the stable configurations of the cell, i.e. those configurations of the cell which can be experimentally observed. Also, since in practice R will be a reduced state-output-dependent machine, $\theta(R_{\overline{L}_0})$ can be replaced by $\#_G(R^S)$ in (6.60). (See Definitions 5.40, 5.41, and 5.42.) As is not difficult to prove for reduced R, we have $\#_G(R^S) = \max\{\theta(R_q) : q \text{ is a state of } R\}$.

(b) We can consider (Range $\overline{\delta}, R^S$). For notation let $Q =$ Range $\overline{\delta} \equiv$ the (stable) states of R and let $S = R^S \equiv$ the *semigroup of the cell.* Now since (Q, S) is a right mapping semigroup, $\#_G(S) \leq |Q| - 1$ by (3.16g).

Thus the maximum $\theta(R_{\overline{L}_0}) = \#_G(S)$ can be is $|Q| - 1$. Also Q is in one-to-one correspondence with the stable configurations of the cell. Then the Evolution-Complexity Relations for the cell assert: the complexity of the cell (i.e. of R) is the maximum it can be, namely, $|Q| - 1$. Fortunately, $|Q|$ has a simple physical meaning, namely, the number of distinct stable configurations of the cell.

(c) Now M represents cell metabolism, G represents the genetic control, and R represents the global input-output behavior of the cell where the inputs are the natural inputs to the cell and the outputs are the resulting stable internal configurations of the cell. Thus we have the very important functor of Figure 19.

$$\boxed{M} \rightleftarrows \boxed{G} \rightsquigarrow R$$

Figure 19. Functor sending M interacting with G to the result R of this interaction.

Now $\#(R)$ will denote the number of states of R. $\#_G(R)$ by definition equals $\#_G(R^S)$, the complexity of the semigroup of R (see Definition 5.40 and Chapter 3). Now by (3.16g),

$$\#_G(R) \leq \#(R) - 1 .$$

Now R is the *result* of the interaction of M and G. Thus $\#(R)$ measures the *contact* between M and G and $\#_G(R)$ measures the *complexity of the contact* between M and G.

With reference to Figure 17, in our case of the cell it is reasonable to define $\#_G(R)$ to be the *complexity of the cell*. Also in our case $\#(R)$ is the number of stable configurations of the cell. Thus we can restate (6.59)–(6.60) as follows.

6.62. Second Version of the Evolution-Complexity Relations for the Cell. The complexity of the cell is the maximum possible, namely,

cell complexity = (number of stable configurations) − 1

or

(6.63) $\#_G(R) = \#(R) - 1 =$ (number of stable configurations) − 1.

We now have the precise statements ((6.59) or (6.62)) of the Evolution-Complexity Relation for the Cell which are complete when supplemented by the Appendix at the end of this Part II which describes the modeling of cell metabolism and cell genetic control by M and G respectively.

But why are the Evolution-Complexity Relations true? We take this question up now. Consider the functor given by Figure 19. We want to define 'perfect balance' between M interacting with G. We do this in terms of the *result of interaction R*. $\#(R)$, the number of states of R, is a reasonable measure of the *contact* between M and G. This is so since $\#(R)$ is the number of stable configurations which can exist between M and G, which is the obvious quantitative measure of contact between M and G. In general only stable configurations can be experimentally observed. Thus $\#(R)$ is the number of distinct relations existing between M and G which can be experimentally observed.

Not only do we wish to measure the contact between M and G but we also wish to measure the *complexity of the contact* between M and G. The most obvious way to define this is to pass to the result R and define the complexity of the contact between M and R as the complexity of the state-output-dependent circuit R. Thus (see (6.61a)) complexity of the contact between M and G by definition equals $\#_G(R^S) = \max\{\theta(R_q) : q$ is a state of $R\}$. This is a reasonable definition since *we have defined the complexity of M interacting with G to be the (minimum) complexity of a machine which can simulate the external behavior of external result* (i.e. R) *of this interaction.*

Similarly, *we have defined the contact between M interacting with G to be the (minimum) number of states of a machine which can simulate the external behavior or external result* (i.e. R) *of this interaction.*

However, see (6.75) for some suggested refinements of these definitions.

6.64. Intuitive proof of the Evolution-Complexity Relations. We
want to argue that $\#_G(R) = \#(R) - 1$. We know $\#_G(R) \leq \#(R) - 1$. Thus suppose $\#_G(R)$ is quite a bit less than $\#(R) - 1$, i.e. $\#_G(R) + \epsilon = \#(R) - 1$ with ϵ large (relatively). This means the contact between the metabolism M and the genetic control G is large but the complexity of the interaction is (relatively) small. Large contact means a large number of stable configurations, which represents a sophisticated mechanism for transferring and using enzymes. However, small complexity (relatively) means that relative to the potential use of this mechanism, the actual use is very limited. It is rather like a novice violinist playing a Stradivarius violin. The Stradivarius violin represents high contact since this violin is capable of many configurations (as opposed to cheap imitations). On the

other hand, the novice player represents low complexity (relatively), low ability to act, low utilization of the possibilities.

In this situation of low cell complexity relative to the contact (or possibilities), the forces of evolution (mutation, natural selection, genetic drift for micro-evolution, and the isolating mechanisms of macro-evolution like spatial isolation and genetic isolation leading to external reproductive isolation and internal reproductive isolation) force either the possible configurations to diminish because of inadequate use or the complexity to rise under a fortuitous mutation having survival value, or both together. In short, our novice violinist becomes much better or we give him a less sophisticated violin or both, i.e. he becomes much better and he is given a quality but not an excellent violin.

In short, low complexity and a high possible number of configurations is not a stable situation under the forces of evolution. Either the organism uses effectively more of the possibilities, or the possible configurations diminish with lack of use.

Thus if an organism has stopped evolving (e.g. the cell) then it must be stable under the forces of evolution and thus the complexity must be close to the possible number of configurations. In other words, the cell achieves a 'perfect balance' where the total number of configurations possible (i.e. $\#(R)$) is matched or balanced by a complete utilization of the possible configurations (i.e. $\#_G(R)$ is the maximum possible: $\#(R) - 1$). Thus we define a cell to be in *perfect balance* or *perfect harmony* when the utilization of the possible configurations is maximum, i.e. when $\#_G(R) - \#(R) - 1$. Then we have

6.65. Third Version of the Evolution-Complexity Relations for the Cell.
Let O be an evolved organism. Then O is in perfect balance or perfect harmony. In particular the cell R given by the interaction of the metabolism M and the genetic control G are in perfect harmony, so the complexity is the maximum possible, namely, one less than the total number of stable configurations.

In short, if an organism is not in perfect balance the forces of evolution will increase the complexity of use of the possible configurations or, failing this, decrease the number of possible configurations due to lack of use. Thus if the organism is stable under the forces of evolution it must be in perfect balance (i.e. $\#_G(R) = \#(R) - 1$).

This completes our heuristic proof.

6.66. We suggest at this point that the reader consult the Appendix at the end of this Part II for the precise method of modeling the metabolism M and the genetic control G by finite state machines.

We next want to consider the *deductions* which can be made from the Evolution-Complexity Relations. We start with a few general comments and then get down to specifics.

First, the power of mathematics comes by considering the global situation in a precise way (e.g. *all* the stable states of the cell) even if this immense amount of data is not yet available or may never be completely available or, if available, is unmanageable. Then by discovering general laws governing this immense data, we have guidelines for reasonable means of organizing the data. If the global, mathematical, approach is ignored no reasonable methods of organization of the data are likely to be found.

The Evolution-Complexity Relations state that the cell is as complicated as it can be, subject to the obvious restraints (i.e. the total number of stable states possible for the cell, which in turn is related to energy considerations, molecular configurations, etc., of the cell). The mathematics is usually successful at the extremes, i.e. very low complexity being the simplest case while highest complexity possible for the situation being the next simplest case (e.g. for groups, $\mathbb{Z}_p$'s are easy, and S_n, the symmetric group on n letters, not too difficult).

Getting down to specifics we will *derive the following statements* (by deduction) *from the Evolution-Complexity Relations.*

6.67. Enzymes catalyze (in almost all cases) specific reactions.

6.68. The genetic properties of the cell, including the possibility of a Mendelian type theory, can be *deduced* from the Evolution-Complexity Relations. In short, a Mendelian genetic theory must hold for all evolved organisms, whether we have observed them directly or not; it follows mathematically from the Evolution-Complexity Relations.

6.69. No simplification of the basic (but immense) data of an evolved organism (e.g. the cell) is possible.

6.70. Long chains (nearly as long as the total number of stable states) of essential dependencies (see Theorem 4.11a) must exist among the stable states for evolved organisms. Otherwise said, evolved organisms have internal relations of great depth.

6.71. In the cell, the control of the genes over metabolism need not be absolutely complete but must be very good.

We now concern ourselves with deducing (6.67)–(6.71) from the Evolution-Complexity Relations, and in so doing we will also clarify these assertions when and where necessary.

Outline of the proof of (6.67). The assertion (6.67) is clear. But one might say that it is simply a matter of experimental observation and it might have turned out to be otherwise, just accidentally true thus far it might be found to be false tomorrow. But this is not the case. We can derive (6.67) from the Evolution-Complexity Relations as follows. (In physics one would like to derive from a general principle why charge always comes in positive or negative multiples of the charge of an electron.)

First reconsider Proposition (3.31). Next hold the model of the Appendix of this Part II firmly in mind. Then by a technical argument (which we omit), if the enzymes were not specific in general (each enzyme catalyzes several reactions), then the semigroup of the resulting machine R would have no elements with very high spectrum (e.g. at most $|X| - \epsilon$ elements in the range with X the collection of stable configurations of the cell) and thus by Proposition (3.31), $\#_G(S) \leq |x| - \epsilon - 1$, violating the Evolution-Complexity Relations.

Outline of the proofs of (6.68) and (6.71). First the assertions of (6.68) and (6.71) are ambiguous and may be unclear. Soon we will clarify them, but now we want to give a heuristic proof.

The Evolution-Complexity Relations state that the cell is as complicated as possible, meaning the complexity is the maximum possible for X states, where X is all possible stable configurations of the metabolism. *Identifying complexity with control* yields (after consideration of the model in the Appendix of this Part II, where G has only one state but is not state-dependent-output) that the control of G over M has a maximum value. However, 'absolutely complete' control would mean the semigroup of the cell consists of all maps of X into itself. We do not assert this, we merely assert that the complexity (of the semigroup) of the cell is the same as if the genes of the cell did indeed have 'absolutely complete' control. If the genes have good control then important manifestations of the metabolism correspond to specific mechanisms of the gene, so 'genetic maps' (associating gene functions on genotypes with manifestation of the metabolism on phenotypes) and a Mendelian view is possible. This gives a heuristic proof of (6.68) and (6.71).

For a precise proof we must first precisely define the assertions (6.68) and (6.71), which we shall do using an old idea of Willard Gibbs from physics, and then through mathematics show we can identify complexity with control. We now proceed to do this.

What precisely do we mean by control? Following Willard Gibbs we consider N identical copies of the cell R (where N is a large integer to be specified later). Now one can consider this in a mathematical way or, alternatively, in a physical way, since large numbers of very nearly identical cells do in fact exist. Let N be the total number of distinct configurations of the cell and let $R_1, R_2, \ldots, R_N$ be the N identical copies of R. Further assume at the present time R_1 is in stable configuration S_1, R_2 is in stable configurations $S_2, \ldots$, and R_N is in stable configuration S_N.

$$R_1$$
$$\widehat{S_1}$$

$$R_2$$
$$\widehat{S_2}$$

.
.
.

$$R_N$$
$$\widehat{S_N}$$

Figure 20. Identical copies of the cell R in all possible stable configurations.

Now the control of the cell will be defined in terms of the effect of the same input sequence t applied simultaneously to all N copies.

For example,

Definition 6.72. The *control* of R is '*absolutely complete*' iff for any function $f : \{1, \ldots, N\} \to \{1, \ldots, N\}$ there exists an input sequence t for R such that t applied to R_j (equals R in stable configurations S_j) leaves R_j

in stable state $S_{f(j)}$, or symbolically,

$$R_j \cdot t = R_{f(j)}$$

for all j with $1 \leq j \leq N$.

Thus the control of R is *'absolutely complete'* iff input sequences can perform arbitrary operations on the possible stable configuration. Of course, the control of R is *'absolutely complete'* iff the semigroup of R is *all* maps of states of R into themselves.[*]

Also we can delimit the other extreme.

Definition 6.73. The *control* of R is *'above the bare minimum'* iff there exists input sequences $t_1, t_2, t_3, \ldots, t_{N-2}$ for R so that if $\alpha_j = t_1 t_2 \ldots t_j$, then applying α_j to R in all possible N stable configurations leaves exactly $N - j$ distinct configurations or, symbolically,

$$|\{R_k \cdot \alpha_j : k = 1, 2, \ldots, N\}| = N - j$$

for $j = 1, \ldots, N - 2$.

The idea of control above the bare minimum is that t_1 kills exactly one stable state (but one does not necessarily have control over which one) then then t_2 kills another (but one does not necessarily have control over which one) until α_{N-2} kills all but two states. Intuitively the sequence α_{N-2} is a very slow death for the cell, making it run through a very large number of states, a Hamiltonian circuit type death. *(A proof that living things can be tortured.)*

Perhaps we should mention here an unsatisfactory idea of control, namely, given (j, k) with $1 \leq j, k \leq N$, there exists input sequence t such that $R_j \cdot t = R_k$. (This of course amounts to saying the semigroup R is transitive on the states of R.) It might be reasonable to include this condition in our definition of control above the minimum, but by itself it does not amount to any significant control. Here is why. Assume we want state R_k. We 'dump' into the cell enough 'stuff' to destroy the workings but force R into R_j independent of the present state, i.e., we apply t such that $R_j \cdot t = R_k$ for *all* j, $1 \leq j \leq N$. But after this the cell is damaged and worthless and can merely be driven from state to state by gross outside sources while its complexity has become zero (namely, all constant

---

[*]A connection between this idea of absolutely complete control and finite simple nonabelian groups (SNAGs) arises at the end of Chapter 5 – see the section "Computing with SNAGs" especially Theorem (5.94) and Corollary (5.95). This again suggests a special role for SNAGs in biological systems. -CLN

maps on the states). Thus this definition of control does not amount to much and should not be taken seriously. Even our weak Definition (6.73) is much stronger, more realistic and useful. (In the case of permutation groups acting, transitive is a useful definition of control but not in the case of semigroups acting.)

Proposition 6.74. *The control of the evolved cell R must be above the bare minimum.*

Proof. Use the Evolution-Complexity Relations and Proposition (3.31). □

Actually, the Evolution-Complexity Relations can show that the control of the cell (by the genes) is quite a bit above the bare minimum. However, because of the large amount of mathematics involved (some not yet in final form) we cannot go into complete details. But to indicate why this is so we detour to consider (6.70). The reader should reconsider (4.11) and (4.11a) together with the long discussion of (4.11a) in Part I included in our computations of the elementary model of the Kreb cycle (see the computations following (6.25), Discussion (6.38) and the long interpretation of chains of essential dependencies which follows (6.38)).

Outline of the proof of (6.70). The definition of essential dependencies of (4.11) can be strengthened so that equality is obtained in Theorem (4.11a) (see [LB I],[LB II] and subsequent papers by Tilson and Rhodes). When these theorems have been proved and written then (6.70) will follow with the strengthened definition of essential dependencies.

However, the definition presented in (4.11) is close enough to the strengthened version so that it is profitable to see how (6.70) holding (with the (4.11) version) implies a form of control greater than the bare minimum.

One idea of *locally good control* of a collection of states $\{q_1, \ldots, q_p\} = Q$ is that a group of permutations of Q acts transitively on Q (i.e., given $q_1, q_2 \in Q$, there exists an input t yielding Π so that $\Pi(Q) = Q$ and $q_1 \cdot \Pi = q_2$) and also for each $q \in Q$ there exists an input t_q such that $q_j \cdot t_q = q$ for all $j = 1, \ldots, p$. The semigroup S on Q under locally good control is a transitive permutation group G on Q together with all the constant maps (or resets) on Q. The control is only *locally* good since whenever a constant map is used (a reset) there is no way for any map but a constant map to occur again (since constants$\cdot S \subseteq S \cdot$ constants $\subseteq$ constants). Clearly how effective a good local control is is inversely proportional to the number of states. Of course if S is in series with other locally good control units, then if each unit just controls a small number of states, the global control can be quite good, so heuristically complexity, relative to the total states, means many locally good control units in series, while low complexity relative to the total states implies few locally good control units in series. In the first case each local unit has few states to manage so the local control is good everywhere, so the global control is good. Similarly in the second case the global control is bad, since locally good control units have so many states that their global control is bad.

Using the precise definition of essential dependencies of (4.11) we can make the above heuristic argument precise.

Let $X_1 \supsetneq X_2 \supsetneq \cdots \supsetneq X_n$ be a chain of essential internal dependencies. Consider the step from k to $k+1$, $X_k \supsetneq X_{k+1}$. Using the notation of (4.11) we have G_k a transitive group on $\mathcal{X}_{k+1} = \{X_{k+1} \cdot g_k : g_k \in G_k\}$. Using some standard semigroup theory (details of which we omit), $\mathcal{I}_{k+1}$ can be shown to have the following property. For each $X'_{k+1} \in \mathcal{X}_{k+1}$ there exists $f \in \mathcal{I}_{k+1}$ so that for *all* $X^*_{k+1} \in \mathcal{I}_{k+1}$ either $X^*_{k+1} \cdot f \equiv f(X^*_{k+1}) = X'_{k+1}$ or $|X^*_{k+1} \cdot f| \equiv |f(X^*_{k+1})| < |X^*_{k+1}| = q_{k+1}$ where q_{k+1} is the number of elements in each member of $\mathcal{X}_{k+1}$. Also, for at least one element of $\mathcal{I}_{k+1}$, the first condition must hold. In other words, f acts like the constant map sending every element of $\mathcal{X}_{k+1}$ to X'_{k+1} or to 'zero' and it sends at least one element to X'_{k+1}. Thus $G_k \cup \mathcal{I}_{k+1}$ acting on $\mathcal{X}_{k+1}$ has locally good control (where our previous definition has been slightly generalized to allow 0-constant maps).*

Thus each essential dependency $X_k \supsetneq X_{k+1}$ leads to a locally good control. Since n is nearly as large as the number of states by the Evolution-Complexity Relations, and a strengthened definition of essential dependency (similar to the above but going into precise details about how many zeroes are allowed in the 0-constant maps and not allowing for many zeroes) and a lot of mathematics, the number of locally good control units is large, so each has only a few states to govern, so the global control is good.

This completes our outline of the deduction of (6.71) from the Evolution-Complexity Relations.

---

*A function is said to be a 0-*constant* map if it takes a constant value except possibly at some points where its value is undefined or outside the range of interest and so reported as 'zero'. In the example here, the application of f induces a 0-constant map from $\mathcal{X}_{k+1}$ to itself. Such maps arise frequently in semigroup theory (e.g. in relation to the wreath product of permutation groups augmented with constant maps — implicitly related to the case discussed here – or in the global action of a semigroup on the local Rees-Sushkevych coordinatizations). -CLN

We now return to considering (6.68). (6.68) can be deduced from (6.11), for if the genes have a good control of the cell then this is exactly what a Mendelian genetic viewpoint means. If the genes have a good control of the metabolism of the cell, then the genes determine the phenotype.[†]

In general the argument goes as follows. Let $\mathcal{O}$ be an *evolved* organism. Break $\mathcal{O}$ into two parts mathematically: control C and activating part A. Since $\mathcal{O}$ does not change under the forces of evolution C and A are in perfect balance. Stating this precisely is the Evolution-Complexity Relations. Deduced from this is that C has a good control of A. Since A determines the phenotype of $\mathcal{O}$ and C has a good control of A, C determines the phenotype of $\mathcal{O}$. This is Mendelian type genetics and this argument holds even if $\mathcal{O}$ has never been observed.

When we observe $\mathcal{O}$ we physically describe C and A and then assert that C determines the phenotype with surprise (?). In short, since we are observing an evolved organism, it must be invariant under the forces of evolution, and thus much is true about it simply from this invariance (e.g. the Evolution-Complexity Relations and a Mendelian type genetics).

To continue the discussion of genetics we must discuss the proof of (6.69). It asserts that a concrete program of relating physical or behavioral traits of the organism to specific genes on the chromosomes *must fail*. The genetic code can be determined but *the map of genotype into phenotype is as complex as possible subject only to the obvious restraints.*

Mendel proposed that a gene for each hereditary trait is given by each parent to each of its offspring. The part of this which asserts that genotype (plus environment) determines phenotype is valid, but the implication that the correspondence of individual genes to individual traits holds, on the main is *not* true. If it were the map genotype $\rightarrow$ phenotype would be of very low complexity. The map genotype $\rightarrow$ phenotype being of high complexity implies most physical and behavioral and emotional traits, etc. are a (complex) function of the entire collection of genes.

Let g denote the map of genotype into phenotype or the map of the genome G into the resulting cell $R = g(G)$; g is the function of finite sets into finite sets. By suitably digitalizing the genes, the resulting finite state machine $g(G) = R$, and defining the complexity of the resulting Boolean function (say, by using the method of Winograd (see (5.34) and the references)), we want to argue that g is as complicated as possible, subject

---

[†]Of course genes do not necessarily determine phenotype uniquely but only determine it in relation and response to the environment, interaction, epigenetic marking, variability, etc. See below. -CLN

only to the obvious restraints (size of genome, energy, configuration of the metabolism, i.e., number of states of R, etc.).

If g were (relatively simple) then as the genome was mutated the resulting metabolism changes are related in a relatively simple fashion to gene mutation. But this implies the relation between the genes and the metabolism is not highly complicated, which contradicts the Evolution-Complexity Relations.

Otherwise said, if all the non-lethal genetic mutations are complexly related to the metabolism, then the map genes → metabolism must be complex.

Now it may be found (as indeed it is) that a few genes control some direct property of the metabolism (e.g., eye color, hair color, etc.) but appealing to (6.70) we know that very long chains of essential dependencies among the (stable) states of the cell exist, and this means that they are strongly interrelated and thus the simple program 'one gene, one trait' must fail. (If 'one gene, one trait' were strictly true the complexity of the cell would be zero!)

Thus if in (6.69) one means by simplification anything like the simple 'one gene, one trait' program, then we have shown that this simplification does not exist. In fact the philosophy we suggest is that to first approximation, no simplification is possible.

Thus our result and philosophy suggest the following program. Determine the basic data of the cell, i.e. the substrates, enzymes, structure of the genes, genetic map (which is essentially just the map G defined in the Appendix following this Part II), the metabolic pathways. Then when this data has been gathered, accept that no important simplifications are possible (meaning everything is related to everything by (6.70)). Then the experimental work stops and the mathematics takes over because then the only thing true about the cell is that which is true mathematically for systems satisfying the Evolution-Complexity Relations. (By analogy, if some wires are placed randomly, experimentally we would determine the average number of red wires, blue wires, etc., and their standard deviation, then work out the details by probability theory, check some details experimentally and, when verified, ignore experiment henceforth and calculate by mathematical probability theory.) Sample assertions would be strengthened forms of (6.74) and more detailed versions of (6.70).

In short, if the organism is evolved (and so stable under the forces of evolution) then examine by experiments and determine the basic data. Then (only) mathematics can give further information *because the evolved organism is as complicated as it can be subject only to the basic data and thus' the only additional things true beyond the basic data are those assertions which can be deduced from the assertion of maximum complexity.*

Complexity and biology are intimately wedded. *Living things are complex.* We have made complexity precise so we can make (as above) this assertion about life precise and useful. But what about *evolving* organisms? We take this up, together with mutations, in the next part.

We close with a few technical remarks.

Remark 6.75.

(a) It is perhaps better to replace $\#(R)$, the number of states of R, by a function f which satisfies Axiom A (5.25) and $f((X, F_R(X))) = n$, i.e. for the machine M whose semigroup is all maps on n letters, $f(M) = n$ (For example something like β of (5.92).) The reason is that $\#(R)$ is supposed to measure the *contact* and we might want to break R into *parallel* pieces and look at the maximum number of states in each parallel piece. For the cell there is probably little parallel breaking so the two functions would not differ for cell. However, when parallel functions occur, $\#(R)$ could be replaced by a new function as suggested.

(b) In the Evolution-Complexity Relations we mean only of course that the complexity is within 5% (say) of the number of stable states, not exact equality, but within 'experimental error'.

(c) An interesting mathematical question is: What percent of the semigroups on n letters have complexity $n - 1$, i.e., what percent of the subsemigroups of $F_R(X_n)$, the semigroup of all maps on n letters acting on the right, have complexity $n - 1$? How many have complexity less than or equal to k for $0 \le k \le n - 1$? Little is known about this important problem. Are there about $1/n$ of each complexity?

(d) To introduce cell division more directly to our model, we might take as the output of the cell not the resulting stable state but merely whether the cell is reproducing (mitosis). Thus we have a map $\delta' : X \to \{0, 1\}$ where X denotes the collection of stable states of the cell and $\delta'(x) = 1$ denotes that the cell is state x reproduces (mitosis), while $\delta'(x) = 0$ denotes that reproduction does not occur in state x. Thus the circuit R is changed by replacing the identity map by δ'.

We conjecture that this new R is reduced if the old R is reduced, and that the number of states, complexity, and the semigroup, etc. remain unchanged. The idea is that the essential thing a cell does is to exist and reproduce!

Appendix to Part II

In this Appendix we model the metabolism M and the genetic control G of a living cell by finite state machines leading to a model of the cell as M interacting with G.

We assume here that the reader is familiar with Parts I and II up to (6.66) and with the basic mechanisms of the regulation of protein synthesis and function, and particular the Jacob and Monod models (see, for example, the paper by F. Jacob and J. Monod and the book by J. Watson *et al.* listed in the references at the end of this Appendix).

Since the most is known about *E. coli* we can assume this is our typical example. We begin by presenting our model of the metabolism of *E. coli* by a finite state, state-output-dependent circuit M. For the basic inputs of M we take all the inducers and co-repressors and enzymes of *E. coli*. (There are an estimated 600 to 800 different enzymes in *E. coli* when glucose is its primary carbon source.) As we will consider other carbon sources as well, the number of enzymes will be even greater.

Let S denote the set of all substrates of *E. coli*. S will include the inducers and co-repressors but not the enzymes. In other words, S will include all the substrates, inducers and co-repressors, that is, any substance changing the state of repressors or taking part as a substrate in metabolic reactions, but S will not include the catalysts. Since the metabolic enzymes are (very nearly) specific in the reactions they catalyze we will àlso include the coenzymes in S. Thus S equals metabolic substrates, inducers, co-repressors and coenzymes, but S does *not* include the enzymes.

The states of M are all subsets of S, denoted subsets(S). Thus the states of M form a very very large but finite set. (However, the stable configurations of M interacting with G will be large but much smaller than $|\text{subsets}(S)| = 2^{|S|}$.)

The outputs of M equal the states of M, and the output is the present state. Thus to complete the definition of M we need only give the next state function.

First consider the case where the enzyme e is entered into M which is in state $X \subseteq S$. Suppose e catalyzes a reaction $a \xrightarrow{e} b$ (assuming all substrates and coenzymes are present). Then

$$\lambda_M(X, e) = X \cdot T_{ab} = \{x \cdot T_{ab} : x \in X\},$$

where $x \cdot T_{ab}$ equals b when $x = a$ and equals x otherwise. Thus enzymes act on S just as in Part I, and the action is extend to acting on subsets(S) in the obvious way.

The action of the inducers and co-repressors is somewhat more complicated. We first define a very simple action and then modify it later. If x is an inducer (e.g. the carbon source lactose) or co-repressor (e.g. histidine) then our first simple definition of the action is $\lambda_M(X, x) = X \cup \{x\}$, where $X \subseteq S$ is a state of M. This tentatively defines M. We will go on to define G and later return to modify M.

Biologically G is the chromosomal DNA of the cell. By the "central dogma" of molecular genetics, DNA serves as a template for the synthesis of RNA and RNA in turn is the template of the protein synthesis of enzymes. The genetic information of the cell is contained in a sequence of nucleotide bases in a DNA (or RNA) molecule. Substrates whose introduction in the growth medium specifically increase the amount of an enzyme are called *inducers*. Biosynthetic enzymes whose amount is reduced by their end products are termed *repressible enzymes*.

Repressors are coded by chromosomal DNA. The genes that code them are called *regulatory genes*. (The DNA is subdivided into functional parts of sequences of nucleotide bases, each part termed a *gene*.) Each repressor blocks the synthesis of a unique set of proteins. Repressor molecules can exist in both active and inactive states, depending on whether they are combined with the specific small molecules, the inducers and co-repressors. Often repressors control more than one protein. The collections of adjacent nucleotides that code one or more messenger RNA (mRNA) molecules, but are under the control of a single repressor, are by definition termed the *operons*. The functioning of an operon is under the direct control of a specific chromosomal region, termed the *operator*.* The regulatory genes always trigger the creation of the operon's specific repressor. If the specific inducer is present, the repressor then is not able to continue to block the operator and the synthesis of the proteins (enzymes) of the operon is triggered.

When an end product (co-repressor) builds up, an inactive repressor can become active and block the operator, preventing further protein (en-

---

*The genetic regulatory model in this Appendix follows the insights of the groundbreaking early Jacob-Monod model [J-M]. The details of genetic regulatory control can be much complex, e.g. involving multiple transcription factors interacting via various possible mechanisms, cf., the latest edition of [Watson], or E. H. Davidson, *The Regulatory Genome: Gene Regulatory Networks in Development and Evolution* (Academic Press, 2006), or M. J. Schilstra and C. L. Nehaniv, "Bio-Logic: Gene Expression and the Laws of Combinatorial Logic", *Artificial Life* (special issue on Systems Biology) 14(1):121-133, 2008. With generally straightforward modifications, the reader can create models in the same spirit as the one presented here taking into account state-of-the-art knowledge and further details on particular genetic regulatory systems and mechanisms. -CLN

zyme) synthesis of the operon. For example, when an *E. coli* cell growing in a minimal glucose medium is suddenly supplied with the amino acid isoleucine, production of the RNA molecules needed to code specific enzymes in isoleucine biosynthesis ceases. In fact, isoleucine inhibits the enzyme threonine deaminase which transforms threonine into α-ketobutyrate.

Thus G is most simply and effectively modeled as a one-state circuit (not state-dependent output (see Definitions 5.3 and 5.4)). The inputs of G are the states of M, namely, subsets(S). The outputs of G are subsets(enzymes). Thus G is simply a map G: subsets(S) $\rightarrow$ subsets (enzymes).

Physically G is defined as follows: $G(X)$ equals those enzymes synthesized via mRNA when X is present in the cell. On the basis of the model developed in the above discussion, we can define G as follows. Let the normally active repressors of the cell be $R_1, \ldots, R_m$. Let $e(R_j)$ be the gene products of the operon(s) repressed by R_j, i.e. the enzymes whose synthesis is normally actively inhibited. Let the normally inactive repressors be $N_1, \ldots, N_n$. Let $e'(N_j)$ be the enzymes that are continually synthesized by the operator associated with N_j, i.e. $e'(N_j)$ is the operon of N_j. Let $E_0 = \bigcup \{ e'(N_j) : j = 1, \ldots, n \}$, i.e., E_0 are those enzymes which are normally always being synthesized. Let $I_1, I_2, \ldots, I_k$ be the inducers. Let $f(I_x) = \bigcup \{ e(R_y) : R_y \text{ is rendered inactive by } I_x \}$. Let $f(I_{x_1}, \ldots, I_{x_p}) = \bigcup \{ f(I_{x_y}) : y = 1, \ldots, p \}$. Thus, $f(A \cup B) = f(A) \cup f(B)$. Similarly let $C_1, \ldots, C_q$ be co-repressors. Let $g(C_x) = \bigcup \{ e'(N_j) : N_j \text{ is rendered active by } C_x \}$. Let $g(C_{x_1}, \ldots, C_{x_p}) = \bigcup \{ g(C_{x_q}) : q = 1, \ldots, p \}$.

Given $X \subseteq S$, $f(X)$ are all those enzymes normally not produced, but produced because of the inducers contained in X. Similarly $g(X)$ are all those enzymes normally produced, but not produced because of co-repressors in X. Then by definition

$$(6.76) \qquad G(X) = (E_0 \cup f(X)) - g(X)$$

where $A - B = \{ a \in A : a \text{ is not in } B \}$. It is interesting to note that G can be realized by a simple McCulloch-Pitts net consisting of one bank of components. See [*Automata Theory* by J. Rhodes, (end of Chapter 1)]. That is, G can be realized from one bank of threshold elements with threshold $+1$ and input lines of value $+1$ or $-\infty$. In fact each repressor/inducer or repressor/co-repressor unit is one such element.

Now we increase the input set of M by allowing subsets(enzymes), inducers and co-repressors. We summarize the model thus far in Figure 21.

machine	inputs	outputs	states
M metabolism, a state-dependent output circuit	inducers, co-repressors and subsets(enzymes)	subsets(S) with S the substrates, inducers, co-repressors and coenzymes	same as outputs of M
G genetic control, a circuit	same as outputs of M	subsets(enzymes)	one element

Figure 21.

We must explain how subsets of enzymes act on subsets of S. E denotes the collection of enzymes. Let $X \subseteq S$ and $F \subseteq E$ be given. Let $r(F) = \{(a, b) : a, b \text{ substrates and } a \xrightarrow{f} b \text{ for some } f \in F\}$. In the reaction $a \xrightarrow{f} b$ we assume all coenzymes and cofactors are present. Thus $r(F)$ is the collection of reactions catalyzed by members of F. Let paths$(X, F) = \{a_k : \text{ there exists substrates } a_0, a_1, \ldots, a_k \text{ such that } a_0 \in X \text{ and } (a_0, a_1), (a_1, a_2), (a_2, a_3), \ldots, (a_{k-1}, a_k) \text{ all belong to } F\}$. Thus paths$(X, F)$ equal those substrates which can be obtained from reaction catalyzed by members of F beginning with members of X (assuming all coenzymes and cofactors are present). Certain members of X will be considered 'distinguished' (denoted dist(X)). Usually dist(X) will be the (relatively) large quantities of external foodstuffs, carbon sources, etc. We will define dist(X) later. But now we define for $X \subseteq S$, $F \subseteq E$,

$$(6.77) \qquad \lambda_M(X, F) = \text{paths}(X, F) \cup \text{dist}(X) .$$

Also if $X \subseteq S$, and x is a co-repressor or inducer (external inputs to the cell),

$$(6.78) \qquad \lambda_M(X, x) = X \cup \{x\} .$$

Now that our model has tentatively been given, we will let M and G interact. Since G has but one state we can omit all references to the states of G and consider only the states of M. Suppose for ease of starting we assume M begins in the empty set. Also, being a carbon source, we denote lactose as a distinguished substrate. Next suppose lactose is introduced to the cell. Then

$$\phi \xrightarrow{L_a} \{L_a\} .$$

Lactose (or β-galactoside) is an inducer, and the β-galactosidase repressor does not block the synthesis of β-galactosidase mRNA.

Thus $G(\{L_a\}) = \beta$-galactosidase. β-galactosidase hydrolytically cleaves the sugar lactose to galactose and glucose.

Thus $\{\beta$-galactosidase $\equiv e\}$ is fed-back into M and so

$$\lambda_M(\{L_a\}, e \equiv \beta\text{-galactosidase}) = \text{paths}(\{L_a\}, e) \cup \text{dist}(\{L_a\})$$
$$= \{\text{galactose, glucose}\} \cup \{L_a\}$$
$$= \{\text{galactose, glucose}, L_a\} \equiv S_1 .$$

S_1 induces the enzymes for the familiar breakdown of glucose, etc., which leading to pyruvate via the Embden-Meyerhof pathway, and, e.g., in eukaryotes, to the Krebs cycle, etc. Notice if $S_0 = \phi$, $S_1 = \{\text{galactose, glucose}, L_a\}$, $S_2, S_3, \ldots$ are given, then $S_0 \subset S_1 \subset S_2 \subset S_3 \subset \ldots$ until for some q, $S_q = S_{q+1} = S_{q+2} = \ldots$ and steady state is reached. $S_j \subseteq S_{j+1}$ holds for all j since $\{L_a\}$ is distinguished. Now the steady state S_q will be the collection of all substrates involved in the conversion of lactose to energy (forming mainly energy-rich phosphate bonds).

Suppose another energy source x is presented to the metabolism M and the supply of lactose is stopped. Then the action of x is to eliminate lactose and replace it with x, and x is distinguished so, assuming M is in S_q,

$$\lambda_M(S_q, x) = (S_q - \{\text{lactose}\}) \cup \{x\} \equiv T_0 .$$

Now x is an inducer and $G(\{x\})$ are the enzymes which begin the breakdown of x. At the same time $G(\{\text{lactose}\}) = \beta$-galactosidase will not be forthcoming so $T_1 = \{x\} \cup G(\{x\}) \cup S_q - \{\text{lactose,galactose,glucose}\} = \{x\} \cup G(\{x\}) \cup (S_q - S_1)$. Thus as the breakdown of x proceeds, the substrates involved in the breakdown of lactose (and not involved in the breakdown of x) slowly disappear, until finally, $T_p = T_{p+1} = \ldots$ appears. Then T_p contains the substrates necessary for the breakdown of x. T_p might be somewhat different depending on whether we began M in ϕ or in S_q.

Also we might allow a mixture of lactose and x to enter M together. In this case both lactose and x would be distinguished and M starting from ϕ would proceed by $\phi \rightarrow \{L_a, x\} \rightarrow \lambda_M(\{L_a, x\}, G(\{L_a, x\})) = \text{paths}(\{L_a, x\}, G(\{L_a, x\})) \cup \{L_a, x\}$. Notice $G(\{L_a, x\})$ might *not* equal $G(\{L_a\}) \cup G(\{x\})$. It might be the case that in the presence of L_a the breakdown of x cannot proceed in exactly the same manner as before.

Thus we see that in our model the inputs to M are the natural inputs to the cell. The process by which M interacting with G comes to a stable loop (which is almost always in a stable state, i.e. loop of length one in the case of a cell) corresponds to setting up and maintaining the basic metabolic pathways. The resulting stable state is all the substrates involved in the various pathways. When another input is introduced the original pathways are disturbed and the sequence of states running to the stable state correspond to breaking down the old pathways and establishing the new ones.

For the Evolution-Complexity Relations to be valid, the inputs to the cell should be those it receives in its natural environment under all conditions faced in a typical cell lifetime. These will include the carbon sources (energy sources), perhaps various amino acids which the growth medium may contain, etc., and also any inputs serving as 'signals' from other cells for this cell to perform special functions (e.g., secreting a substance when properly 'signaled').

When the input is lactose, say, instructions may be given for the application. For example, feeding in lactose in normal amounts, normal manner and normal concentrations for a length of time adequate to establish normal digestion. Or, feeding a poor dilute lactose mixed with certain concentrations of various amino acids, etc.

The same substances administered in a different way will be considered to be different inputs and will be denoted by distinct symbols.

Let α denote an inducer or co-repressor $x = x(\alpha)$ being applied in a specific manner. Now $\text{dist}_\alpha(X \cup \{x\})$ must be defined for each α (see (6.77)). The idea is that if lactose (say) has been fed in normal amounts to the cell leading to the stable state S_q and $x = x(\alpha)$ is then introduced, α might denote that lactose is suddenly replaced by the new carbon source x, so $\text{dist}_\alpha(S_q \cup \{x\}) = x$. On the other hand, β might denote that lactose is supplemented with a mild solution of $x = x(\beta)$ so $\text{dist}_\alpha(S_q \cup \{x\}) = \{L_a, x\}$, etc.

Thus we have the inputs, outputs and states of M and G given by Figure 21 and G defined by (6.76). For each external input α to the cell we have a co-repressor or inducer $x(\alpha) = x$ defined. Also for each $X \subseteq S$ and $Y \subseteq Y \cap$ (inducers $\cup$ co-repressors) we have

$$(6.79) \qquad \text{dist}_\alpha(X \cup \{x\}, Y) \subseteq X \cup \{x\}$$

defined.

Then for $X \subseteq S$ and given $\text{dist}(X) = Y$, by definition

$$(6.80) \qquad \lambda_M((X, \text{dist}(X)), \alpha) = X \cup \{x(\alpha)\}$$

and

$$(6.81) \qquad \text{dist}(X \cup \{x(\alpha)\}) = \text{dist}_\alpha(X \cup \{x(\alpha)\}, \text{dist}(X)) \,.$$

Finally for $F \subseteq E =$ all enzymes and $X \subseteq S$, and assuming dist(X) defined, we let

(6.82) $$\lambda_M(X, F) = \text{paths}(X, F) \cup \text{dist}(X) .$$

This completes the definition of M and G.

In closing we note some salient features of our modeling. First the *stable states* obtained by M interacting with G *correspond with* the various *metabolic pathways* of the cell. Trajectories through transient states leading to the resulting stable state correspond to setting up the metabolic pathways. The problem of rates has been avoided by the discrete concept of distinguished state. Distinguished states are those substrates present in (relatively) large amounts. Thus even though enzymes are present which convert them into new substrates the distinguished states persist (because of their quantity) until the stable state is reached. However, if a distinguished state then becomes non-distinguished (i.e. more is not provided from external sources), it can be destroyed by conversion into another substrate provided the proper enzymes are produced by G. In this way introduction of the exact rates and concentration of reaction can be avoided.

The Evolution-Complexity Relations simply assert that the complexity of the machine R which results from M interacting with G is (approximately) the total number of distinct stable states of the cell (minus one).

The modeling presented here should be looked at merely as guidelines of how to model the cell with finite state machines. Once the general approach is grasped and the relevant data is available the relevant model can be generated by the reader.

Bibliography

On Molecular Biology, Control of Protein Synthesis, Enzyme Systems, etc.

[J-M] F. Jacob and J. Monod. Genetics regulatory mechanisms in the synthesis of proteins. *Journal of Molecular Biology*, 3:318-356, 1961. (Review of mRNA and the control of protein synthesis).

[Rose] S. Rose. *The Chemistry of Life*. Pelican Books, 1966. (Elementary background book). [Fourth revised edition, published by Penguin Books, Ltd., 1999.]

[SciAm] R. H. Haynes and P. C. Hanawalt, editors. *The Molecular Basis of Life: Readings from* Scientific American, W. H. Freeman, 1968. (Many excellent articles).

[Watson] J. D. Watson. *Molecular Biology of the Gene.* Benjamin, N.Y., 1965. (Excellent text). [Sixth edition by J. D. Watson, T. A. Baker, S. P. Bell, A. Gann, M. Levine, and R. Losick, published by Benjamin Cummings, 2008.]

On Gibbs' Ideas in Physics

[Schrödinger] Erwin Schrödinger. *Statistical Thermodynamics.* Cambridge University Press, 1967. New edition, Dover, 1989. (Excellent. See chapter 1 on exposition of Gibbs' ideas.)

Mathematics

[LB I] J. Rhodes and B. R. Tilson. Lower bounds for complexity of finite semigroups. *Journal of Pure & Applied Algebra,* 1(1):79–95, 1971.

[LB II] J. Rhodes and B. R. Tilson. Improved lower bounds for complexity of finite semigroups. *Journal of Pure & Applied Algebra,* 2:13–71, 1972.

[Wino] S. Winograd. On the time required to perform multiplication. *Journal of the Association for Computing Machinery,* 14(4):793–802, 1967.

See also the end of Chapter 5 and additional references there.

On Evolution[†]

[Darwin] Charles Darwin. *The Origin of the Species and the Descent of Man.* Modern Library (Random House), New York. [no date. First edition of "Origin of Species" published 1859, Sixth edtion 1872. First edition of "The Descent of Man" published 1871, second edition 1882. (See 'The Complete Work of Charles Darwin Online' http://darwin-online.org.uk/ .]

[Evolution] Jay M. Savage. *Evolution,* 2nd edition. Holt, Rinehart & Winston, New York, 1969.

[Haldane] J. B. S. Haldane. *The Causes of Evolution.* Cornell University Press, Ithaca, NY, 1966. [Originally published by Longmans, Green & Co., Limited, 1932. Also by Princeton University Press, 1990.]

[Simpson65] George Gaylord Simpson. *The Meaning of Evolution.* Yale University Press, New Haven, Connecticut, 1967. [Revised edition. Original edition 1949.]

[†]Editor's note. See also:

[Davidson] Eric H. Davidson, *The Regulatory Genome: Gene Regulatory Networks in Development and Evolution,* Academic Press, 2006.

[Ridley1] Mark Ridley, *Evolution,* 3rd edition, Blackwell Science, Ltd., Oxford, U.K., 2004.

[Ridley2] Mark Ridley, editor, *Evolution* (Oxford Readers), 2nd edition, Oxford Universty Press Inc., New York, 2004. [Selection of important classical and modern papers on evolution.]

Part III. The Lagrangian of Life

A. The Laws of Growing and Evolving Organisms

> *"The purpose of Life is the evolution of consciousness"*
> — from mystical folklore

In the previous section (Part II) we considered the conditions which evolved organisms must satisfy. In this section (Part III(A)) we consider the laws which evolving organisms must satisfy. By the very general concept of evolving organism we include both the evolution of a species and also the evolution of the individual member of the species from birth (individual existence) to death (individual destruction).*,† By the standard biological concept‡ that "ontogeny recapitulates phylogeny" (the life of the individual retraces the path of evolution of the species, e.g. the baby in the womb a passes through an early "fish-like" state, etc.) these two concepts of evolution are intimately wedded.

The purpose of this section is to determine some laws which hold for all evolving organisms. Our laws will take the form of defining a function

---

*Here the term "evolution" is formulated in such a manner that includes, but is not limited to, Darwin-Wallace evolution. In particular it also includes the behavior and development of organisms in evolving populations. However this is not merely 'time evolution' or 'dynamics' (the sense in which some physicists often use the term "evolution", and which completely misses the major insights of modern evolutionary theory), but is specifically constrained by the 'Principle of Evolution' formulated by the author in what follows. Thus 'evolution' in the sense of Rhodes encompasses Darwin-Wallace evolution, as well as instances of ontogeny, growth, morphogenesis, development, learning, and interactive behavior, where the dynamics is described by the maximization of a specific 'Lagrangian operator' (construed in a wide sense as made mathematically precise in the subsequent text below). -CLN

†For an introduction to fundamental background on evolution, see Mark Ridley, *Evolution*, Blackwell Science (3rd edition), 2003. For modern evolutionary theories also involving particular attention to scope over multiple levels, *cf.* Eva Jablonka and Marion J. Lamb, *Evolution in Four Dimensions: Genetic, Epigenetic, Behavioral, and Symbolic Variation in the History of Life*, Bradford Books/MIT Press, 2005, or, Eörs Szathmáry and John Maynard Smith, *The Major Transitions in Evolution*, W. H. Freeman, 1995. -CLN

‡This concept has been variously revised and appears in various guises in biology from the relatively naïve to increasingly sophisticated forms beginning from Ernst Haeckel in 1866 to Stephen Jay Gould in *Ontogeny and Phylogeny* (Harvard University Press, 1977) and throughout modern 'Evo-Devo', e.g., Brian K. Hall, *Evolutionary Developmental Biology* (2nd ed., Springer, 1999). -CLN

H which for each possible relation R of the organism to itself and its environment (i.e., for each possible complete description R of the organism, its environment, and their relations). H will yield a positive integer $H(R)$. Then the laws will be of the form that the organism evolves or grows in such a way as to *maximize H*. We call H the Lagrangian of the organism because of the similarity to the Lagrangian formulation of Newtonian mechanics.[1]

We believe the *principle* on which the definition of H is based is universally valid. However to obtain precise and quantitative results by the methods presented herein it is necessary to model the environment, the organisms, etc., by finite state machines. In many important cases this poses no major obstacles; for example, in the Appendix to Part II we modeled the cell by finite state machines.

The modeling usually proceeds by determining the basic discrete inputs to the organisms as they naturally occur in the environment. The continuous inputs (if any) are either suitably digitalized or considered in natural divisions (e.g. normal sunlight of mean intensity x for a period of time y) represented by a single symbol. Relevant stable states of the organism are listed again with the continuous state parameters (if any) suitably digitalized or considered in natural divisions. Then the machine is defined 'naturally' by applying a sequence of inputs to the organism (or plant, etc.) and experimentally observing the final state (or deducing from theory the final resulting state). For example, for a flower (i.e. a flowering plant), say, the inputs would include sunlight (e.g. α_1 denotes normal intensity for one hour, α_2 high intensity for half-hour, etc.), water, pollination by a bee (e.g. β_y denotes pollination by a bee with the exact location/quantity/quality, etc. of the pollen given), entrance of specified harmful parasites, etc.

The state space of the flower would include the basic metabolism (e.g.

---

[1]We wish to warn the mathematically sophisticated reader that by a Lagrangian we do *not* mean the precise definitions as formulated in Classical Mechanics and in modern treatments of Classical Mechanics (e.g. Global Analysis). We mean that there is a function H defined on the collection of possible paths (subject to energy restraints, etc.) an organism might possibly pass through in space-time, and that the path *actually followed* by the organism has a maximal H value. We assume no continuity, differentiability, etc., so that the maxima are not necessarily unique. We consistently use the term Lagrangian in this wide sense. So potentials give rise to Lagrangians in our sense, but in the technical sense, only trivial spaces (one point or less) could have both a potential and a Lagrangian. We feel the wide use of this term is justified (e.g. Freud's Lagrangian, Marx's Lagrangian) by the importance of the concept of maximizing a function (i.e. Lagrangian) on a function space and we should not be required to accept the narrow condition of the term.

the photosynthetic carbon cycle, etc.) which could be modeled as in Part I and the Appendix to Part II. Also the present topology of the flower (say its differentiable or piecewise linear structure) could be conveniently digitalized or coded by, for example, breaking cartesian three-space into small cubes and denoting which cubes contain any part of the flower where we include the roots, stalk, petals, etc. Alternatively we could have finite sequences of symbols with certain symbols denoting leaf locations and the following symbols indicating shape and size of the previously denoted leaf, etc. The status of any reproductive part of the flower would also be included in the state space.

To determine the resulting machine for a flower the botany and chemical theories could be utilized together with a motion picture camera run over several days (say) to determine the change in form caused by the inputs, etc. Of course a large amount must be known about the flower, both theoretically and practically, to effectively determine the resulting machine. However the actual task of modeling is a very constructive way to be forced to organize the known facts and theories of the flower and can by itself, even when not completely successful, increase one's understanding of the global organism.*

The Principle of Evolution is the following: *An evolving organism transforms itself in such a manner so as to maximize the contact with the complete environment, subject to reasonable control and understanding of the contacted environment.*

Later we will make this principle of evolution mathematically precise. But first we give a heuristic discussion.

Contact with the environment means (a monotonic function of) the memory space required to simulate the behavior on a computer or (a monotonic function of) the number of flip-flops required to build a machine to simulate the behavior of the organism, etc.

---

*Systems biologists have now begun to use computational approaches modeling from the cellular to the whole organism level to simulate its development, see e.g. H. Kitano, "Systems Biology: A Brief Overview", *Science* 295(5560):1662–1664, 2002, or B. E. Shapiro, A. Levchenko, E. M. Meyerowitz, B. J. Wold, E. D. Mjolsness, "Cellerator: Extending a Computer Algebra System to include Biochemical Arrows for Signal Transduction Simulations", *Bioinformatics* 19(5):677–8, 2003, or J. T. Kim, "**transsys**: A Generic Formalism for Modelling Regulatory Networks in Morphogenesis", *Advances in Artificial Life*, Springer Lecture Notes in Computer Science, vol. 2159, 242–251, 2001, where the latter describes such simulation system applied using the ABC model of flowering plant development for angiosperms suchas *Arabidopsis thaliana*. -CLN

The complete environment means all forms of contact physical and mental, i.e. to measure the contact of an artist with his environment we would have to model his thinking (which we haven't the slightest idea how to do at present) and then simulate this model on a computer (say). One would expect that Van Gogh's contact would be multiples larger than Norman Rockwell's contact. We repeat that we believe this principle of evolution is universally valid, but in areas where our knowledge is very dim or nonexistent the results cannot be made numerically precise and contain only heuristic value (e.g. in the area of artistic thinking).

So much for contact. Next what do we mean by "reasonable control and understanding" of the contacted environment? From the many arguments presented in Part II we mean that the complexity of the resulting interaction of the organism and its environment is very high (relatively) or is approximate in numerical value to the numerical value of its present contact. Thus we *identify* the following statements: reasonable control of the present contacted environment plus reasonable understanding of the present contacted environment; the complexity of the present organism/environment is the maximum value possible given the value of the present contact.

The arguments in Part II have shown that increased control implies increased complexity. Clearly increased understanding implies increased complexity. Further, since in Chapter 4 we found that complexity is deeply related to understanding (complexity being the minimal number of coordinates necessary to understand a process) an organism operating at high (relative) complexity in its interaction with its environment implies high (relative) understanding and ability to act.

Thus, at least heuristically, the statement of this principle of evolution is clear. But why is it true? Intuitively we can argue as follows. One of the main goals of the organism is just to avoid destruction (stay alive and intact). There are two reasons why one dies. First, something happens which the organism doesn't know about and it kills it. For example, an anti-science English professor walks into an area of high radioactivity and dies, or a mathematics professor involved in his research does not read the newspapers and goes out into the streets during a *coup d'état* and is struck down by a bullet, or a baby eats some poison, etc. In all these cases the organism dies because some element is *suddenly* presented to its environment (e.g. radiation, bullets, poison) and kills the organism. In short, the *contact* was too limited and, as a direct result, the organism dies. It should have been aware of the existence of radioactivity, revolution or poison.

At the other extreme, the contact may be very high but the complexity too low; things get "out of hand" and one dies. For example, a spy disguised in an enemy uniform is shot and killed by his own men at night as he returns from enemy territory with vital information; or two psychoanalysts develop a new theory of madness requiring them to spend many hours alone with previously violent individuals. One of the psychoanalysts is nearly strangled to death but lives and recovers, but the other becomes very morose after this violent incident and apparent setback to the theory, eventually committing suicide. In these two cases the contact with the environment was high but the complexity, control and understanding of the situation was far enough behind to lead to death. A flower may bend toward the sun but then become visible from the footpath and be picked by a passing schoolgirl. The contact with the sunlight was heightened but the control to avoid being picked after being seen was not existent. The sly deer who waters at the river bend may, by understanding the hunter's methods, escape by his sense of smell and rapid running but die from the polluted water. Here excellent control and understanding was no defense against the sudden new element of the modern environment — water poisoned with industrial waste.

But if the organism can increase contact and increase control — understanding and complexity in nearly the same amount — it is in all ways better off. Hence our principle of evolution.

We now proceed to give a precise formulation using the construction previously introduced in Part II.

Let E_t be a finite state circuit representing the environment at time t and let O_t be a finite state circuit representing the organism at time t. Let R_t be the result of E_t interacting with O_t as we precisely and carefully defined in Part II. Let $C_t = \#(R_t)$ denote the number of states of R_t and let $c_t \equiv \theta(R_t) + 1$, which is the complexity of R_t plus one. Then C_t is the *basic contact number* at time t and c_t is the *basic complexity number* (the ordinary complexity plus one) at time t. Thus $1 \leq c_t \leq C_t$ for all t.*

Let $\mathbb{N}^+$ denote the positive integers, $\mathbb{N}^+ = \{1, 2, 3, \ldots\}$. Let h map pairs of positive integers into the real numbers $\mathbb{R}^1$, i.e., $h : \mathbb{N}^+ \times \mathbb{N}^+ \to \mathbb{R}^1$. If $a, b, c \in \mathbb{N}^+$ and $a < c$ and $b < c$, we then require $h(a, b) < h(c, c)$. The

---

*This follows since the number of states plus one is an upper bound to complexity (3.16g). To see this more directly, consider the complexity axioms and apply them to decomposition of the full transformation semigroup on n states in the alternating wreath product of direct products of flip-flops and full symmetric groups on k states (for each k with $1 < k \leq n$) — see discussion following Principle I (4.10) above. Since there are $n - 1$ group levels in this decomposition, the complexity in the case of n states can be no more than $n - 1$. -CLN

function h is called the *culture function*. Let $O = \{O_t : t \in \mathbb{R}^1\}$ and $E = \{E_t : t \in \mathbb{R}^1\}$.

Definition 6.83. The Lagrangian H for O in environment E with respect to culture function h is by definition

$$(6.84) \qquad\qquad H(t) \equiv H(O_t, E_t) = h(C_t, c_t) \ .$$

We will explain the significance of the culture function after a few examples. In many cases E_t will not depend on time so $E_t = E$ for all t.

Now let us return to our previous example of the flower to illustrate the use of the Lagrangian (6.84). The inputs to the flower are sunlight, water, bees and parasites. The states (also equals outputs) of the flower include descriptions of the metabolism, form, topology, position, and status of the reproductive parts. The inputs of the environment E will be the states of the flower. The states (also equals outputs) of E will be the inputs of the flower. The finite state machine E will be defined so that if E is in state $\alpha = $ "rainfall of intensity x from direction $\vec{v}$ for time t", then independent of the output of flower O_t this input will continue for time t. On the other hand, the flower at time t, O_t, by changing its position can greatly affect the result of this output on itself, but in this case it cannot directly affect the environment by stopping or increasing the water (as distinct from the animal who can get in or out of the rain or get a drink of water or seed clouds). With a limited number of exceptions, the flower cannot directly affect its environment but can, by changing its state, greatly change the effects of this environment on itself. Thus a flower adapts to its environment and has limited feedback which changes the environment. Thus the environment E for a flower can usually be modeled by a very simple finite machine similar to the case of the genes G in the Appendix of Part II or omitted altogether. In this last case we assume that the sunlight, water, wind, bees, parasites, etc. are not at all affected by whatever state the flower is in. If we model the environment like the genes in the Appendix to Part II then we assume the effects of the state of the flower on the bees (say) is direct (the output of the environment depends only on the plant, the environment has no memory). If the environment has memory then we must model E as a finite state circuit with more than one state, etc.

At any rate, at time t, $E = E_t$ and O_t (equals the finite state machine representing the flower at time t) yield the result R_t. Intuitively R_t is given by placing a basic input into the environment E and then letting E and O_t feed back on one another until a stable configuration is locally formed. In the simplest case let us assume that we omit E altogether by assuming no state of the flower affects the environment. In this case $R_t = O_t$. Then the Lagrangian H of (6.84) is used as follows. Consider all functions $f : \mathbb{R}^1 \to M =$ all finite state machines with inputs and outputs suitable for the flower. By setting $f(t) = O_t$ we obtain a possible *growth function* for the flower. Now, by known botany, chemistry and physics consider those growth functions not excluded by these theories (e.g. f has compact support, before existence a zero-state machine and after decay a one state machine, that is, all flowers grow from a seed and finally die and decay). Let us denote this collection of possible growth functions as *Poss*. If *Poss* consisted of only one element then this is the way all flowers would grow. But *Poss* will contain in general a large number of functions. But we assert if $O_t \equiv O(t)$ is the way the flower in fact grows, then

$$(6.85) \qquad \int_{\mathbb{R}^1} h(C(O(t)), c(O(t))) \geq \int_{\mathbb{R}^1} h(C(f(t)), c(f(t)))$$

for all $f \in Poss$. Here $C(f(t))$ is the basic contact number of $f(t)$, etc. Of course (6.85) merely says that the flower grows to maximize the Lagrangian H.*

Another interesting example is the fetus forming and growing in the mother's womb. Here O_t is the fetus (or baby) modeled as a finite state machine at time t. E is the womb modeled as a finite state machine, etc.

We next want to explain the purpose of the culture function. Intuitively *the purpose of the organism is to keep c_t close to C_t and then maximize the two nearly equal quantities.* But what strategy should the organism adopt if C_t becomes quite a bit larger (relatively) than c_t? Should it immediately attempt to lower C_t to c_t or attempt to bring them both to a midway point, or do nothing? Of course that the attempt is made to do any of these things does not necessarily imply that this is what will happen. However, the culture function h is what determines the attempt, the *local strategy*, when $C_t - c_t$ is (relatively) large.

For example, let us consider a society like the English society where a (relatively) high premium is placed on social propriety, emotional control, duty, etc., but withstanding this a fairly strong force for personal advancement exists. We might represent this by the culture function $h_E : \mathbb{N}^+ \times \mathbb{N}^+ \to \mathbb{R}^1$ with

$$(6.86) \qquad h_E(a, b) = a^3 - e^{a-b} .$$

---

*Thus (6.85) expresses the 'Principle of Evolution' in the course of an organism's lifetime with respect to a particular strategy (i.e. Lagrangian determined by culture function h). -CLN

Here the negative term e^{a-b} represents a severe penalty for $C_t - c_t$ being large while a^3 represents some incentive to maximize contact C_t. In short, h_E represents a culture which is repressed and productive.

Alternatively we might consider a society like the American society which sacrifices all other considerations to economic considerations. It encourages inventiveness and enterprise and discourages activities not directly related to economic productivity (e.g. art, music, deep emotional relations, sex, etc.) in the same bland but frantic manner. Thus we might represent this by the culture function

$$(6.87) \qquad h_A(a,b) = a^2 + b^2 - (a - b)^2 = 2ab .$$

Here contact (e.g. enterprise) and understanding (e.g. inventiveness) are rewarded while contact minus understanding (uncontrolled activities) are penalized similarly. In short h_A represents a bland, immature but productive culture.

We might also consider the French society. Here the culture transmits enjoy yourself, manipulate the environment for your pleasure, etc. This might be represented by

$$(6.88) \qquad h_F(a,b) = b .$$

There is no penalty for a (relatively) large contact minus understanding (uncontrolled activities). The culture function h_F indicates that always one should increase control, understanding, manipulation, etc. In short h_F represents a society full of love, sex, talk and personal strivings idealizing beauty and intellectuals, and losing a war to the Germans in six weeks!

The idea behind the English culture function (6.86) is that there is a culture force to control yourself and then advance on symbolically, first $\longrightarrow\!\!\!\times\!\!\!-$, then $\uparrow$. The idea behind the American culture function is "don't rock the boat", or restrain what should be restrained a little and increase what should be increased a little. The idea behind the French culture function is: don't worry, live — wine, women and song — and — since this takes money, sexual attractiveness, etc., — manipulate.

Thus in h is contained the strategy which the organism will attempt during its growth. Exactly what form of h must be determined experimentally for each organism. In some cases the culture function will not radically affect (6.85). This will be the case where the environment has direct and relatively strong effects on the organism so the exact strategy attempted by the organism has limited effects. However, as we consider organisms higher on the evolutionary scale the importance of the culture function h becomes proportionately greater. At any rate, it must be determined by theoretical intuition and experiment.

In Part II we stated the Evolution-Complexity Relations which hold for *evolved* organisms. We can now state that *evolved organisms satisfy the Evolution-Complexity Relations* (see Part II) *and evolving organisms tend toward these Evolution-Complexity Relations in a direct and maximal manner.*

The results of this section as presented here are somewhat vague, since to incorporate precision it would be necessary to model the organism and its environment in detail by finite state machines, necessitating a mass of specific detail. This we will attempt to do in later papers. Here we must content ourselves with statements of more generality. However, in this vein, we would like to discuss what *deductions or suggestions* can be made from the principle of evolution in its 'Lagrangian of Life' formulation as given by (6.83)–(6.85). Below we consider six possibilities, but the list could be lengthened.

6.89. Application to Thom's program for studying morphogenesis. We have reference to the methods of R. Thom as presented in *Topological Models in Biology* and his *Stabilité Structurelle et Morphogenèse*. Thom has suggested a method or an 'art of models' for use in organizing the immense data of present-day biology with particular emphasis on the problem of morphogenesis, i.e. the origin, development, and evolution of biological structures.

Implicit in Thom's model is that the growth and behavior of the organism or phenomenon is determined by some implicit physical variables *minimizing* some potential (or *maximizing* the negative of this potential). Without this potential or (in some wide sense) Lagrangian lurking in the background the method cannot even begin.

Two questions immediately arise. First, does the potential exist in concrete situations and, if so, how is its mathematical form discovered? The second question is how is the problem of morphogenesis handled assuming this potential exists? An answer to the second question emerges from Thom's work, namely, "... there appears to be a striking analogy between this fundamental problem of theoretical biology (morphogenesis) and the main problem considered by the mathematical theory of topology, which is to reconstruct a global form, a topological space, out of all its local properties. More precisely, a new mathematical theory, the theory of *structural stability* — inspired from Qualitative Dynamics and Differential Topology — seems to offer far reaching possibilities to attack the problem of the stability of self-reproducing structures, like the living beings. But ... this type of dynamic description exceeds by far the biological realm, and may be applied to all morphological processes — whether animate or inanimate — where discontinuities prohibit the use of classical qualitative models."

The mathematical heart of Thom's theory is the complete geometric classification of those discontinuities which can arise from a system minimizing some potential and where the discontinuities have only a finite number of types of state in a neighborhood of the discontinuity. Such discontinuities are called *elementary catastrophes.* Up to diffeomorphism there are seven elementary catastrophes possible in space-time. See the above references for more details.

For an application of Thom's method to the breaking of water waves, see [Zeeman]. For an excellent review of the mathematical theory of structural stability, see S. Smale [Review].

However, regarding the first question of identifying the physical variables and the potential, Thom apparently offers no solution. It is with this first problem that we believe the principle of evolution, as formulated earlier in this section, can be of service.

We suggest starting with some of the more elementary plants and modeling the plant and its environment as simply as is possible by finite state machines and attempting to determine the basic contact number and the basic complexity numbers for the growth of an actual plant. From this determine a culture function leading to a wide-sense Lagrangian. Then use Thom's method (when the hypotheses apply) to determine the time-space location of the catastrophes (e.g. initial appearance of a new leaf, blossoming, etc.). The finite state machine descriptions of the plant will include the topology of its form (e.g. an abstract simplicial complex coded digitally) so catastrophes will be intimately related to the states of the plant machine. Intuitively it is completely clear when the catastrophes occur in space-time for the plant. Thus if C_t and c_t can be determined, very few h's will exist such that the potential $h(C_t, c_t) = H(t)$ will yield the correct type of catastrophes at the correct point in space-time. In fact, C_t and the space-time locations of the catastrophes can probably be relatively easily obtained. Then slight information on general properties of h and c_t may very well uniquely determine h and c_t.

This program is very ambitious. One of the main technical points is to relate the discrete descriptions of the plant's structure (by finite state machines) to differentiable models of the plant's form. The functions $t \mapsto C_t, c_t$ are integer valued and Thom works with differentiable functions. But there are many technical methods available. For example if the topology of the form is carried by the machine as a digitalized abstract simplicial complex, then changes in the form can be taken to be simplicial transformations. Then by working the piecewise linear setup we can obtain piecewise linear transformations, etc. Just as in our modeling of the cell in the Appendix to Part II each locally stable configuration reached from the organism and the environment interacting is the result of feedback between the environment and the organism and corresponds in the differential category to a transformation in a dynamical system leading to local stability.

Also, a large difficulty is that probably the existence of a potential is not satisfied very often for physical biological systems or, as Smale says "many systems jiggle", which is not a phenomenon possibly guided by a potential. Thus we will probably need a detailed theory of dynamical systems (and not just those guided by a potential). But the central program of Global Analysis is to do just this. Maybe Smale's theorem, that every diffeomorphism (on a compact manifold) can be moved (i.e. is differentiably isotopic) to one whose non-wandering points are well behaved and has the no-cycle condition, can be used.

In short, I believe (in the wise sense) the Lagrangian indicated here is the function that living organisms are attempting to maximize. But its definition requires modeling organisms with finite state machines plus a lot of algebra. When this is done then it might be defined on some function space and might be differential but it won't be very nice, i.e., it won't be given by a potential or classical Lagrangian.

So we will need Thom's detailed results extended to arbitrary diffeomorphisms (or at least to almost all diffeomorphisms), plus detailed modeling of organisms by finite state machines, plus algebraic theory of complexity, plus how to define a differential wide sense Lagrangian from the algebra, etc.

One interesting thing appears on both the finite side and on the differentiable side, namely, *Lie groups*. Of course Lie groups occur continuously (e.g. Hilbert's fifth problem) but they also appear on the finite side since finite state machines reduce to finite semigroups which reduce to finite groups which reduce to SNAGs (= simple non-abelian groups). But via Thompson's and others work, most (all?) SNAGs come from Lie algebras (plus some "twists") via the Chevalley method. Thus the two approximations to reality (continuous and finite) both lead to Lie groups (the building block of reality?).

We note in closing the similarity of many of Thom's general ideas on morphogenesis to those of Goethe's as expressed in his *Metamorphosis of Plants* published in 1790.

6.90. The Complexity of Plants. Model the plant P by a finite state machine in the manner previously described. Here we choose those plants for which (in the sense previously discussed) the environment E can be ignored. Similarly model the plant P'. Then define P to be less complex than P' iff $\theta(M(P)) < \theta(M(P'))$ where $M(P)$ is the machine model of P and $\theta(M(P))$ is the complexity of the machine $M(P)$. Then compare this means of classification of the complexity of plants with the naive and intuitive conception of the complexity of plants. Also verify that more highly evolved plants are more complex. Also obtain directed graphs from the plant structure in a natural way (e.g. veins and edges of the leaf). Then using the complexity of graphs (as introduced at the end of Part I), compare the complexity of graphs associated with P (denoted $G(P)$) with the complexities of P, P', $G(P')$, etc.

6.91. Mutations. With reference to our discussion of the map, g, of genotype into phenotype for the cell (say *E. coli*), we suggest the following hypothesis for verification. *The lethal mutations drive the complexity of the resulting cell down while the non-lethal mutations drive the complexity of the cell up.* Actually we should consider the value of $\int h(C_t, c_t)$ over the lifetime of the organism, but since $C_t = c_t$ (by the evolution-complexity relations) and mutations are small (but perceptible) changes, the above is a good enough approximation (up to ordering).

6.92. Continuity and Discontinuity of Evolution — Genius, Stupidity and Fragmentation. Our principle of evolution provides some insight into the important (and much discussed) question of whether discontinuities have appeared in the evolution of (say) man. See M. J. Adler [Adler]. The current opinion of the biologist is that the elemental forces of mutation, natural selection and genetic drift produce relatively small population changes, the so-called *micro-evolution*. At the next level speciation occurs on the fragmentation and development of new populations. At another level *macro-evolution* or *adaptive radiation* occurs which is characterized by the splitting of groups into many new subgroups, invasion of numerous new environment situations, and diversification of structure and biology. Special adaptation tends to be produced by micro-evolution and speciation, while macro-evolution generally develops from a general adaptation and special new adaptations derived form the general adaptation. Rarely, a new biological organization of general adaptation occurs, called *mega-evolution*. The important question of how the elementary forces of micro-evolution give rise to the origin of new populations and eventually new species has been grappled with since Lamarck and Darwin. The interaction of ecologic isolation and differential micro-evolution leads to new reproductively isolated species but the origins of the isolating mechanisms are obscure. For more details see J. Savage [Evolution] and the references contained there. We return to discussions of macro-evolution and mega-evolution in (6.96).

However here we would like to discuss the *continuity* or *discontinuity* of evolution. One way to make this precise is the following. Let $\ldots, O_{x_1}, O_{x_2}, O_{x_3}, \ldots$ be the sequence of organisms where O_{x_i} is the 'mother' of $O_{x_{i+1}}$. Assume the complexity $\theta(O_{x_t})$ is defined for each t and consider the mapping $t \mapsto \theta(O_{x_t})$ of a subset of the integers into a subset of the integers. We expect that as t becomes larger so does $\theta(O_{x_t})$ or, more precisely, eventually the average

$$\sum_{\alpha=-\infty}^{k} \theta(O_{x_\alpha}) \Big/ \sum_{\alpha=-\infty}^{k} 1 \;=\; A(k)$$

is a monotonic function (there exists N so that if $k_1 \geq k_2 \geq N$, then $A(k_1) \geq A(k_2)$). But in discussing continuity or discontinuity we are interested in *how large* $|\theta(O_{x+1}) - \theta(O_x)|$ *can be.*

We cannot determine the sequences $\ldots, O_{x_1}, O_{x_2}, \ldots$ very far back in the past so it is interesting to consider mathematically whether continuity or discontinuity (by precise definition given below) is *mathematically possible.*

Continuity can be defined as an *evolutionary jump in complexity never exceeds 1*, the minimum possible since complexity is integer-valued, i.e. $|\theta(O_{x+1}) - \theta(O_x)| \leq 1$. We further assume O_{x_j} can be modeled by a finite state machine with semigroup S_{x_j}, so $\theta(O_{x_j}) = \#_G(S_{x_j})$. Now choose time 0 in an arbitrary but suitable way and consider S_0 and S_T where T is large. Under evolution S_T evolved from S_0. We do not know the exact path so we ask what are the mathematical possibilities of S_0 'evolving' to S_T

in steps increasing in complexity by no more than one. We can assume S_0 divides S_T (by taking S_0 to be very close to $\{1\}$ or by using cases where the evolved organism contains the activities of the older generations as proper subsets, i.e. O_0 divides O_T as machines).

Thus we are led to considering the *possibilities* of finding $S^{(1)}, S^{(2)}, \ldots, S^{(c)}$ so that

(6.93)
$$S_0 = S^{(1)} \mid S^{(2)} \mid \ldots \mid S^{(c)} = S_T \quad \text{and}$$
$$|\#_G(S^{(j+1)}) - \#_G(S^{(j)})| \leq 1 .$$

But we consider these theorems in Chapter 3, see (3.28), (3.29a) and (3.30).

Now if $S_0 = \{1\}$ then (6.93) can be solved. In fact, solved in the stronger sense with divides ($\mid$) replaced by surmorphism ($\leftarrow$) by using (3.28), or with divides ($\mid$) replaced by subsemigroup ($\leq$), or even by divides replaced by 'is an ideal of' by using (3.30).

However, taking $S_0 = \{1\}$ is not very realistic. By using the very new results of (3.29a) it appears to be the case that there exists S and T such that

(6.94)
$$S \mid T$$
$$S \mid R \mid T \quad \text{implies} \quad R = S \text{ or } R = T$$
$$\#_G(S) = k$$
$$\#_G(T) = 2k + 1$$

and

(6.95)
If $S \mid T$ and $S \mid R \mid T$ implies $R = S$ or $R = T$, then
$$\#_G(S) = k \leq \#_G(T) \leq 2k + 1 .$$

Thus we can say that starting from nothing (e.g. $S_0 = \{1\}$) a chain of continuous evolution is possible. However it might happen that in fact $S_{x_j} = S$ is reached at time j and at the next generation $S_{x_{j+1}} = T$ is obtained with $S \mid T$ but no in-between organism is possible (i.e. $S \mid R \mid T$ implies $R = S$ or $R = T$), so-called *irreducible jumps*. In this case the complexity of T can be double the complexity of S (plus one) but no more than this by (6.95). On the other hand, by (6.94) the jump of double plus one is mathematically possible.

Passing from complexity to its logarithm base 2, $\log_2$, we see the irreducible jumps in complexity are by at most $1 + \epsilon$.[*]

---

[*]This can be considered as relating to a mathematically precise formulation of "punctuated equilibrium" including bounds on the magnitude of any complexity jumps in evolution. See also C. L. Nehaniv and J. L. Rhodes, "On the Manner in which Biological Complexity May Grow", in *Mathematical & Computational Biology*, Lectures in the Life Sciences, Vol. 26, American Mathematical Society, pp. 93–102, 1999. -CLN

Summing up, *for any organism there are irreducible jumps of complexity zero or one, starting from* {1}, *which eventually yield the organism. However, if the actual path of evolution of the organism is refined into irreducible jumps, some jumps may double the complexity plus one but never increase it more than this. In short, the logarithm* $\log_2$ *of the complexity increases in jumps of at most* $1 + \epsilon$ (where ϵ is a very small real number).

It is very interesting to consider those examples S and T satisfying (6.94) to be the 'genius jumps of evolution'. Without going into technical details of semigroup theory let us say that one class of examples are of the form $T \leq S$ and T breaks into the union of two disjoint subsemigroups T_1, T_2, each with complexity k (in fact T_1 is an ideal of S and T_2 is a subsemigroup which is a union of groups) and S is obtained by adding a single link between T_1 and T_2 (T_1 is a chain of $\mathcal{J}$-classes, T_2 is a chain of $\mathcal{J}$-classes and $T_1 \cup J \cup T_2 = S$ is also a chain of $\mathcal{J}$-classes with J linking the two chains being the middle non-zero $\mathcal{J}$-class of S) driving the complexity of S to $2k + 1$ while the complexity of T_1, T_2 and T are all k. Further, this class of examples *seems typical* of the '$2k + 1$ irreducible jumps'.

This has an interesting interpretation for the 'genius irreducible jumps'. Namely, two distinct structures must be developed (e.g. theories, biological organization, etc.) represented by T_1 and T_2 which thus far have not been related (e.g. $\#_G(T) = \#_G(T_1) = \#_G(T_2)$) but in S they become related for the first time. Thus the 'genius jumps' depend on the existence of developed but previously unrelated structures.[2] Thus genius depends on fragmentation and then in turn genius destroys fragmentation. Also T is rather stupid in development since its complexity could have been $2k$ (if each new $\mathcal{J}$-class had raised the complexity by 1, since t has $2k$ $\mathcal{J}$-classes). Thus if 'non-stupid' is defined as "the complexity equals the length of the longest nest of ideals (i.e. the number of non-zero $\mathcal{J}$-classes in a chain)" then 'genius' depends on 'stupidity' since if S is 'non-stupid', then each irreducible jump from it can only be an increase of zero or 1 in complexity.

Thus genius depends on stupidity and fragmentation and remedies both. Of course one genius could unite T_1 and T_2 via T but then later another genius unites T and T', etc., so one genius destroys the old fragmentation and creates even larger parts in fragmentation, etc.

---

[2] Galileo first stated the three laws of motion in non-mathematical terms. Descartes developed analytic geometry. Newton learned analytic geometry as an undergraduate at Cambridge (which he found difficult), then learned the laws of Galileo's physics and synthesized both by developing the calculus during the time that Cambridge was closed because of the Great Plague in 1665-66 (when Newton was 24).

Finally, you will not meet anyone more than twice (plus ϵ) as intelligent (i.e. complex) as you, assuming you have slightly above average intelligence.

6.96. Evolution above the Species Level and levels of Complexity of Biological Organization. Quoting from *Evolution* by Savage:

"Macro-evolution and mega-evolution are both incredibly complex in their dimensions. They involve, in addition to the elementary forces and isolation, the following characteristics, all related to natural selection and environmental relations: -

(1) The taking on of new general adaptation or occupancy of a new adaptive zone.
(2) The breakthrough into new zones or subzones within the new adaptive zone by development of special adaptations.
(3) The loss of evolutionary flexibility, and channelization into greater and greater specialization within subzones.
(4) The ecologic reinvasion of a zone or subzone when it becomes partially unoccupied because its original occupiers are now specially adapted (ecologic replacement).
(5) The irreversibility of evolution. Since each step is dependent upon the previous progressive changes, once a group is on the adaptive road, it is usually trapped in an adaptive zone and cannot reverse evolution against the direction of selection."

Macro-evolution and mega-evolution are not well understood processes. Using our principle of evolution can we obtain any insight into these processes?

The possibility (6.96)(1) above clearly corresponds to changing the environment to a very different finite state machine E. (6.96)(2) corresponds to radical changes in the organism O so as to maximize the (wide-sense) Lagrangian, but now with respect to the interaction with the new E. The result is (6.96)(3) because the maximum (often) lies in more specialization of structure. (6.96)(5) follows since we are maximizing a function.

We believe that once some experience and success is obtained in working out the Lagrangian for actual plants and other organisms, then the development of the plant or other organism in a new hypothetical environment E can be determined in some marked degree. This then would be a powerful tool for studying mega- and macro-evolution.

Bibliography

On Evolution

[Adler] Mortimer J. Adler. *The Difference of Man and the Difference It Makes.* Holt, Rinehart & Winston, New York, 1967.
[Evolution] Jay M. Savage. *Evolution.* Holt, Rinehart & Winston, New York, 2nd edition, 1969.

(See additional references at the end of Chapter 6, Part II.)

Morphogenesis

[DArcy] D'Arcy Thompson. *On Growth and Form.* Cambridge University Press, 1961. Abridged edition edited by J. T. Bonner.

Mathematics

[Review] S. Smale. Differentiable dynamical systems. *Bulletin of the American Mathematical Society*, 73(6):747–817, 1967.
[Smale] S. Smale. Stability and isotopy in discrete dynamical systems. In M. M. Peixoto, editor, *Dynamical Systems.* Academic Press, New York, 1973.
[Thom1] R. Thom. Topological models in biology. *Topology*, 8:313–335, 1969.
[Thom2] R. Thom. *Stabilité Structurelle et Morphogenèse: Essai d'une Théorie Générale des Modèles.* W. A. Benjamin, Reading, MA, 1972.
[Zeeman] E. C. Zeeman. Breaking of waves. In *Proceedings of the Symposium on Differential Equations and Dynamical Systems*, Springer Lecture Notes in Mathematics, vol. 2006, pp. 2–6, 1971.

B. Complexity, Emotion, Neurosis and Schizophrenia

"The human race is a myriad of refractive surfaces staining the white
radiance of eternity. Each surface abstracts the refraction of refractions of
refractions. Each self refracts the refractions of others' refractions of
self's refractions of others' refractions ... "
— R. D. Laing

"Life is boredom interrupted by pain."
—The author in his unhappy youth

In Part III(A) of this section we introduced a Lagrangian guiding the growth of an organism or a species. The principle from which this previous Lagrangian was derived is also valid for the emotion growth of an individual during his or her lifetime. We would like to introduce this "Lagrangian of individual emotional development" and discuss its relation to so-called "mental illness", namely, neurosis and schizophrenia, and so-called normal or adjusted individuals. On the way we will also give a *precise definition of emotion* (in a restricted context).

Assuming the reader is already familiar with the arguments presented in Part III(A), let us simply say that *the Lagrangian (true goal) of emotional life for the individual is to contact the environment maximally subject to reasonable understanding and ability to act.* For those situations which can be modeled (or reasonably approximated) by finite state machines, the contents of Part III(A) will provide a completely precise formulation of this Lagrangian together with methods yielding exact numbers. However in almost all situations the meanings will be relatively clear although there might not exist precise definitions or techniques to obtain exact relations and/or numbers. In any case, deductions can be derived from the assumption that emotional development is governed by this Lagrangian and in fact this is the purpose of this section.

In those situations where those aspects of the environment and the individual considered important can be modeled by finite state machines we obtain I (a finite state model of relevant aspects of the individual) interacting with E (a finite state model of relevant aspects of the environment) in the precise sense of Part III(A). The interaction of I and E yields the (finite state machine) result R in the precise sense of Part III(A). Let $\#(R)$ denote the number of states of the (reduced) circuit R. Let $\theta(R)$ denote the complexity of the (reduced) circuit R. Then, as we have seen from Part III(A), $\#(R)$ measures the contact between the individual I and his or her environment E, while $\theta(R)$ measures the understanding and ability of I to act with the important aspects of the environment E which I has contacted.

In this situation we propose to define $e(R) \equiv \#(R) - \theta(R)$ to be the *emotion* of I in the interaction with E. Similarly, if (X, S) is a right mapping semigroup we define $e(X, S)$, read 'the emotion of (X, S)', by $e(X, S) = |X| - \#_G(S)$. Since $\theta(R) \leq \#(R) - 1$ and $\#_G(S) \leq |X| - 1$ it follows that $e(R)$ and $e(X, S)$ are always positive integers.

The intuition behind this definition is that emotion arises when new elements from the environment appear to the individual which are not well understood and/or not well controlled. For example, a gun suddenly being leveled at one, or a man finding his wife in bed with the newly hired attractive female maid he is having an affair with, etc.

Since $e(R) = 1$ iff $\#(R) - 1 = \theta(R)$ and $\theta(R) \leq \#(R) - 1$ we can reformulate the Lagrangian of emotional life as follows: *Maximize contact with the environment subject to ensuring emotional stability.* Here emotional stability means emotions of small magnitude, or $e(R) = 1$, or experimentally $e(R)$ is relatively small. Of course an entirely equivalent formulation is: Maximize control (ability to act) and understanding of the environment subject to insuring emotional stability.

The mathematical properties of $e(R)$ and $e(X, S)$ can be investigated but we will not produce the results here. It is also desirable to replace $\#(R)$ (the number of states of R) and $|X|$ by closely related functions which, in addition, satisfy Axiom A of Chapter 5, Part II, similar to $\beta(S)$ of (5.92).

Now, assuming that emotional development of the individual is governed by the Lagrangian previously introduced we would like to obtain some deductions concerning so-called "mental illness".

For each individual I_t and environment E_t at time t we obtain the resulting interaction R_t and thus also $\#(R_t)$ and $\theta(R_t)$. We will denote $\#(R_t)$ by C_t, read 'the *contact* of I with the environment at time t', and we will denote $\theta(R_t)$ by c_t, read 'the *understanding* or *control (ability to act)* possessed by I at time t'. Now, $C_t - c_t = e_t$, read 'the *emotion* of I at time t', and $e_t \geq 1$.

The goal of individual mental development is to make C_t and c_t large and their difference $C_t - c_t = e_t$ small. Roughly there are four main extreme cases, namely:

(6.97)

Case α : C_t high and c_t high and $C_t - c_t$ low

Case β : C_t high and c_t low and $C_t - c_t$ low

Case γ : C_t low and c_t low and $C_t - c_t$ low

Case δ : C_t low and c_t low and $C_t - c_t$ somewhat high

Case α corresponds to high contact with the environment, high ability to act with understanding and emotional stability. In short, the active, full, happy, creative, stable, competent, intelligent individual.

We next want to show that Case β corresponds to the neurotic individual, and Case γ corresponds to the schizophrenic individual.

For the *theory* of mental behavior of human beings the only good references are early (non-philosophical) Freud — see *The Ego and the Id*, *Three Essays on the Theory of Sexuality*, "Analysis of a phobia in a five-year-old boy", and the other references listed for S. Freud at the end of this section (*The Ego and the Id* and the case histories as written by Freud are the best) — and R. D. Laing's *The Divided Self*; *Sanity, Madness and the Family*; and *The Self and Others* (but not *The Politics of Experience* or *Knots*). See also (the partly good, partly bad) *Interpersonal Perception* by Laing, Phillipson and Lee [IP], which we will comment on later. In the following we assume the reader is familiar with these references.

Now case β is the situation of the individual with (relatively) high contact but (relatively) low understanding and control (ability to act) of the environment which is being extensively experienced (or contacted). In this situation the individual's emotions ($e_t = C_t - c_t$) are understandably high. We are going to argue that this situation, Case β, can be identified with the classical neuroses as studied by Freud.

Freud discovered and made conscious certain classes of experiences of the individual which could not be forgotten even many years later. In our theory this corresponds to C_t "being stuck at some high value". Roughly, some important members of these classes of experiences are

(6.98)

(a) Sexual advances of a mother to a son.

(b) Sexual desires of a son to a mother.

(c) Anger of a father to a son because of (a) and (b).

(d) Anger of a son to a father because of (a), (b) or (c).

Of course a similar formulation holds when "mother" is interchanged with "father" and "son" interchanged with "daughter".

To be more graphic and clear, let us suppose that a husband and wife have a very cold sexual relationship (e.g. bad love making once or twice a month); the husband is a sea captain who often is away from home. Assume the wife, in a misdirected attempt to obtain adult affection and gratification, gives the son excessive affection (often lies nude, half covered on the bed and asks her thirteen year old son to "rub her back"). Suppose that evening her husband returns from a two-month sea voyage, the wife is cold toward him but displays much affection toward the son. Suppose during dinner the son becomes "nervous" and knocks a candle off the dining table and burns a small hole in the dining room rug. The father becomes very angry and after verbal abuse toward the son, pushes him somewhat roughly and orders him to bed. The wife then runs from the table to the son's room, comforts him, and later locks herself in her separate bedroom for the night. The husband drinks himself into a stupor and falls asleep on the couch while the son dreams of a young knight saving the queen from a terrible death.

In this illustration, all the elements (a)–(d) are clear. *A very important aspect of these experiences is that they cannot be forgotten.* (Man can forget many things, but not making love to his mother.) This is an empirical fact that has been verifying that experiences of the class (a)–(d) cannot be forgotten. The detailed observations of Freud and later observers also give ample evidence for this empirical assertion.

Now *repression* can be defined as the strategy of emotional development which attempts to lower contact (i.e. lower C_t) by forgetting certain past experiences and leaving ability to act and the understanding (i.e. c_t) at the same level. Another very important empirical fact uncovered by Freud is that for experiences of the class (a)–(d), *repression won't work.* This seems mainly to stem from the fact that the experiences keep reappearing in the adult nightly dreamlife and this cannot be simply erased by the waking conscious mind.

However, society appears to recommend repression, probably stemming from its abhorrence of emotion as displayed by its individuals. Now, if the contact could be lowered then demanding emotional stability (i.e. no emotion) would result in C_t falling to c_t, giving $C_t - c_t = 1$, i.e., emotional stability. However, since experiences of the classes (a)–(d) cannot be erased the pressure from society cannot lower C_t (insofar as it depends on (a)–(d) and in fact c_t merely becomes lower which exacerbates the problem for both the individual and for the society).

Freud brilliantly grasped the situation and saw that the "cure" for both the individual and the society was to increase the understanding of experiences of class (a)–(d), increase the ability of the individual to act in society (e.g., to be loving) and to discourage fruitless attempts at repression. The detailed methods of psychoanalysis as developed by Freud are well known.

Roughly, the individual must assimilate his sexual desires for his mother (e.g., realize that many of his dreams are making love to his mother. A few months ago I dreamed I was making love to my ex-wife in my mother's bedroom and the ex-wife suddenly changed in appearance below me into a colleague's unattractive wife, similar in appearance to my mother.) The son must realize that the father's anger toward him was displaced anger because the relationship of the mother and father was defective and realize that he has very strong anger toward the father because of his unjust punishment (my father often beat me after unnatural goading from my mother) and his desire to be rid of the father so as to have all the mother's affection, etc. This assimilation leads to understanding. Also, as time passes, he is no longer the son, he is the adult who can control his life and destiny, etc. This leads to the ability to act (like relating deeply to a wife, girlfriend, work, etc.).

Thus, all in all, Freud's classical analysis and suggested therapy of neuroses fit into Case β.

We next want to consider Case γ and show that it can be identified with the schizophrenic individual or at least with the individual using a (relatively) larger number of "schizophrenic defenses".

Case γ is the situation of the individual with (relatively) low contact with the environment but (relatively) high understanding and control (subject to this low contact) of the situation. Thus the individual's emotions $(C_t - c_t)$ are low.

For the theory of schizophrenia the reader is referred to R. D. Laing's *The Self and Others* and *Sanity, Madness, and the Family*. Laing is to schizophrenia what Freud is to neurosis. Since, as we have seen, neurosis involves certain experiences which cannot be forgotten, the "analysis" involves determining the exact experiences, and the "cure" involves assimilating by non-repressing understanding, etc., this whole process can be carried out in a room with the so-called "analyst" and the so-called "patient". Basically this is possible because the "patient" contains (in nonerasable form) all the information necessary (written in his "conscious" or "subconscious").

This is not the case with schizophrenia. This "mental disease" is the result of a certain type of *interaction* of two or more persons (e.g. mother *and* daughter). If only one member of this interaction is observed the cause of the "illness" will totally escape the observer. (This is how the absurd "physical basis theory" of schizophrenia developed.) In short, *neurosis is a malfunction in the feedback between an individual's conscious and "subconscious", while schizophrenia is a malfunction in the feedback or interaction between the "patient" and the family members.* All the information necessary for the analysis and cure of neurosis lies within the single individual but the information necessary for even the analysis of the (so-called) schizophrenia patient lies in the whole family and its interaction. In short, you cannot even analyze the situation of the so-called schizophrenic until you interview the family; see [Sanity, Madness and the Family].

Before we portrayed a graphic family situation which led to a son's neurosis. We now give a generic family situation which can lead to a daughter's schizophrenia. (See [Sanity, Madness and the Family] for actual cases.)

As a child just beginning school, a family rule is that the daughter will go to bed at 8:00 p.m. However, as children are prone to do, the child attempts to stay up longer. The family does not enforce this rule but merely pretends that the child does *not exist* after 8:00 p.m., by not noticing, talking, acknowledging, etc. the child after 8:00 p.m.!

Further, until 18 years of age the daughter *is never alone in public* (picked up at school, driven to the movie and picked up afterwards, etc.) without an accompanying older member of the family. At about 16 years of age the daughter develops a great passion for religious and mystical philosophy and spends many hours alone in her room reading. Simultaneously her school record declines slightly and she becomes more "distant" from the family. With no discussion of these matters and no previous sign of disapproval by the parents, the parents simply take all the books from the daughter's room, in her absence at school, and destroy them. No one ever talks about this to each other. This is also done on two future occasions. Eventually the daughter stops reading and becomes "lethargic" and her school work worsens. She stops talking much with the family.

On dispensing the daughter's books, the mother secretly reads one on ESP (extra-sensory perception) and wonders if her daughter posseses this ability. The father and mother on Sunday drive in the family car and both adopt a secret plan to think about popcorn to see if the daughter can pick up their "thought waves". While driving the parents frequently look at the daughter wondering if the "popcorn" signals are being received. The daughter asks, "Do you want something from me?" and the mother replies, "Oh, no, dear, I was just looking out the window." Similar experiments (unknown and denied to the daughter) are tried on several occasions.

With declining performance in school and non-communicativeness with the family, and on the advice of school authorities, the daughter sees a psychiatrist as an out-patient at the local state mental institution once a week. She "improves" somewhat and also begins to date another patient she has met in the psychiatrist's waiting room, a young auto mechanic with whom she falls in love and wants to marry.

The parents have a secret meeting and send the daughter to her aunt and uncle in the country for a week. Then when the boyfriend comes to the house to see the daughter they tell him the daughter told them to tell him she does not want to see him anymore (which was not true). The boyfriend never calls again and they tell the daughter he came by in her absence and said he was leaving town and wouldn't be calling again (which is also not true). The daughter is heartbroken, retreats to her room and, after a few weeks, refuses to eat and is taken to the local state mental hospital by her parents, where she is diagnosed with "schizophrenia with congenital physical base" and given extensive shock treatments.[3] Improvement is slight.

Now the low contact in the schizophrenic situation, as illustrated above, is easy to see. The daughter is never alone in public until she is 18; there is no discussion of the demands that as a child she must go to bed at 8:00 p.m.; she was simply ignored after 8:00. Similarly, there is no discussion of her reading; the books were simply taken in her absence. There was gross interference in the daughter's love life in such a manner that she believed the man she loved simply left her, not loving her, not even caring enough to say farewell in person (which was not the case, but only what the mendacious parents said). These experiences led to a withdrawal from everyday life, leading to less concentration in school, less conversations, less eating, etc. No friends, no lovers, no studying: less and less life. In short, the contact with the environment C_t is (relatively) small in the schizophrenic situation as illustrated above and presented by Laing in [Self and Others] and [Sanity, Madness and the Family].

Next we want to argue that the complexity or understanding of the situation, c_t, is high (relatively) subject to this low contact, or precisely

---

[3]Shock treatments are the snakes and state mental hospitals are the pits of the modern military-industrial societies.

$c_t + 1 = C_t$ or $e_t = 1$ (or, experimentally, e_t is small relative to C_t and c_t). The idea behind this assertion is the following. Suppose the daughter is riding in the car with her parents a Sunday afternoon when the parents are playing the game of thinking "popcorn" to test if daughter has the ESP ability. Suppose they pass near an amusement park and mother asks, "Are you hungry dear?" Now the parents have never discussed this "game" they are playing with daughter and in fact they always deny they are doing anything at all, and reply, "Why do you ask dear? Is anything wrong? Are you feeling alright?" The daughter has been through this many times, so although not knowing explicitly about the ESP-popcorn game she does know that each time they stop for refreshments her parents seem to act strange and watch her intensely and accuse her of acting strangely if she asks them if anything is wrong. On the other hand, if they don't stop to eat the trip goes on and on, while after stopping they generally go directly home.

Thus, in replying to mother's "Are you hungry dear?", the daughter *thinks*: "Actually I am hungry for a hamburger and tired of riding but if I say I'm hungry the funny scene goes on when we stop, which I really don't like. On the other hand, if we stop at Uncle Bill's I could get a snack and get back the book I left there two weeks ago. But mother doesn't want me to read the book, also she doesn't like for us to eat at Uncle Bill's. On the other hand, Aunt Jean is a nut about home remedies and gives me these awful tasting medicines for my cold which mother insists I take. Thus, if I say I am tired with a cold and need some of Aunt Jean's medicine, we could go there, be offered a snack and I could secretly get the book. However, if I say I want to get the medicine from Aunt Jean right now, mother will say, "well, let's go home", just so I don't get to do what I want, but if she thinks Aunt Jean's medicine is her idea it will probably work out." Thus the reply is,

"No, I am not hungry. I'm tired; I got the sniffles."

"Have you been taking your medicine?"

"I'm out."

"Are you sure?"

"Yes."

"Maybe we should get some more from Aunt Jean."

"No, my cold's not too bad" — while making a face as if tasting bitter stuff.

"Well, I think we should stop by, if you are sure you're not hungry."

In short, the daughter is forced as a rational human being to consider and plan each simple utterance, each simple act, in order to obtain the most basic elements necessary to emotional life. *This is a complex procedure, requiring much understanding and much ability.*

Quoting from *Interpersonal Perception* (denoted [IP]), "... We were specifically concerned with the relationship between young adult schizophrenics and their parents." Laing was anxious to find a way of testing whether the schizophrenic was more "autistic" than the other family members. If part of autism was inability to see the other person's point of view, then his hunch was that, comparing the schizophrenic with his or her mother, and often with a sibling, the schizophrenic was *more* able to see the other's point of view than the other was to see his, or that of a third party. More fully stated (in relation to the mother) Laing's guesses were as follows:

"The schizophrenic sees the mother's point of view better than the mother sees the schizophrenic's. The schizophrenic realizes that the mother does not realize that he sees her point of view, and that she thinks she sees his point of view, and that she does not realize that she fails to do so. The mother on the other hand think she sees the schizophrenic's point of view, and that the schizophrenic fails to see hers, and is unaware that the schizophrenic knows that this is what she thinks, and that she is unaware he knows."

Thus we conclude that Case γ corresponds to the individual using schizophrenic defenses.

Case δ is the situation of Case γ (schizophrenia) worsening with elements of Case β being added (neurosis) leading to the most intolerable emotional conditions for the individual.

It is interesting to compare the neurotic individual (Case β) with the individual employing schizophrenic defenses (Case γ). The neurotic individual is less pleasant to be with locally since he is anxious and "nervous" (anxiety and nervousness are all various monotonic functions of the emotion e_t of the individual at time t) which disturbs and "pricks" others "a little bit all the time". Also, often the neurotic employs great activity and energy with little thought (e.g. getting up from the café every ten minutes to run to the mailbox while knowing the mailman almost never arrives for another hour). This stems from the high contact but lessened understanding. However, unless the emotion becomes very high leading to dangerous theatrics (e.g. having an auto accident so that your lost love will come see you at the hospital), the main drawbacks to neurosis are wasted energy, local unpleasantness to others due to "nervousness", the individual pain of anxiety, inability to carry out certain tasks, etc. However, since the contact is "stuck on high", a reasonable condition in life and the understanding which comes from just living so many days can often make things tolerable-to-OK. Also, the energy, though often misdirected, is greater and often in the long run some of it may be channeled toward an end.

On the other hand, schizophrenics are locally pleasant to be with (since their e_t is low). People say they are "very nice" after a first meeting. Their activities, though often not full with vigor, seem purposeful and well thought out. However, in the long term it seems to be very difficult to "contact" them. It is like peeling layers from the onion. One never reaches the center. They think over the smallest details infinitely (e.g., on being asked "Do you have the stamps?", the schizophrenic, before answering "I'm not sure", *thinks*: I was supposed to get stamps. I forgot. If I say this he will get mad. Also, I spent the money on a mystical book. He won't like this. Therefore I can't say 'No' or he will want the money. I can't say 'Yes' or he will want the stamps. Therefore I will say 'I'm not sure'. If he persists I can say I think they are in the purse I left with the babysitter because he forgot his keys. If I chide him about forgetting the keys he will defend himself and may forget about the stamps. Also I think there are some stamps in the drawer at home but I'm not sure. If I find them I can give these to him."). Lots of thought, little action. No contact emotionally.

However, schizophrenia is potentially very dangerous because of the following phenomenon. If a neurotic eight-year-old boy must walk through a dark forest at sunset to reach home he will fear wolves are chasing him and run home screaming and crying into his mother's arms and be filled with anxiety and fear (while the father says, "Don't coddle the boy so much"). However, when the schizophrenic eleven-year-old boy must walk through the dark forest at sunset to reach home he will *pretend* he is not in the forest. Under this pretense he will calmly walk through the forest and upon reaching the other side will call his name aloud to the sky five or six times so he can return (to himself?) and continue on.

In short, the strategy to handle a problem P is to *pretend* you are something which in fact you are not (e.g. pretending you are not in the forest, pretending you don't know that you didn't buy stamps, etc.) This *may* (but often will not) solve the immediate problem, but it definitely leads to the much bigger problem of what and who you are if you are often pretending, even to yourself, to be something you are not, i.e., it leads to the identity problem. In other words, the schizophrenic plays a very complex "pretense game", with less and less contact over the long run until finally the pretense game even confuses himself and he falls from Case γ (low contact-highly complex pretense game) into even lower contact and the pretense game becomes so personally confusing that inactivity sets in (Case γ, e.g., catatonic state).

For example, suppose a man joins a bank as a window-teller with the purpose of eventually robbing the bank and fleeing to South America. He tells himself he is just pretending to be honest. He knows he is really a crook but he is fooling everyone. Suppose he works hard so as to avoid suspicion. Also he bides his time so as to obtain the combination to the safe. Then he decides that working up to Assistant Manager is the best way to get the safe combination. He does this over three years, thinking every day, "They think I'm honest, but I know I'm one smart crook who is fooling them. I am just pretending to be honest." He also obtains a secret Swiss bank account, buys an inexpensive house in the hills of Bolivia under an assumed name, and plans the get-away route. Then suddenly, a week before he is to obtain the combination to the safe, the manager dies in a plane crash. He is made manager with a very good salary. At the same time he receives a bank award for his "honest and conscientious service". He becomes a little confused and puts the robbery off for a few months. He keeps asking himself, "Am I an honest banker or a clever crook?"

However, during this time a newer employee at the bank, unbeknownst to him, is embezzling funds in such a way as to cast doubt on him, the manager. He decides to put the robbery off another six months. The next day the bank brings charges of embezzling against him. He has an utter breakdown which is not relieved when the charges are dropped six weeks later. He is committed to an institution where he often maintains he is Jesse James in disguise.

The contact is somewhat limited in the beginning: the job, the thoughts and planning of the robbery. Nothing else — no wife, no worries, no hobbies, no spending. However the complexity is high relative to this limited contact. The job, plus the secret planning, plus always being careful to fool everyone that he is honest ("I must be careful to protect my honest image").

The sudden promotion does not increase contact too much but does confuse the complex "honest-but-really-a-crook game". Then after some deliberation he decides to "be a bank manager for a while" and "do the robbery later, if at all". "Was I really serious?" "Yes, I must eventually do it, but later." "Maybe I am honest, I don't know."

Thus the promotion increases the contact a little (travel, meetings, socializing with the other managers, etc.) but the complexity of the reaction goes up too ("Am I really a very smart crook or a hard working bank manager?") The unexpected charges of embezzling lower the contact to a hospital bed and utterly destroy the complex game.

In short, neurosis can lead to pain (e.g., anxiety), dangerous theatrics (e.g., faking suicide, misjudging, and dying) and waste, but schizophrenic can lead to the land of no return of playing more and more complex games subject to less and less contact into which an unpleasant intrusion of fact can snap the game and lead to Case δ (madness).

Summarizing this section thus far, we find that we have defined a Lagrangian of individual emotional development and consider the four extreme cases, Case α, β, γ and δ. We have seen that Case α corresponds to excellent mental health, Case β to neurosis, Case γ to schizophrenia, and Case δ to madness. Given this viewpoint, what deductions or suggestions for therapy or research on new viewpoints, etc., can be obtained? We will devote the remainder of this part to listing some of the results which can be deduced.

Improvements to the Interpersonal Perception Method (IPM) Test introduced in *Interpersonal Perception*, denoted [IP].

In the following we assume the reader is familiar with [IP]. First, let us hasten to add that the IPM (Interpersonal Perception Method) has some excellent merit but we would like to improve it by an addition. To show the necessity of this addition we will first reformulate the viewpoint espoused in IPM in a form closer to the viewpoint of the previous part of this section.

IPM studies "the experiences, perceptions and actions that occur when two human beings are engaged in a meaningful encounter." Let the two interacting individuals be Peter, denoted Pe, and Paul, denoted Pa. We next introduce some definitions. Let Peter's *behavior* at time t, $B_t(Pe)$, be a physically complete description of his actions (with no interpretation placed on these actions) at time t. For example, suppose Peter and Paul are talking on the telephone. Then $B_t(Pe)$, $t \in [a, b]$ is a recording of exactly what Peter said from time a to time b, including coughing, stammering, the words, the pauses, etc. In short, the behavior of Peter is the strictly materialistic description of Peter's actions, words and movements as could be recorded for example on a recording, camera, etc.

We next consider Paul's *experience* of Peter at time t, $E_t(Pa)$. Paul's experience of Peter is a function of the behavior of Peter and the *present state* of Paul $(q_t(Pa))$ or $E_t(Pa) = \lambda(q_t(Pa), B_t(Pe))$. The present state of Paul "entails all the constitutional and culturally conditioned learned structures of perception that contribute to the ways Paul construes his world." Quoting from [IP],

> "Experience in all cases entails the perception of the act *and* the interpretation of it. Within the issue of perception is the issue of selection and reception. From the many things that we see and hear of the other we select a few to remember. Acts highly significant to us may be trivial to others. We happen not to have been paying attention at that moment; we missed what to the other was his most significant gesture or statement. But, even if the acts selected for interpretation are the same, even if each individual perceives these acts as the same act, the interpretation of the identical act may be very different. She winks at him in friendly complicity, and he sees it as seductive. The act is the same, the interpretation and hence the experience of it disjunctive. She refuses to kiss him goodnight out of "self-respect", but he sees it as rejection of him, and so on.
>
> A child who is told by his mother to wear a sweater may resent her as coddling him, but to her it may seem to be simply a mark of natural concern."

Thus for us $q_t(Pa)$, the state of Paul at time t, will include the experience of Paul of Peter's behavior, the "constitutional and culturally conditioned learned structures of perception that contribute to the ways Paul construes his world." In short, the state of Paul at time t, denoted $q_t(Pa)$, is all the information necessary to determine (predict) Paul's behavior (at time $t + \epsilon$) given Peter's behavior at time t, $B_t(Pe)$. Thus assume time is suitably divided into convenient subintervals t, then

$$(6.99a) \qquad q_{t+\epsilon}(Pa) = \lambda_{Pa}(q_t(Pa), B_t(Pe))$$

or, in words, the state of Paul at time $t + \epsilon$ is determined by the state of Paul at time t and the behavior of Peter at time t. Further, the state of Paul at time $t + \epsilon$ includes the experiences of Paul at time $t + \epsilon$ (and in addition the "constitutional and culturally conditioned learned structures", etc.).

Finally, Paul's behavior at time $t + \epsilon$, $B_{t+\epsilon}(Pa)$ is a function of Paul's state of time t, $q_t(Pa)$, and Peter's behavior at time t, $B_t(Pe)$, or

$$(6.99b) \qquad B_{t+\epsilon}(Pa) = \delta_{Pa}(q_t(Pa), B_t(Pe)) \, .$$

Thus the dyadic relationship between Peter and Paul can be represented by the following two circuits Pe (for Peter) and Pa (for Paul) interacting where

$$(6.100a) \qquad Pe = (B(Pa), B(Pe), Q(Pe), \lambda_{Pe}, \delta_{Pe})$$

(6.100b) $Pa = (B(Pe), B(Pa), Q(Pa), \lambda_{Pa}, \delta_{Pa})$

where $B(Pa) = \{B_t(Pa)$ for all $t\}$ = the possible behaviors of Paul, $Q(Pe) = \{q_t(Pe)$ for all $t\}$ = the possible states of Peter, etc. Then λ_{Pa} is defined by (6.99a), δ_{Pa} is defined by (6.99b), etc.

In actual practice, at the present level of development, the behavior of Peter must be reduced to some simple parameter (e.g. "yes" or "no" to a question). Thus most of the behavior is ignored and only some small part which can easily be quantized is used.

The state of Paul at time t, $q_t(Pa)$, is nearly mystical since at the present state of knowledge we haven't the slightest idea what information might be needed (or even that it exists) to predict the behavior of Paul given the sequence of inputs into Paul. However in restricted situations with restricted behavior parameters, we can determine the input/output response of Paul (usually by having Paul answer questions on a written or oral test) and thus determine the circuit from the input/output data via Chapter 5. Of course this just gives the states (or information required) relative to the limited behavior parameters previously chosen.

To get back to Peter and Paul interacting, given Pe and Pa defined by (6.100) we can, by the precise definitions of Part II (see Definition (6.58) and take $M = Pe$, $G = Pa$, A = inputs of Pe and h the identity map) define the *result* R of the interaction of Peter and Paul.*

As before, let $\#(R)$ denote the states of R and let $\theta(R)$ denote the complexity of R. Then, for reasons which have been belabored many times before in this volume, it is reasonable to define $\#(R)$ as the *contact* existing between Peter and Paul, and to define $\theta(R)$ as the measure of understanding and realization of understanding, cohesion, awareness, etc. existing in the interaction of Peter and Paul. Otherwise said, $\#(R) - \theta(R)$ measures the *alienation* (measured in absolute units) between Peter and Paul in their interaction.

Now, the Lagrangian (read 'goal') of Peter and Paul in their interaction is to *minimize alienation and maximize contact*. In our opinion the IPM test in [IP] attempts to measure the alienation which exists between Peter and Paul (or husband and wife). For details of the test we refer the reader to [IP]. However, to give the flavor of the test, it consists of sixty questions (say for the husband to take), the first of which is

---

*Here the result of human interaction can be analyzed just as in Part II of Chapter 6 for the case of metabolism and genome interaction, as the author's mathematical abstraction as just formulated here now allows one to apply the same automata- and semigroup-theoretic techniques. -CLN

1.A. How true do you think the following are?

1. She understands me.
2. I understand her.
3. She understands herself.
4. I understand myself.

B. How would SHE answer the following?

1. "I understand him."
2. "He understands me."
3. "I understand myself."
4. "He understands himself."

C. How would SHE think you have answered the following?

1. She understands me.
2. I understand her.
3. She understands herself.
4. I understand myself.

At this time we cannot go into a detailed analysis and critique of the IPM or present a detailed examination of whether IPM does or does not successfully measure alienation. So let us just assume that IPM does measure alienation in a way as reasonable as any test devised thus far.

However, in our opinion, the weak part of [IP] is in the applications, specifically the application of the IPM to "disturbed and nondisturbed marriages". Roughly, chapter VI of [IP] argues that IPM measures more alienation in "disturbed marriages" than in "nondisturbed marriages". (Here "disturbed marriages" is operationally defined as those couples seeking help from psychologists, while "undisturbed marriages" is operationally defined as those couples satisfied with their marriages!)

The main defect in all the applications is in assuming that lack of alienation is the entire goal (or Lagrangian) in the relationship.   This is false. We want to *maximize contact and minimize alienation*. Thus minimizing alienation is one goal but maximizing contact is another equally important goal. Thus for applications IPM should be supplemented by an additional test which measures *contact* between the individuals (e.g. husband and wife).

To make the necessity of this addition clearer we give a graphic generic example. Suppose one couple, Jack and Jill, rate a little lower on the IMP than Charles and Prudence. On the other hand, suppose Jack and Jill are in a nude encounter group, Jack is an underground revolutionary who sabotages power installations, Jill is deep into Women's Liberation, and both are writing books: one on revolution and the other on sexism in society. Further, Jack and Jill have two children ages 2 and 4 whom they plan to educate themselves.

In contrast Charles and Prudence were childhood playmates, both from the most distinguished families of the town. Charles' father is a lawyer and it had been planned from his birth that he would take over his father's law firm. Prudence's father is a doctor and she, after going to the same eastern college as her mother, married Charles. They have two children, 2 and 4. They live in their home town and live lives very similar to those of their parents 30 years before.

Now Charles and Prudence have little alienation but they also do not have the contact that Jack and Jill have. Jack and Jill often have quiet evenings at home but sometimes all hell breaks loose. When a heavy criminal charge comes down on Jack, Jack and Jill have intense arguments about why it happened, or when Jill turns onto another guy in the nude encounter group, etc.

Now if Jack and Jill could develop so that their level of alienation equaled that of Charles and Prudence then, since Jack and Jill had higher contact (and equal alienation) we should clinically judge Jack and Jill's to be the "better" relationship.

In devising a test to measure the contact between a couple, the organizing idea is to measure (Jill's contact with the environment including Jack) + (Jack's contact with the environment including Jill) - (Jill's contact she does not communicate to Jack) - (Jack's contact he does not communicate to Jill). Thus questions like, "Do you flirt with other women?", "If you flirt with other women do you tell your wife?", "If an attractive woman tries to seduce you by asking you to sleep with her 10 minutes after meeting her, would you?", "Would you tell your wife if you did?", "If you didn't?" In other direction, "Would you torture people as part of your job as a member of the C.I.A.?", "If you tortured someone for the C.I.A. who dies, would you tell your wife?". Finally, "If you were talking to your buddies while slightly drunk in a bar about your wife and someone recorded the conversation unbeknownst to you, would you allow (if asked) the tape to be played for your wife?"

Thus for each couple we should obtain not only the IMP score but the "contact" score also. Now for couple 1, scores $(IMP)_1$ and $(C)_1$ are obtained, while for couple 2, $(IMP)_2$ and $(C)_2$ are obtained. Then if

$$(IMP)_1 \leq (IMP)_2 \quad \text{and} \quad (C)_1 \leq (C)_2$$

hold, couple 2 is judged "clinically better". In other cases what judgment is to be rendered? Usually couples with less contact will be able to concentrate on this smaller amount of mutual activity and break down alienation. On the other hand, high contact leads to so much happening (acid trips, bombings, jail terms, Ph.D.'s, books being published, new homosexual experiences) that often it cannot be mutually assimilated (e.g., "if you're going to fuck around and bomb things, I'm leaving!") leading to alienation or rupture of the relationship.

We note the very basis in [IP] of operationally defining "disturbed marriages" as those seeking professional advice is using a parameter of *contact to define disturbed!* Clearly seeking advice increases the contact in a relationship.

Now, just as in (6.97) we can distinguish four extreme types of relationships between individuals

		contact	alienation
	case α	high	low
(6.101)	**case** β	high	high
	case γ	low	low
	case δ	low	high

Case α corresponds to a very good relationship, case β corresponds to a high contact with alienation usually produced by neurotic individuals, case γ corresponds to a low contact but unalienated relationship often coming from a "good" relationship between individuals having many schizophrenic defenses, while case δ is total failure as a human relationship.

The modern university (e.g. Berkeley) often produces relationships between students like case β since there is high contact (cramming for courses together, political actions, sex) but often high alienation in the "multiversity". Thus the modern university often produces neurotic relationships. On the other hand, conservative businesses (e.g. IBM) often produce relationships between employees like case γ. The contact is kept low (work 9–5 Monday through Friday in suit and tie, live in selected suburbia, etc.) but subject to this small contact there is little alienation.

In general, society favors unalienated relationships, but when they become alienated *society* (almost) *always wants the alienation to be diminished by diminishing contact* (i.e. by repression of experience). One of Freud's great societal contributions is to be a social force for release of alienation without repressing contact or experience. In fact, the reason why a "contact test" has not been implemented before is that by simply taking such a test the individual is increasing his or her contact, or will tend to increase the contact in the near future (e.g. asking one if one is in a nude encounter group makes the probability that one will explore the possibilities of this activity much higher).

In short, society (up to now) favors low contact, low alienation (case γ of (6.101)), but the individual desires high contact and low alienation (case α of (6.101)) while neurotic relationships fall between these two (case β of (6.101)).

Given two scores of $(IMP)_1$, $(C)_1$ and $(IMP)_2$, $(C)_2$ for couples 1 and 2, we could define $(c)_1 = (C)_1 - (IMP)_1$ and $(c)_2 = (C)_2 - (IMP)_2$, and then compare couples 1 and 2 by computing $h((C)_1, (c)_1)$ and $h((C)_2, (c)_2)$ where h is a culture function as was considered in Part III (see (6.86–6.88)).

However, probably the best thing is merely to report the two parameters and suggest where these scores are best received (i.e. (30, 25) is well received at IBM in upstate New York, while (100, 60) is well received in the student culture of Berkeley). Also the couple would know where they were strong and where they were weak with respect to the Lagrangian.

In summary, "normal marriage" or "nondisturbed marriage" does not mean anything. The purpose of relationships is to maximize contact and minimize alienation. Thus the tests should rate both contact and alienation. If a couple is "satisfied" with their marriage, but contact is low they should strive to up the contact. If the contact is high and the alienation is high, they should strive to lower the alienation. "Normal", "nondisturbed", etc. are stupid concepts; increasing unalienated contact is the goal. Of course when the contact is L and the alienation is very low (zero) then the Lagrangian has value L (independent of any culture function) and the relationship has L units of love (measured in absolute units). The goal is to maximize love, the hell with normality, restrictive experience or alienation. The Lagrangian of personal relations is love.*

6.102. Suicide. An old idea that goes back to Freud and further is that emotional health for the individual depends on having a large number of emotional ties to others and with the environment of various kinds and intensities. The idea being that if one has few ties or one major tie, when

---

*Note the author has tacitly formalized a measure for *love* as amount of understanding of the contact, rigorously defined as the Krohn-Rhodes complexity of the interaction (corresponding to understanding and ability to act and interact). It is maximized when contact is maximal and at the same time this complexity matches the amount of contact (which bounds its maximal possible value). -CLN

this is strengthened or weakened the individual oscillates from heights of ecstasy to depths of depression. In contrast if one has a deep relationship with a woman, several close friends, creative activity in writing, a passion for skiing, etc., then when some of the ties break or are strengthened the remaining ties hold relatively stable (e.g. if one breaks his leg skiing, he can plunge into writing books).

From this it can be argued that suicide often occurs when individuals have just a few major emotional ties and these, by circumstances, are broken or shattered (by death, economic factors, divorce, etc.).

Now "emotional ties" (by proceeding as a physicist) can be related to the previous concepts of contact C and emotion or lack of emotional stability e by noting that "emotional ties" are directly proportional to the contact C_t of the individual (at time t) and inversely proportional to the lack of emotional stability e_t of the individual (at time t). If one doubles his contact, leaving his emotional stability intact, his emotional ties double. If his contact remains constant but his emotional stability halves (e.g. by developing neurotic behavior and avoiding all bald-headed men, which includes his best friend) then his emotional ties half, etc. Thus

$$(6.103) \qquad (ET)_t = \frac{k_1\, C_t}{k_2\, e_t}$$

where $(ET)_t$ = the emotional ties at time t, C_t is the contact at time t, and e_t is the lack of emotional stability at time t, and k_1 and k_2 are constants. Now the probability of suicide is inversely proportional to the emotional ties, or,

$$(6.104) \qquad (S)_t = \frac{1}{(ET)_t} = \frac{k_2\, e_t}{k_1\, C_t}$$

where $(S)_t$ is the (unnormalized) probability of committing suicide.

We suggest a study to attempt to determine $(S)_t$ for those who have committed suicide and for those who have not to see if the above analysis in fact holds. By using IPM and the "contact" test we could determine C_t and e_t, respectively.

6.105. Personal evolution (mental ontogeny) and mental illness.

An individual is born into a certain social setting, in our western culture, usually into a family. The individual adapts to the surroundings as a baby and young child to survive first physically then mentally. If your mother is getting a Ph.D. and leaves you with babysitters a lot you respond differently than if your mother beats you because she is unhappy or if your mother "loves" you so much you are never apart from her all your waking life for 10 years.

By puberty you have evolved to become a marvelous individual able to survive in this specialized social setting, which often has numerous unlivable and unbearable elements which you as a child could not change. Then suddenly you are thrust into a new world (e.g. run away from home, go to college, get married) where the situation is entirely different. Often some of your responses should be changed to the opposite of what they were a few years before. For example, if you had a very domineering mother you spent your childhood resisting her as far as possible, but now you are married and your wife is pregnant. You should allow her to depend on you but you go on resisting her dependence (which was appropriate with your mother years ago but is not appropriate now).

The tragedy is that often the human emotional response changes too slowly. As mature adults we respond as adolescents, as older adults we respond as young adults, etc. Most mental illness is based on responding to the current situation in a manner appropriate to the (distant) past. This lowers both the contact, the understanding, and the ability to act.

Every individual is reborn anew every day. The Lagrangian of life says live and understand. Don't be chained by the past. Don't slavishly follow the ways of yesterday. The Lagrangian of life gives the purpose of life as living and loving. Experience, understand, act and so *love*.

Bibliography

[Freud00] S. Freud. *The Interpretation of Dreams*. Allen & Unwin, London, 1955 [1900]. [New edition published by Kessinger Publishing 2004].

[Freud01] S. Freud. *The Psychopathology of Everyday Life* (The Standard Edition of the Complete Psychological Works of Sigmund Freud, volume VI). Hogarth Press, London, 1960 [1901]. [New edition published by Vintage Books, 2001].

[Freud05a] S. Freud. *Three Essays on the Theory of Sexuality*. Hogarth Press, London, 1962 [1905]. [Revised edition published by Basic Books, 2000].

[Freud05b] S. Freud. *Jokes and Their Relation to the Unconscious*. Routledge and Kegan Paul, London, 1960 [1905]. [New edition published by Vintage Books, 2001].

[Freud09a] S. Freud. Analysis of a phobia in a five-year old boy. In *The Standard Edition of the Complete Psychological Works of Sigmund Freud*, volume X.

Hogarth Press, London, 1955 [1909 + 1922]. Republished in *Two Case Histories* by Vintage Books, 2001. (Previously two papers, printed together. Also in *The Sexual Enlightenment of Children*, Collier Books, New York, New York, 1963).

[Freud09b] S. Freud. *Two Case Histories* (Standard Edition, volume X (1909)). Hogarth Press, London, 1955 [1909]. pages 259–260 ("The Case of the Rat Man"). [New edition published by Vintage Books, 2001].

[Freud18] S. Freud. *An Infantile Neurosis and Other Works* (Standard Edition, volume XVII (1917–19)). Hogarth Press, London, 1955 [1918]. page 99 ("The Case of the Wolf Man"). [New edition published by Vintage Books, 2001].

[Freud23] S. Freud. *The Ego and the Id.* Hogarth Press, London, 1962 [1923]. [New edition published by Vintage Books, 2001].

[Divided Self] R. D. Laing. *The Divided Self: An Existential Study in Sanity and Madness.* Penguin Books, Harmondsworth, 1960. [Reprint edition published by Routledge, 1998].

[Self and Others] R. D. Laing. *The Self and Others.* Tavistock, London, 1961. [Reprint edition published by Routledge, 1998].

[Sanity, Madness and the Family] R. D. Laing and A. Esterton. *Sanity, Madness and the Family.* Tavistock, London, 1964. [Reprint edition published by Routledge, 1998].

[IP] R. D. Laing, H. Phillipson, and A. R. Lee. *Interpersonal Perception.* Tavistock, London, 1966. [Reprint edition published by Routledge, 1998].

Part IV. Complexity of Games

> *One does not have to be Levi-Strauss to realize that games are conflict
> situations abstracted from reality in which we can play, experience and
> learn without paying the dreadful penalties demanded by loss and defeat
> in the Great Games of Life.* — John Rhodes

We assume the reader is familiar with the rudiments of von Neumann's
Theory of Games [Games]. For simplicity of exposition we restrict ourselves
here to two-person games with perfect information and no chance moves
(e.g. Chess, Checkers, Go). Our theory can be extended to more general
games and we briefly consider these extensions at the end of this Part.

The *idea* of how to define the complexity of a game is very simple.
Let μ be the von Neumann value of the game G (μ is the amount you can
expect to win if your opponent plays perfectly and then you play in the best
manner under the assumption that your opponent always plays perfectly).
Let $S_1, S_2, \ldots$ be the set of (pure) strategies which achieve the value μ.
Let $M_1, M_2, \ldots$ be reduced finite state machines such that M_k "plays the
pure strategy" S_k. Then, by definition, $\theta(G)$, *the complexity of the game*
G, equals

$$\min\{\theta(M_j) : j = 1, 2, \ldots\}.$$

In short, the complexity of a game G is the minimum complexity of any
finite state machine which plays (a pure strategy of) G at the von Neumann
value of G. By convention the minimum of the empty set is $+\infty$.

In almost all cases of interest every pure strategy can be realized by a
unique finite state sequential machine. Thus in almost all cases of interest
the complexity is a non-negative integer. Again, *the complexity of a game
G is the minimum complexity of any finite state machine which can play
the game G perfectly* (in the sense of von Neumann).

We now sketch a (fairly) precise definition of "a machine which can play
the game" and later illustrate the concepts with specific well known games.
We assume for the purposes of the following that G is a board game for
two people with a finite (but perhaps large) number of bound positions,
e.g. Chess or Checkers.

6.106. The machine C_{q_0} *can play the game* G *with strategy* S if $C =
(A, B, Q, \lambda, \delta)$ is a circuit (see Chapter 5), so $\lambda : Q \times A \rightarrow Q$ and $\delta :
Q \times A \rightarrow B$ with the following conditions.

(a) A is the set of plays which the opponent of C can make. For example
in Chess, A includes PK4, BQ4, PQB5 check, PK5 checkmate, etc.

(b) B is the set of plays which C can make.

(c) Q is the set of board positions for G as they could appear to C's opponent.

(d) $\lambda(q, a)$ and $\delta(q, a)$ are determined as follows: if q is a board position which can appear to the opponent of C, and a is a move which can be made by the opponent of C, let $q \cdot a$ be the board position after the play a is made on the board position q. Let $S(q \cdot a) = b$ be the move for C given by strategy S when the board position presented to C is $q \cdot a$. Define $\delta(q, a) = b = S(q \cdot a)$ and define $\lambda(q, a) = (q \cdot a) \cdot b = (q \cdot a) \cdot S(q \cdot a).$*

(e) Let q_0 be the initial board position of the game, assuming that C goes second.

(f) Replace $C = (A, B, Q, \lambda, \delta)$ as defined by (a)–(d) by its reduction C^* (see Chapter 5) and replace q_0 by q_0^* such that $C_{q_0} = C_{q_0^*}^*$.

We illustrate the above with the following very simple example. Consider the following game tree given in Figure 22.

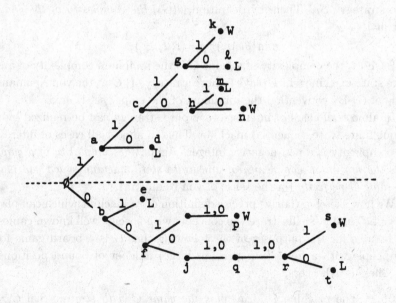

Figure 22.   A game of complexity two.

---

*In other words, δ gives the response move b to make given that C's opponent just made move a in state q, while λ gives the new state after C's response is played.   -CLN

In this game each player plays zero or one successively and the opponent of C goes first initially, then the loser goes first in future plays. A play of the game has at most six turns (split 3 and 3). W denotes a win for the player entering a dot marked W, similarly L denotes a loss for the player entering a dot marked L. The payoff is $+1$ for W and -1 for L. Clearly the von Neumann value for this game is $+1$ for the player who goes second.

Assuming C goes second, the strategies achieving the von Neumann value of $+1$ are listed in Figure 23.

Position	S
$\emptyset$	arbitrary
a	1
b	0
c	arbitrary
d	arbitrary
e	arbitrary
f	arbitrary
g	1
h	0
i	arbitrary
j	arbitrary
k	arbitrary
ℓ	arbitrary
m	arbitrary
n	arbitrary
p	arbitrary
q	arbitrary
r	1
s	arbitrary
t	arbitrary

Figure 23. Strategy S of second player giving appropriate move in each position to achieve von Neumann value of the game in Figure 22.

Now let $C = (A, B, Q, \lambda, \delta)$ with $\lambda : Q \times A \to Q$ and $\delta : Q \times A \to B$ be defined as follows:

$$A = \{0, 1, (0, W), (0, L)\}$$
$$B = A$$
$$Q = \{c, f, k, n, p, q, s, \emptyset\}$$

	$\lambda(\cdot, 0)$	$\lambda(\cdot, 1)$	$\delta(\cdot, 0)$	$\delta(\cdot, 1)$
c	n	k	(0,W)	(1,W)
f	q	p	0 or 1	(1,W) or (0,W)
k	f	c	0	1
n	f	c	0	1
p	f	c	0	1
q	s	s	(1,W)	(1,W)
s	f	c	0	1
$\emptyset$	f	c	0	1

and $\lambda(x, (0, W)) = \lambda(x, (0, L)) = \lambda(x, 0)$, etc. Here an entry 0 or 1 means one of 0 or 1 is chosen in an arbitrary but fixed way, etc.

C is not reduced since clearly k, n, p, s and $\emptyset$ may be identified, yielding

C^*	$\lambda^*(\cdot, 0)$	$\lambda^*(\cdot, 1)$	$\delta^*(\cdot, 0)$	$\delta^*(\cdot, 1)$
c	$\emptyset'$	$\emptyset'$	(0,W)	(1,W)
f	q	$\emptyset'$	0 or 1	(1,W) or (0,W)
$\emptyset'$	f	c	0	1
q	$\emptyset'$	$\emptyset'$	(1,W)	(1,W)

Thus $C^* = (A, B, \{c, f, \emptyset', q\}, \lambda^*, \delta^*)$ is a reduced circuit with $C_{\emptyset'}^*$ playing at value $+1$ (i.e. always winning if it initially plays second).

Now the complexity of C^* (or $C_{\emptyset'}^*$) is clearly independent of the choice of $\delta^*(f, 0)$ and $\delta^*(f, 1)$ and is in fact (see Chapter 5, Part II) the complexity of the subsemigroup S of $F_R(\{c, f, \emptyset', q\})$ generated by

$$x = \begin{array}{c} c \to \emptyset' \\ f \to q \\ \emptyset' \to f \\ q \to \emptyset' \end{array} \quad \text{and} \quad \begin{array}{c} c \to \emptyset' \\ f \to \emptyset' \\ \emptyset' \to c \\ q \to \emptyset' \end{array} = y \ .$$

We next show S has complexity 2 *so the game of Figure 22 has complexity 2.*

The upper bound for the complexity follows from the following elementary

Proposition 6.107. *Let S be a semigroup of mappings on the finite set X. Let r be the maximum range (or fixed points) of any idempotent $e = e^2 \in S$. Then $\#_G(S) \le r - 1$.*

Proof. Let I be the ideal generated by the idempotents of S. Then S/I is combinatorial and $I \le \{f : X \to X : |f(X)| \le r\} = I_R$. Further, I_k, $k = 1, \ldots, n$ are the ideals of $F_R(X)$, the semigroup of all maps of X into X. Then using [2-J], Section 2, we can show $\#_G(S) = \#_G(I)$ and $\#_G(I_r) \le r - 1$. Thus $\#_G(S) = \#_G(I) \le \#_G(I_r) \le r - 1$. This proves (6.107). $\qquad\square$

If $f : X \to X$, let $\mathrm{rank}(f) = |f(X)|$. Clearly, $\mathrm{rank}(fg) \le \min\{\mathrm{rank}(f), \mathrm{rank}(g)\}$ and $\mathrm{rank}(e^2 = e) =$ the number of fixed points of e. Since in our example S is generated by x, y and $\mathrm{rank}(x) = 3$, $\mathrm{rank}(y) = 2$, we have $r \le 3$ for S and thus $\#_G(S) \le 3 - 1 = 2$ by (6.107).

To show $\#_G(S) \ge 2$ we use Theorem (4.11a). Let $z = yx$. Then replacing q by 1, $\emptyset'$ by 2, and f by 3, yields

$$x = \begin{matrix} 1 \to 2 \\ 2 \to 3 \\ 3 \to 1 \\ c \to 2 \end{matrix} \qquad y = \begin{matrix} 1 \to 2 \\ 2 \to c \\ 3 \to 2 \\ c \to 2 \end{matrix} \qquad z = yx = \begin{matrix} 1 \to 3 \\ 2 \to 2 \\ 3 \to 3 \\ c \to 3 \end{matrix} .$$

Let α be x restricted to $\{1, 2, 3\}$ and let β be z restricted to $\{1, 2, 3\}$ so

$$\alpha = \begin{matrix} 1 \to 2 \\ 2 \to 3 \\ 3 \to 1 \end{matrix} \qquad \beta = \begin{matrix} 1 \to 3 \\ 2 \to 2 \\ 3 \to 3 \end{matrix} .$$

Then T, the subsemigroup of $F_R(\{1, 2, 3\})$ generated by α and β, divides S. Thus $\#_G(T) \le \#_G(S)$, so it suffices to show $\#_G(T) \ge 2$.

Using the notation of (4.11) we let $e_1 = \alpha^3$ and $G_1 = \{\alpha, \alpha^2, \alpha^3 = e_1\}$ and we let $e_2 = (\beta\alpha^2)^2$ and $G_2 = \{\beta\alpha^2, (\beta\alpha^2)^2 = e_2\}$. Thus

$$\beta\alpha^2 = \begin{matrix} 1 \to 2 \\ 2 \to 1 \\ 3 \to 2 \end{matrix} \qquad (\beta\alpha^2)^2 = \begin{matrix} 1 \to 1 \\ 2 \to 2 \\ 3 \to 1 \end{matrix} .$$

Thus $X_1 \equiv X(G_1) = \{1, 2, 3\}$ and $X_2 \equiv X(G_2) = \{1, 2\}$. Now $(\alpha\beta\alpha)^2$, $(\alpha^2\beta)^2$ and $(\beta\alpha^2)^2$ are all idempotents since

$$(\alpha\beta\alpha)^2 = \begin{matrix} 1 \to 1 \\ 2 \to 3 \\ 3 \to 3 \end{matrix} \qquad (\alpha^2\beta)^2 = \begin{matrix} 1 \to 2 \\ 2 \to 2 \\ 3 \to 3 \end{matrix} \qquad (\beta\alpha^2)^2 = \begin{matrix} 1 \to 1 \\ 2 \to 2 \\ 3 \to 1 \end{matrix} .$$

But if $(\beta\alpha^2)^2(\alpha\beta\alpha)^2(\alpha^2\beta)^2(\beta\alpha^2)^2 = \gamma$ then

$$\gamma = \begin{array}{c} 1 \to 2 \\ 2 \to 1 \\ 3 \to 2 \end{array}$$

so $\gamma = \beta\alpha^2$.

Further, $\text{Range}((\alpha\beta\alpha)^2) = \{1,3\}$, $\text{Range}((\alpha^2\beta)^2) = \{2,3\}$, and $\text{Range}((\beta\alpha^2)^2) = \{1,2\}$. Thus all these ranges lie in $\{\text{Range}(e_2) \cdot \alpha^j : j = 1,2,3\} = \{\{1,2\} \cdot \alpha^j : j = 1,2,3\} = \{\{1,2\},\{2,3\},\{3,1\}\}$. Thus $X_1 \supsetneq X_2$ is an essential dependency and so $2 \leq \#_G(T)$ by Theorem (4.11a). Thus the game of Figure 22 has complexity two.

Roughly, *the complexity of a game G measures the amount the loops intermesh in the best program which plays G*. For example, the program which computes the best strategy of the game of Figure 22 is the following:

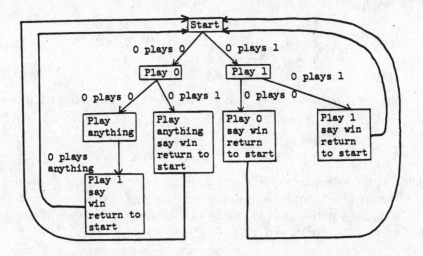

Figure 24. A best program for the game of Figure 22. Labeled arrows indicate what the opponent O plays leading to boxes indicating correct response move for optimal play.

Now a loop with three opponent plays intermeshes with a loop with two opponent plays, yielding a program of complexity two.

Do games of every complexity, $0,1,2,3,\ldots$, exist? The answer is yes by the following trivial argument. Let f be a machine of complexity n with inputs and outputs zero and one. Consider the game where O_1 plays either zero or one each turn and O_2 must play $f(a_1, \ldots, a_k) \in \{0, 1\}$ (if $a_1, \ldots, a_k$ have been the first k moves of O_1) to tie and otherwise O_2 loses. Thus the game has value 0 but goes on infinitely. Since the value zero consists of computing $f(a_1, \ldots, a_n)$ the complexity of the game must equal the complexity of f, namely n.

Another less trivial method of constructing games of arbitrary large complexity is by constructing *wreath (or cascade) products of games*. Let $G_1, G_2, \ldots, G_n$ be two-person board games. O_1 makes a play in $G_1, G_2, \ldots$ and G_n and then O_2 makes a play in $G_1, G_2, \ldots, G_n$, etc. However the payoffs, exact rules, etc. of the game G_i depend on *the board positions of games* $G_1, \ldots, G_{i-1}$. Since wreath products of games can easily be constructed so that the optimal strategy is given by a circuit which is the wreath product of the optimal strategies of the games G_i, we can construct games of arbitrary large complexity by using (3.16(i)).

We could also generalize the ideas used previously for Figure 22 and construct a game tree with complexity n but we suppress the details here as they become a bit involved. However, as the $n \times n$ board size tends to ∞ the complexity of $n \times n$ Go (Wei-Ch'i) tends to ∞ (see (6.119)). However these details are far beyond the scope of section.

Any game for which an optimal strategy depends only on the opponent's last move is clearly a game of complexity zero. In fact any such strategy can be realized by a one-state machine. For example consider the game of placing coins (of the same size) flat on a round table, with the loser being the first not to be able to place a coin. The first player can always win by placing his first coin in the center of the table and thereafter playing symmetrically to his opponent. Thus this game has complexity zero.

We next state an important theorem giving an upper bound for the complexity of some games.

Theorem 6.108. *(a) Let G be a two-person board game played on an $n \times m$ board with rules of the form that O_1 plays a piece in a square, then O_2 plays a piece in an unoccupied square, then O_1 plays a piece in an unoccupied square, etc., and the game stops at least by the time the board is filled. Also assume O_1 always goes first, independent of who won the preceding game (e.g. Hex with O_1 always going first). Then the complexity of G is less than or equal to half the area of the board, or*

$$\theta(G) \leq \frac{nm}{2}$$

(b) Let G be a game which reverts to the initial position independent of the past plays with less than or equal to n moves for each player. Then the complexity of G is less than or equal to $n - 1$, or

$$\theta(G) \leq n - 1 .$$

Proof. Clearly (a) follows from (b) since the game of (a) has $\leq \frac{mn}{2} + 1$ moves per player, so it suffices to prove (b).

To prove (b), let $M = (A, B, Q, \lambda, \delta)$ be a reduced circuit such that for the initial state q_0, M_{q_0} yields a strategy for G achieving its von Neumann value. Then given $a_1, \ldots, a_n$ with $a_j \in A$ and given $q \in Q$, one of $q \cdot a_1, q \cdot a_1 a_2, \ldots, q \cdot a_1 \ldots a_n$ *must be* q_0. This follows since each $q \in Q$ can be written as $q \cdot a'_1 \ldots a'_m$ for suitable $a'_1, \ldots, a'_m \in A$ (because M is reduced) and by the hypothesis of (b) the game returns to initial position in $\leq n$ moves.

Thus let $x_1, \ldots, x_m = t \in A^+$ represent an idempotent of the semigroup $S = M^S$ of M. Thus

(6.109) $\qquad\qquad q' \cdot t = q' \cdot tt \quad$ for all $\quad q' \in Q$.

Let q belong to the range of $\cdot t$ or, equivalently

(6.110) $\qquad\qquad\qquad\qquad q \cdot t = q$.

Then (6.109) and (6.110) imply

(6.111) $\qquad\qquad\qquad q \cdot t^j = q \quad$ all $\quad j \geq 1$.

Choose j so that $|t^j| = mj > n$. Set $\beta_0 = t^j$ so $q \cdot \beta_0 = q$. Then if $\beta_0 = y_1 y_2 \ldots$ and $|\beta_0| > n$, q_0 must appear among $q \cdot y_1, q \cdot y_1 y_2, \ldots$ so we can write $\beta_0 = \alpha_1 \beta_1$ with $0 < |\alpha_1| \leq n$, $0 < |\beta_1| < |\beta_0|$ and

$$q \cdot \alpha_1 = q_0 \quad \text{so} \quad q_0 \cdot \beta_1 = q .$$

Now if $|\beta_1| > n$ we can continue writing $\beta_1 = \alpha_2 \beta_2$ with $q_0 \cdot \alpha_2 = q_0$, $q_0 \cdot \beta_2 = q$ until we obtain

$$q_0 \cdot \beta = q$$

(6.112) $\qquad\qquad\qquad\qquad \alpha\beta = t^j$

$$|\beta| \leq n .$$

From (6.112) we find $\beta = x_a x_{a+1} \ldots x_m$ and $|m - a| + 1 \leq n$ and $q = q_0 \cdot x_a x_{a+1} \ldots x_m$. Thus the range of the idempotent $\cdot t$ is a subset of $\{q_0 \cdot x_a \ldots x_m : t = x_1 \ldots x_m, |x_a \ldots x_m| = |m - a| + 1 \leq n\}$. Thus the range of the idempotent $\cdot t$ contains $\leq n$ elements and (b) follows from Proposition (6.107). $\qquad\qquad\qquad\qquad\qquad\qquad\qquad\qquad\square$

6.113. Remarks on the Statement, Proof and Applications of Theorem (6.108). (a) The most important thing to observe is that the upper bounds not *not* depend on the total possible number of (board) positions, but instead depend on the length of the game. The hypothesis of Theorem (6.108) can be varied to cover more situations and similar upper bounds for the complexity to be derived. Also, in an honest sense, the conclusions of Theorem (6.108) are not the best possible. However we have presented Theorem (6.108) here to demonstrate methods by which upper bounds to the complexity of games can be obtained. Of course by applying Theorem (3.24) instead of Proposition (6.107) improvement in the upper bounds can be made.

(b) For some games Theorem (6.108) gives useful upper bounds. For example, 7×7 *Hex with* O_1 *always going first* (and hence always winning, by the theorem of John Nash (see [Gardner], chapter 8)) *has complexity* ≤ 24. Similarly, Tic-Tac-Toe has complexity ≤ 4 but we will see later Tic-Tac-Toe has complexity 1.

Go is played on a 19×19 board with 361 intersections and 180 white and 181 black stones. (See [Go]). However, for various reasons (e.g. "Ko" or Repetitious Situation) Theorem (6.108) cannot be applied directly to yield the upper bound 181. But by using the basic idea of the proof of Theorem (6.108) one can obtain that *the complexity of Go* ≤ 200.

6.114. Important Comments on the Proper Modeling of Games and Strategies. (a) We first give an illustration of how *not* to model a game, say Chess. There is an astronomically large but finite number of possible distinct board positions m. Since, by the rules, the play of the game is over when the same board positions occurs three times, no play of the game could be longer than $2m + 1$ moves long. Thus (in our stupid model) we might demand that each play of the game is $2m + 1$ moves long by having the symbol x played p times after the normal end of a play of the game of r moves, until $r + p = 2m + 1$. Then all plays of the game have a uniform length. Then any strategy S of the game can be realized by the finite state machine which simply stores the last k moves where $k \equiv p$ $\mod(2m + 1)$ and m is the total length of the input string to the machine. Thus the k moves stored are simply the k moves of the present play of the game. Then the output function is the map of the k moves (which together with the strategy S determine the present board position B) into the output which S demands from B.

Since the machine must store a large number of sequences of board positions of length $\leq 2m + 1$ the number of states is very very large, roughly exponentially greater than $2m+1$, which is itself astronomically large. However, the machine itself has complexity one, being the parallel combination of a $2m+1$ counter and delay machines of delay $1, 2, \ldots, 2m+1$ respectively. The reason, of course, for this low complexity is that all the "complex computation" is being done in the output function which is possible because of the stupid convention by which we made all plays of the game the same length. (The Winograd complexity of this output function will be high.)

One of the reasons for presenting this stupid model is to point out to the reader the importance of choosing natural and correct models for standard games so as to obtain the complexities of these games.

The model for Chess was stupid for several reasons. First, it did not take into account in a realistic way the fact that plays of Chess end in ties, wins or losses at various number of moves from the beginning play of the game and that upon termination of a play in the game by tie, win or loss the machine must immediately be ready to play again. Also no consideration was given as to what to do in the case of an opponent making an illegal move. Finally, an absurdly large number of states were required.

(b) Now let us lay down principles by which we may obtain reasonable models for games. One does not have to be Levi-Strauss to realize that games are conflict situations abstracted from reality in which we can play, experience and learn without paying the dreadful penalties demanded by loss and defeat in the Great Games of Life. Thus games are already (essentially mathematically) abstracted problems (of conflict between individuals, individuals and environment, etc.). For board games (say) the real-life concept of conflict is translated into mathematics as the 5-tuple $(A_1, B, A_2, \lambda_1, \lambda_2)$ where A_1 are the possible moves of O_1, A_2 are the possible moves of O_2, B is the set of all possible board positions, and $X_1 \subseteq B \times A_1$ consists of all those $(b, a_1) \in B \times A_1$ such that a_1 is a legal move by O_1 when the board is in position b and $\lambda_1(b, a_1)$ is the resulting board position obtained by playing a_1 on b. Thus $\lambda_1 : B \times A_1 \supseteq X_1 \to B$. Also λ_2 is defined in a similar way. Thus a game gives rise to a "set of game positions B with two sets A_1 and A_2 acting as partially defined functions on B." (This is a standard idea in artificial intelligence. For a reference see (the rather limited book) [Banerji].)

Thus the first principle in modeling games is to remember they are abstracted conflict situations and model in accordance to this. For example, our convention of playing x many times at the end of short Chess games is stupid for in real-life situations no such x appears. Also in Chess tournaments no such x occurs. Furthermore it is essential that the machine be in position at the end of a play of the game to immediately commence the next play of the game.

Also the problem of an illegal move by an opponent must be carefully considered. One convention is that any illegal move is considered a loss for the errant player. Another convention is that the illegal move is so named by the other opponent and another move is made. However there must be some bound on the number of errors permitted per play of the game without incurring a loss. Thus, up to such considerations these two conventions (almost always) lead to the same definition of complexity.

Also it is very important that the machines can in fact play the game legally against all opponents (both those who can play legally and those who cannot).

6.115. An Example of Modeling a Two-Person Board Game.

The game gives rise to the 5-tuple $(A_1, B, A_2, \lambda_1, \lambda_2)$ defined in (6.114b). Suppose α is a strategy of O_2. Then the circuit realizing α, $C(\alpha) = (A_2, A_1, B, \lambda, \delta)$ with $\lambda : B \times A_2 \to B$ and $\delta : B \times A_2 \to A_1$ is defined as follows. Using the notation of (6.114b) for $(b, a_2) \in X_2$ (i.e. when $\lambda_2(b, a_2)$ is defined or equivalently when a_2 is a legal move for the board position b), let

$$(6.116) \qquad \lambda(b, a_2) = \lambda_1(\lambda_2(b, a_2), \alpha(\lambda_2(b, a_2)))$$
$$\delta(b, a_2) = \alpha(\lambda_2(b, a_2)).$$

Notice, by definition of a strategy, if $(b, a_2) \in X_2$, then $(\lambda_2(b, a_2), \alpha(\lambda_2(b, a_2))) \in X_1$.

If $(b, a_2) \notin X_2$ (i.e. when $\lambda_2(b, a_2)$ is not defined or when a_2 is an illegal move for the board position b, let $\lambda(b, a_2)$ and $\delta(b, a_2)$ be defined by the convention of the game obtained when errors are committed. For example, $\lambda(b, a_2) = q_0$, the initial board position, and $\delta(b, a_2) = E$ (where E stands for error).

Now $(M(\alpha)^*)^S$, the semigroup of the reduction of $M(\alpha)$ (with starting state q_0), is, by definition, *the semigroup of the strategy of* α, denoted $S(\alpha)$. Thus $\theta(G) = \min\{\#_G(S(\alpha)) : \alpha$ achieves von Neumann value for $G\}$.

If $M(\alpha)$ is already reduced (which holds in most cases of interest) then $S(\alpha)$ is generated as follows: for $a_2 \in A_2$ and $b \in B$ and $(b, a_2) \in X_2$, $b \cdot a_2 = \lambda(b, a_2)$ with $\lambda(b, a_2)$ defined by (6.116) and for $(b, a_2) \notin X_2$,

$b \cdot a_2 = q_0$ where q_0 is the initial position of the game. Thus $\{\cdot a_2 : a_2 \in A\}$ is a set of mappings on B, and the functions obtained by taking products of these functions (the semigroup generated) is $S(\alpha)$.

6.117. Further Remarks on Modeling of Games and Strategies.

(a) Since the complexity of the game reflects the "complexity of the intermeshing" of the loops of the program implementing the optimal strategies, no conventions should be employed to artificially lessen the number of program loops as was stupidly done in (6.114a). When ties, wins or losses occur the programs return to the correct initial position. A reasonable error convention should be employed. In fact one might define the *eloquence* of a game to be the minimal complexity of a machine which can play the game legally starting from an arbitrary but possible game position. Thus the eloquence is the minimal complexity to play the game legally. Clearly the eloquence of a game is less than or equal to the complexity of a game. Certainly any reasonable strategy can play legally. So all legal strategies can play legally, and all legal strategies have complexity greater than or equal to the eloquence of the game.

In almost all games of interest, ties, wins or losses can occur at various times. Further, errors can be made and it is not entirely trivial just to play the game legally (e.g. there are books on Go and Chess which just explain the rules). Wins, ties, errors, etc. set the program back to some original positions which yield program loops.

(b) To simplify matters let's assume the game is given by a game tree (e.g. Figure 22) of which each player plays either a zero or a one and so no errors are possible. The loops comes from ties, wins and losses which set the final position back to some initial position. We have seen games of any complexity arise this way.

Suppose we further restrain the game by making all plays the same length, so the first win, tie or loss is the same distance from any possible initial position. Still it is possible to obtain games of any complexity of this form since the final position can determine the next starting position (including who goes first) and thus loops appear.

However, assume we further restrain the game by demanding O_1 always go first and the game always returns to the same initial position after exactly m moves. Thus we are considering game trees with inputs 0 and 1 and every path of length m from the starting position is a tie, win or loss. All the ties, wins, and losses occur only at these positions, and after m moves the game returns to the initial position. Thus the game is at the initial position at exactly $0, m, 2m, \ldots$ moves.

Since no errors are possible and the game returns to the initial position q_0 at multiples of m, all loops must pass through q_0 *so the complexity is exactly 1* (an m counter to count moves modulo m plus memory). This is not difficult to prove rigorously. Almost *no* games of interest have these properties. For example, usually the winner goes second, ties and wins can occur at differing distances from the start, errors are possible, etc.

However, if desirable and necessary (which in actual cases will be very infrequent) we can introduce the *complexity of the game relative to* $\mathcal{M}$ where $\mathcal{M}$ is some collection of finite state machines. For example, let $\mathcal{M} = \mathcal{M}_n = \{M :$ the set of machines which can be realized by a circuit with $\leq n$ states$\}$. Then $\theta_{\mathcal{M}}(G)$, read "the complexity of the game G relative to $\mathcal{M}$", is defined in the obvious way by adding a new rule to the game G yielding $G_{\mathcal{M}}$, which allows as permissible strategies only those which *are strategies of G and can be realized by a machine $M \in \mathcal{M}$*. Then by definition, $\theta_{\mathcal{M}}(G) = \theta(G_{\mathcal{M}})$.

When $\mathcal{M} = \mathcal{M}_n$ we write $\theta_n(G)$ for $\theta_{\mathcal{M}_n}(G)$. Thus $\theta_n(G)$ can be computed for reasonable values of n. Here of course n is the maximum amount of memory allowed (or $\log_2 n$ is the number of bits of memory allowed) in any strategy. In many cases it is interesting to plot $n \mapsto \theta_n(G)$. Clearly for large values of n, $\theta_n(G) = \theta(G)$.

(c) Continuing in a similar vein we can define the *complexity necessary to overcome the strategy S of the game G*, denoted $O(S)$ as follows. Change the rules of G so that one opponent must play S and denote this new game by G_S. Then define $O(S) = \theta(G_S)$. By the min-max theorem, if S is an optimal strategy, the von Neumann value of G and G_S are identical. However, even if S is an optimal strategy we can have $O(S) < \theta(G)$. For example, consider Hex where O_1 always goes first. Since in Hex the first player can always win, the second player can play any legal strategy and achieve the value -1. So choose a simple minded legal strategy S of complexity zero for O_2 such that it can be beaten by an elementary strategy of complexity zero. The n $O(S) = 0 < \theta(\text{Hex})$. The idea is that even if O_2 can always be beaten, some strategies are much harder to beat than others. This is reflected in the values of $O(S)$.

It is a *very interesting problem* to determine the set of strategies of G such that $O(S) = \theta(G)$. These strategies "put up the best possible fight".

When they exist they are the better strategies. In general if S_1 and S_2 yield the same von Neumann value, but $O(S_1) < O(S_2)$, it is better to choose S_2.

The original theory of games by von Neumann and Morgenstern was motivated by possible applications to economic behavior and economics. However, for Chess, Go, Hex, etc. the analysis given was not very deep. On the other hand, the theory of games has wide applicability.

The theory of complexity for games leads to deeper results but requires a great effort for even relatively elementary games (e.g. Chess instead of the entire economic system). If one is not too familiar with a game and attempts to compute the complexity of the game one invariably finds that one must become an expert on the game to do this. In short, computing the complexity leads to a critical analysis of strategies of this game. This is just what a mathematical theory of complexity of games should do!

Also the von Neumann value is defined "conservatively" (minimize your possible losses, not maximize your possible gains). However, by determining the strategies such that $O(S) = \theta(G)$, we overcome these "conservative" assumptions to an extent.

In short, classical game theory is a general analysis of a wide class of games which often cannot go too deep. In contrast, the complexity of games is a deep analysis limited in scope by the greater difficulty of the computations and methods.

6.118.   Evolution-Complexity Relations for Evolved or Well Known Games. Taking into account the argument in Part II, Complexity of Evolved Organisms, and the initial parts of (6.114b) and Theorem (6.108b), we *conjecture* the following as worthy of being tested:

6.119.   The complexity of an evolved or well known game is approximately equal to the number of moves per game among masters of the game.

Some comments on the meaning of the terms of (6.119) is in order. By an evolved or well known game we mean one which has existed in its present form for 300 years or so, e.g. Chess, Go, but not 3-dimensional Tic-Tac-Toe. See [Falkener].

By the moves per game we mean until the point where play in fact naturally stops (i.e. one resigns, or in Go when further plays will not result in profit or loss for either opponent).

The heuristic argument for the validity of (6.119) is the following. By Theorem (6.108b) the number of moves per game is an upper bound. Since the games have not changed much historically, if this upper bound was not approximately a lower bound, either the rules of the game or the board or pieces would have been changed so that it was shorter or more complicated. In short, we always like games to be as complicated as possible subject to a simple format. If the format allows more complexity than our existing game we either leave the format alone and introduce another rule, piece, convention, etc. (driving the complexity up with the same format) or we simplify the format (e.g. making the board a little smaller) leaving the complexity nearly intact.

The result for 300-year-old games is thus, heuristically, (6.119). At the present time we have the following table of popular games:

Game	Complexity Conjecture	Evidence
Tic-Tac-Toe	1	proof
Nim	1	proof
Hex	≤ 24	proof
	~ 18	(6.119)
Checkers	~ 25	(6.119)
Chess	~ 50	(6.119)
Go	≤ 200	proof
	~ 140	(6.119)

Figure 25.

For those games whose rules apply in an obvious way to larger boards (e.g. Go and Hex but not Chess) it is very interesting to ask if the complexity tends to infinity as the board size tends to infinity. We conjecture

Conjecture 6.120. *(a) The complexity of Go tends to ∞ as the $n \times n$ board size tends to ∞.*

(b) The complexity of Hex tends to ∞ as the $n \times n$ board size tends to ∞.

(c) The complexity of 3-dimensional Tic-Tac-Toe does not tend to ∞ as the board size tends to ∞.

The proofs follow by applying the upper and lower bounds of complexity exposited in Chapter 3.

We close this section with a few brief comments which extend this theory to n-person games, mixed strategies, and games with random elements. In

n-person games with perfect information and no chance moves we extend the definition of complexity by minimizing the maximum complexity of any set of (pure) strategies, each realizing the von Neumann value for each player.

When mixed strategies are required we consider probabilistic combination of pure machines $\sum \lambda_i M_i$, $\lambda_i \in [0,1]$, $\sum \lambda_i = 1$ and define $\theta(\sum \lambda_i M_i) = \sum \lambda_i \theta(M_i)$. If one wants to be fancy let u be a measure on the pure machines and let $\theta(u) = \int \theta \, du$ be the complexity of the measure. Then we proceed as before!

However complexity is most important for two-person zero-sum games with no random elements and perfect information, e.g. Go.

Bibliography

[2-J] B. R. Tilson. Complexity of two J-class semigroups. *Advances in Mathematics*, 11:215–237, 1973.

[Banerji] R. B. Banerji. *Theory of Problem Solving*. Elsevier, New York, 1969.

[Boorman] S. A. Boorman. *The Protracted Game, a Wei-ch'i Interpretation of Maoist Revolutionary Strategy*. Oxford, New York, 1969.

[Falkener] E. Falkener. *Games Ancient and Oriental and How to Play Them*. Dover, 1961.

[Games] J. von Neumann and O. Morgenstern. *Theory of Games and Economic Behavior*. Wiley, 1967.

[Gardner] M. Gardner. *Mathematical Puzzles and Diversions*. Penguin, 1959.

[Go] S. Takagawa. *How to Play Go*. Nihon Ki-in and Japan Go Association, 1969. 17th printing.

Index